Adsorption of Inorganics at Solid-Liquid Interfaces

Adsorption of Inorganics at Solid-Liquid Interfaces

Edited by
Marc A. Anderson
Alan J. Rubin

ANN ARBOR SCIENCE
PUBLISHERS INC / THE BUTTERWORTH GROUP

Copyright © 1981 by Ann Arbor Science Publishers, Inc.
230 Collingwood, P. O. Box 1425, Ann Arbor, Michigan 48106

Library of Congress Catalog Card No. 77-85090
ISBN 0-250-40226-2

Manufactured in the United States of America
All Rights Reserved

PREFACE

Early adsorption modeling was simply a matter of describing the partitioning of an ion or molecule between solution and solid phases. In practice, the concern was with the capacity of a particular substrate to adsorb a given solute. Simple models such as the Langmuir and Freundlich equations were often used quite successfully. In fact, in systems where many variables are essentially constant, the added sophistication imposed by more elaborate and elegant models may be unnecessary. Soils are a good example in many applications, because their pH and ionic strength are usually constant. Even ionic adsorbates have been modeled using the Freundlich and Langmuir equations with some degree of success.

In more complicated systems, however, simplistic models of adsorption have limited utility. Indeed, even before questions of pH and ionic strength came in vogue, there were indications that these simple models were unable to describe systems even in situations where adsorbate concentration was the only apparent variable. In many cases, it became necessary to account for different types of surface sites, each with a different energy of adsorption. Research also began to question the ability of existing models to describe the data over the extremely large concentration ranges often employed in these types of experiments. Were the variable partition (adsorption) coefficients caused by an alteration of the electrostatic environment at the surface by the adsorbate, were they from lateral interactions with the adsorbates themselves, or were they possibly caused by surface precipitation? Additional questions pertaining to the effects of pH on the adsorption of hydrolyzable cations and protolyzable anions soon became obvious as more data were gathered. Finally, the ionic strength dependence of adsorption reactions came into question as a natural extension of double layer theory and because of the dependence of chemical equilibrium constants on activity coefficients.

At this stage in our knowledge of the adsorption of ions at solid-liquid interfaces, we have progressed from simple and somewhat empirical descriptions of adsorption as a function of dissolved adsorbate concentration to a very complicated thermodynamic language that describes the process as a function not only of concentration, but also of pH and ionic strength. It was the intention of the editors of this book to locate, in one volume, most of the current theories on the adsorption of ions so that the reader might better be able to compare models and thus intelligently choose a particular model for a given application. It was also our intent to point out some of the assumptions and shortcomings of these models, and thus provide some suggestions for future research.

The chapters in this text are not organized from a historical view of adsorption modeling, but rather in a fashion that assists the reader in getting used to the vocabulary of adsorption chemistry. Schindler introduces this new vocabulary in Chapter 1 by addressing both cation and anion adsorption reactions from a mechanistic, complexation point of view. This chapter is also an introduction to some of the constraints imposed by the electrostatic double layer. The author in systematic fashion also describes the effects of concentration, pH and ionic strength in chemical format.

The literature of anion and cation adsorption is covered in the next two chapters. Hingston, in Chapter 2, presents a review of equilibrium anion adsorption and the all-too-often neglected problem of adsorption kinetics. In Chapter 3, Kinniburgh exhaustively reviews cation adsorption including some of his work on hydrolyzable cations. Whereas most of the text is addressed to the adsorption of ions on oxy-hydroxide surfaces, Kinniburgh discusses clay adsorbents as well.

Corey, in Chapter 4, discusses the very important concept of surface precipitation and the differences between adsorption and precipitation reactions. This is not only an important consideration for the experimentalist, but also presents a problem to the modeler who attempts to use some sets of previously generated adsorption data where concentrations of adsorbate were high. In Chapter 5, Huang discusses the potentiometric titration which is utilized in many models not only to determine pK values for surface ionization but also to measure surface charge σ_s, which is a input variable for the more sophisticated models. Together, Chapters 4 and 5 illustrate some of the experimental constraints connected with adsorption modeling.

Chapter 6 reintroduces the surface complexation model, but this time James adds the sophistication imposed by double layer constraints. The result is an extremely sophisticated model which is then applied principally to cation adsorption reactions. In this chapter, James also discusses double extrapolation techniques which are used to determine intrinsic stability constants in the absence of ionic strength and surface potential effects. An excellent summary of adsorption models and of the assumptions used in modeling can be found in Chapter 7 by Morel. This chapter not only describes the modeling work of the author, but also serves as a review of models presented in the previous chapters and helps the reader to organize the relationships between these models.

In Chapter 8, Rubin and Mercer discuss adsorption modeling of ions on an organic substrate, activated carbon. Differences between the Freundlich and Langmuir equations are examined in connection with predictability in modeling. Finally, in Chapter 9, Anderson and his co-authors discuss the expectations and limitations of aqueous adsorption chemistry. This chapter reviews the problems, assumptions and potential dangers in the field today. New methods of analysis and alternative techniques for studying adsorption reactions are examined.

Serious questions in the area of adsorption modeling remain to be answered. Much more experimental verification is needed if truly universal predictability is to be achieved. The complex thermodynamic vocabulary that is being employed to describe these heterogeneous reactions now demands equally complex and sophisticated experimental support.

Marc A. Anderson
Alan J. Rubin

Marc A. Anderson is Assistant Professor, Water Chemistry Program, University of Wisconsin, Madison. He received his PhD in Environmental Engineering and his MA in Chemistry from Johns Hopkins University, and his BS in Chemistry from the University of Wisconsin. He has published many research papers and technical reports and has presented papers at conferences and seminars.

Alan J. Rubin is Professor of Civil Engineering and is associated with the Water Resources Center at The Ohio State University. He received a PhD in Environmental Chemistry and a MS in Sanitary Engineering from the University of North Carolina, and his BS in Civil Engineering from the University of Miami. The author of many publications and papers, Dr. Rubin is also the editor of three other Ann Arbor Science publications.

CONTENTS

1. Surface Complexes at Oxide-Water Interfaces 1
   P. W. Schindler

2. A Review of Anion Adsorption 51
   F. J. Hingston

3. Cation Adsorption by Hydrous Metal Oxides and Clay 91
   D. G. Kinniburgh

4. Adsorption vs Precipitation 161
   R. B. Corey

5. The Surface Acidity of Hydrous Solids 183
   C. P. Huang

6. Surface Ionization and Complexation at the Colloid/Aqueous
   Electrolyte Interface 219
   R. O. James

7. Adsorption Models: A Mathematical Analysis in the Framework
   of General Equilibrium Calculations 263
   F. M. M. Morel

8. Adsorption of Free and Complexed Metals from Solution
   by Activated Carbon 295
   A. J. Rubin and D. L. Mercer

9. Expectations and Limitations for Aqueous Adsorption
   Chemistry 327
   M. Anderson, C. Bauer, D. Hansmann, N. Loux
   and R. Stanforth

General Index 349

Index to Elements, Ions and Minerals 355

CHAPTER 1

SURFACE COMPLEXES AT OXIDE-WATER INTERFACES

Paul W. Schindler
Department of Inorganic Chemistry
University of Bern
3012 Bern, Switzerland

INTRODUCTION

Atoms, ions and molecules located at surfaces suffer from an imbalance of the chemical forces. This imbalance may be expressed in terms of surface energy. A finely dispersed solid phase thus tends to lower its surface energy, either by reducing its surface area or by adsorbing molecules and ions from adjacent phases.

This chapter discusses adsorption from the point of view of coordination chemistry, with the abovementioned imbalance of chemical forces understood as coordinative unsaturation. Whereas this approach is widely used in the field of heterogeneous catalysis, adsorption from aqueous solution at oxide-water interfaces has been discussed mostly on the basis of simple and extended double layer models. In the past few years, however, some attempts have been made to explain adsorption of metal ions at oxide-water interfaces, assuming the formation of surface complexes [1-5]. In the same period, coordination chemistry models for the adsorption of anions at oxide surfaces have been presented [6-10]. Two recent papers [11, 12] explain the simultaneous adsorption of metal ions and small ligands by formation of ternary surface complexes. The subsequent paragraphs attempt to summarize and to review these coordination chemistry models, with special emphasis on the presentation of an unified model and on examination of its predictive capabilities.

COORDINATION PHENOMENA AT OXIDE-WATER INTERFACES

Acquiring the Missing Ligands

Figure 1 is a schematic cross section of the surface layer of a metal oxide. The metal ions in the surface layer have a reduced coordination number, thus behaving as Lewis acids. The presence of Lewis acid sites at the surface of dry oxide samples has been demonstrated by numerous investigators. In the presence of water, the surface metal ions may first tend to coordinate H_2O molecules [13] (Figure 1b). For most of the oxides, dissociative chemisorption of water molecules (Figure 1c) seems energetically favored. Infrared (IR) spectroscopy has revealed the presence of surface hydroxyl groups at various oxide surfaces. The concentrations of surface hydroxyl groups have been estimated from weight loss on heating; from BET treatment of water vapor adsorption; from D_2O exchange; and from reactions with various reagents such as $HClO_4$, H_3PO_4, NaOH and CH_3MgI. Representative results are presented in Table I. Hence, oxide surfaces typically carry 4 to 10 hydroxyl groups per 100 (Å)^2 [2]. The actual numbers depend on the geometry of the crystal structure, the nature of the cleavage planes and on the pretreatment of the sample, since the individual oxides differ with respect to the rates of hydroxylation and dehydroxylation. Hydroxylation of SiO_2 is known to be a slow process, whereas the hydroxyl groups on α-Fe_2O_3, γ-Al_2O_3 and ThO_2 are rapidly formed [18].

It would appear from Figure 1c that the surface carries two different types of groups: (1) hydroxyl groups bound to one metal ion ("a"-type); and (2) hydroxyl groups bound to two (or more) metal ions ("b"-type). This point of view was first introduced by Boehm [14], who assumed the b-type groups to be acidic as compared with the basic a-type groups. His model was based on the observation (Table I) that (with TiO_2) D_2O exchange (assumed to involve both a- and b-type groups) yields roughly twice the number of basic groups that react with H_3PO_4. Further, IR spectroscopy has offered evidence for different types of hydroxyl groups as in the case of γ-Al_2O_3, where five different groups were found [19]. The small differences observed in the frequencies of the valence vibrations were attributed to differences of adjacent oxygen ions [20]. On the other hand, Schindler et al. [21, 22], Stumm et al. [1, 2], Yates [23] and Davis et al. [24] were able to explain the protolytic behavior of surface hydroxyl groups on SiO_2, TiO_2, γ-Al_2O_3, α-FeOOH and $Fe(OH)_3$ on the basis of a polyelectrolyte model, assuming identical groups. One might conclude, therefore, that the differences in chemical environment evidenced by the vibrational frequencies are not reflected in the acid-base properties of the surface hydroxyl groups. Moreover, it will be shown here that the different behavior of D_2O and H_3PO_4 is in full accordance with the coordination chemistry model. Despite these arguments, it is felt that the

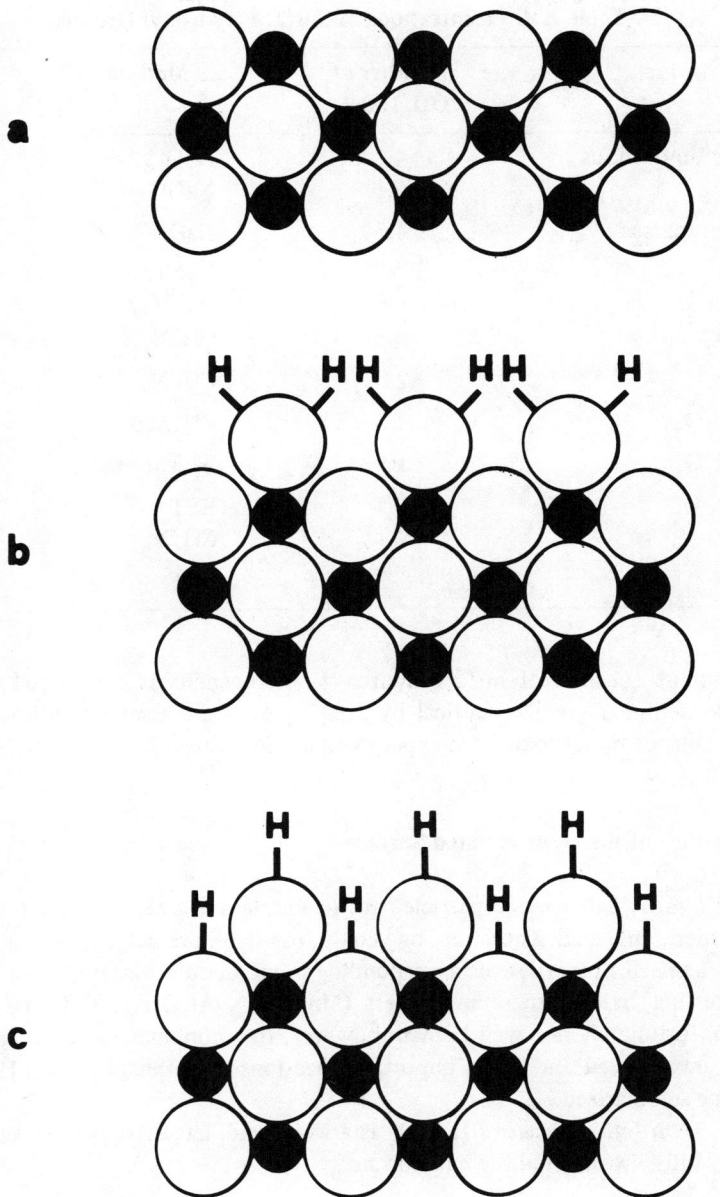

Figure 1. Cross section of the surface layer of a metal oxide: Metal ions, ●; oxide ions, ○:
(a) Surface ions are coordinatively unsaturated.
(b) In the presence of water, the surface metal ions may coordinate H_2O molecules.
(c) Dissociative chemisorption leads to a hydroxylated surface.

Table I. Concentrations of Surface Hydroxyl Groups

Oxide	Number of OH/100 Å^2	Method	Ref.
SiO$_2$ amorphous	4.8	D$_2$O	13
	5.1	NaOH	4
TiO$_2$ anatase	4.5	NaOH	14
	4.9	D$_2$O	14
	2.8	H$_3$PO$_4$	14
CeO$_2$	4.3	CH$_2$N$_2$	14
SnO$_2$	2.0	CH$_2$N$_2$	14
η-Al$_2$O$_3$	4.8	CH$_3$MgI	14
γ-Al$_2$O$_3$	10	Weight loss	15
α-Fe$_2$O$_3$	5.5	BET	16
	9.1	CH$_2$N$_2$	14
ZnO	6.8-7.5	BET	17

important question of surface hydroxyl heterogeneity is not settled. In the subsequent paragraphs identical hydroxyl groups are assumed, although for convenience rather than from experimental evidence.

Reactions of the Hydroxylated Surface

A hydroxylated oxide particle can, to a certain degree, be understood as a polymeric oxo acid. Therefore, one could, partially, predict its reactions from the chemical properties of corresponding monomeric acids. For most of the interesting oxides, these monomers (Mn(OH)$_4^0$, Al(OH)$_3^0$, Fe(OH)$_3^0$, etc.) are unfortunately not well known; however, the monomer of silica has been well investigated and some important reactions of orthosilicic acid H$_4$SiO$_4$ can be summarized:

1. *Protolytic behavior*: H$_4$SiO$_4$ is a weak acid. Lagerström [25] has carefully investigated the equilibrium:

$$H_4SiO_4 \rightleftharpoons H^+ + H_3SiO_4^-$$

From his value of the macroscopic acidity constant (K$_a$ = 3.72 x 10^{-10}, 3 M NaClO$_4$, 25°), one obtains for the microscopic acidity constant (I = 0) a value of $k_a = 9.3 \times 10^{-11}$. In aqueous solution, no cationic Si(IV) species have been observed. This means that $H_5SiO_4^+$ and $-\{SiOH_2^+\}$ must be a very strong acid.

2. *Complex formation*: H_4SiO_4 forms complexes with divalent and trivalent metal ions. Weber and Stumm [26] reported the formation of $FeOSi(OH)_3^{2+}$. Their results were confirmed by Olson and O'Melia [27] who concluded that $FeOSi(OH)_3^{2+}$ is an inner sphere complex. Santschi and Schindler [28] found that Ca^{2+} and Mg^{2+} form complexes of the composition $MOSi(OH)_3^+$. With Ca^{2+} and Mg^{2+} a second species is formed that may be a chelate $MO_2Si(OH)_2^0$ or $M[OSi(OH)_3]_2^0$ or a mixture of both. For the similar Mg^{2+}-H_2CO_3 system, Riesen et al. [29] reported the simultaneous formation of both $MgCO_3^0$ and $Mg(HCO_3)_2^0$, the former species being by far more important. The stability constants of the silicato complexes (Table II) follow the same order as the stability constants of the corresponding hydroxo complexes: $Fe^{3+} \gg Mg^{2+} > Ca^{2+}$.

3. *Ligand exchange*: H_4SiO_4 is known to react with F^- according to the following equation:

$$Si(OH)_4 + 6F^- \rightleftharpoons SiF_6^{2-} + 4OH^-$$

The equilibrium constant is log K = 24.9 ± 0.1 (25°, I = 0.5 M $NaClO_4$) [30]. Hence, the hard Lewis acid Si^{4+} tends to exchange the general ligand [31] OH against the hard Lewis base F^-. Attempts to identify mixed complexes $Si(OH)_x F_y^{(4-x-y)+}$ so far have failed [30].

Table II. Stability Constants of Silicato Complexes at 25°

Equilibrium			log K	I	Ref.
$Fe^{3+} + H_3SiO_4^-$	$\rightleftharpoons$	$FeOSi(OH)_3^{2+}$	9.75[a]	0.1	27
$Mg^{2+} + H_3SiO_4^-$	$\rightleftharpoons$	$MgOSi(OH)_3^+$	0.64	1	28
$Ca^{2+} + H_3SiO_4^-$	$\rightleftharpoons$	$CaOSi(OH)_3^+$	0.39	1	28
$Mg^{2+} + H_2SiO_4^{2-}$	$\rightleftharpoons$	$MgO_2Si(OH)_2^0$	4.17	1	28
$Ca^{2+} + H_2SiO_4^{2-}$	$\rightleftharpoons$	$CaO_2Si(OH)_2^0$	3.09	1	28
or:					
$Mg^{2+} + 2H_3SiO_4^-$	$\rightleftharpoons$	$Mg[OSi(OH)_3]_2^0$	3.82	1	28
$Ca^{2+} + 2H_3SiO_4^-$	$\rightleftharpoons$	$Ca[OSi(OH)_3]_2^0$	2.89	1	28

[a] Calculated from Olson and O'Melia [27] using $\log K_a (H_4SiO_4) = -9.5$ [26].

On the basis of these observations, we may now draft a tentative and speculative scheme of the various possible reactions of a hydroxylated surface (Figure 2). The individual reactions are discussed in the subsequent paragraphs.

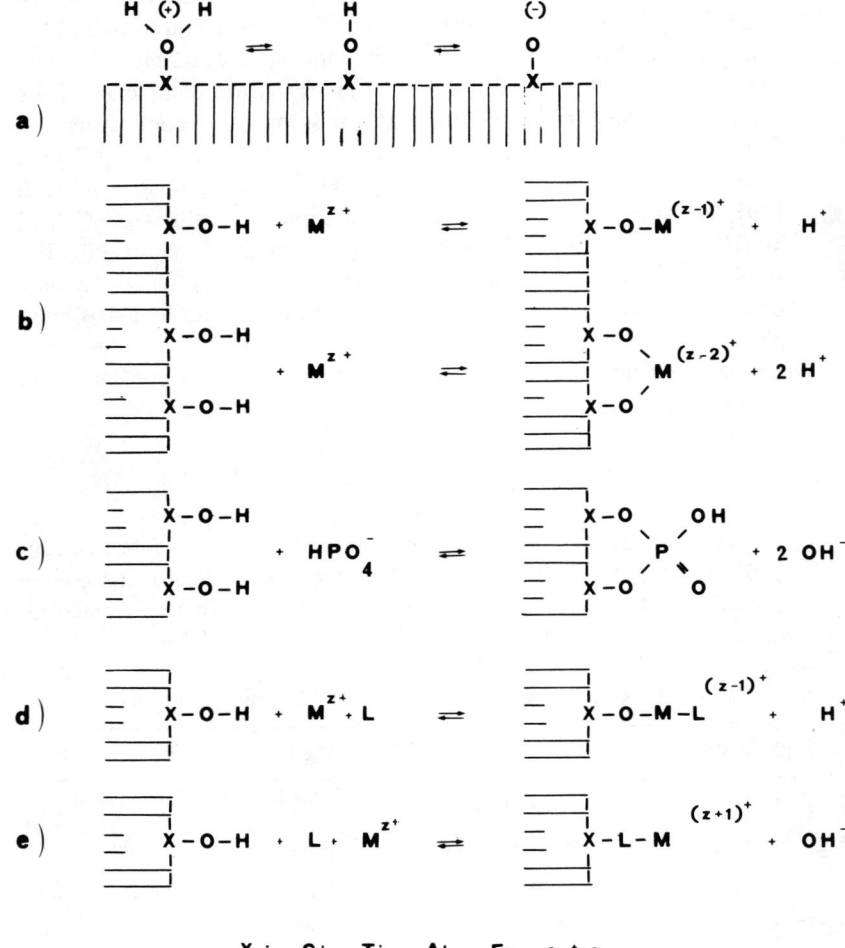

Figure 2. Coordination phenomena at oxide-water interfaces:
(a) Acid-base reactions of surface hydroxyl groups.
(b) Deprotonated surface hydroxyls coordinate with dissolved metal ions. Several species may be formed simultaneously.
(c) Surface hydroxyls are replaced by dissolved ligands.
(d) A dissolved metal ion coordinates with both deprotonated surface hydroxyls and dissolved ligands.
(e) A dissolved multidentate ligand coordinates with both X and the dissolved metal ion M.

Protolytic behavior of surface hydroxyl groups

A hydroxylated oxide surface is known to specifically adsorb H^+ and OH^-. This adsorption can be ascribed to protonation and deprotonation of surface hydroxyl groups. Figure 2a suggests the definition of two acidity constants:

$$K_{a1(int)}^s = \frac{[H^+]\{XOH\}}{\{XOH_2^+\}} \times \frac{\gamma_{H^+} \gamma_0}{\gamma_+} \exp(-F\psi_{H^+}/RT) \quad (1)$$

$$K_{a2(int)}^s = \frac{[H^+]\{XO^-\}}{\{XOH\}} \times \frac{\gamma_{H^+} \gamma_-}{\gamma_0} \exp(-F\psi_{H^+}/RT) \quad (2)$$

where [i] = concentrations of the species i in the aqueous solution [mol/dm^3], and

i = concentrations of surface species [mol/kg of suspended solid oxide],

$\gamma_{H^+}, \gamma_+, \gamma_0, \gamma_-$ = the activity coefficients of H^+, XOH_2^+, XOH, XO^- respectively, and

ψ_{H^+} = the potential at the location of H^+ on the surface.

In a medium of constant ionic strength, γ_{H^+} is a constant. Moreover, Chan et al. [32] have given some arguments for the assumption $\gamma_+ = \gamma_0 = \gamma_-$. From titrations in a constant ionic medium, one can thus obtain the following experimental quantities:

$$Q_{a1(I)}^s = \frac{[H^+]\{XOH\}}{\{XOH_2^+\}_e} \quad , \quad Q_{a2(I)}^s = \frac{[H^+]\{XO^-\}_e}{\{XOH\}}$$

The index (I) was a reminder that the quotients are related to a particular ionic medium. The acidity constants

$$K_{a1(I)}^s = Q_{a1(I)}^s \exp(-F\psi_{H^+}/RT) \quad (3)$$

$$K_{a2(I)}^s = Q_{a2(I)}^s \exp(-F\psi_{H^+}/RT) \quad (4)$$

can be obtained from extrapolations of $Q_{a1(I)}^s$ and $Q_{a2(I)}^s$ to zero charge condition. These experimentally available and thermodynamically sound constants are not identical with the intrinsic constants defined by Equations 1 and 2. As already emphasized by Rossotti and Rossotti [33], measurements in ionic media reveal no information about the interaction between the reacting hydroxyl groups and medium ions. Hence, in an ionic medium consisting of the cations C^+ and the anions A^-, the experimental quantities $[XOH_2^+]_e$ and $[XO^-]_e$ are composed of

$$\{XOH_2^+\}_e = \{XOH_2^+\} + \{XOH_2^+ A^-\} \quad (5)$$

$$\{XO^-\}_e = \{XO^-\} + \{XO^- C^+\} \quad (6)$$

The formation of the surface species $XOH_2^+ A^-$ and $XO^- C^+$ can be described by the intrinsic stability constants

$$K_{A(int)}^s = \frac{\{XOH_2^+ A^-\}}{[A^-]\ XOH_2^+} \times \frac{\gamma_{+-}}{\gamma_{A^-}\ \gamma_+} \ \exp(-F\psi_{A^-}/RT) \quad (7)$$

$$K_{C(int)}^s = \frac{\{XO^- C^+\}}{[C^+]\ XO^-} \times \frac{\gamma_{+-}}{\gamma_{C^+}\ \gamma_-} \ \exp(F\psi_{C^+}/RT) \quad (8)$$

where ψ_{A^-}, ψ_{C^+} = potentials at the locations of A^- and C^+ in the double layer, and

γ_{+-}, γ_{-+} = activity coefficients of $XOH_2^+ A^-$ and $XO^- A^+$. As before, the assumption $\gamma_+ = \gamma_- = \gamma_0 = \gamma_{+-} = \gamma_{-+}$ will be used.

Combining Equations 1, 3, 5 and 7 one obtains the following:

$$K_{a1(int)}^s = K_{a1(I)}^s\ [1 + K_{A(int)}^s\ [A^-]\ \exp(F\psi_{A^-}/RT)\gamma_{A^-}]\gamma_{H^+} \quad (9)$$

And, similarly, from Equations 2, 4, 6 and 8 comes the following:

$$K_{a2(int)}^s = K_{a2(I)}^s\ [1 + K_{C(int)}^s\ [C^+]\ \exp(-F\psi_{C^+}/RT)\gamma_{C^+}]^{-1}\gamma_{H^+} \quad (10)$$

There is no obvious way to evaluate the intrinsic constants ($K_{a1(int)}^s$, $K_{A(int)}^s$, $K_{a2(int)}^s$ and $K_{C(int)}^s$) from a set of ($K_{a1(I)}^s, K_{a2(I)}^s$) values from different ionic media. Approximate values for these intrinsic constants were estimated by Davis et al. [24] using the simplifications:

$$1 \ll K_{A(int)}^s\ [A^-]\ \exp(F\psi_{A^-}/RT)\gamma_{A^-}$$

$$1 \ll K^S_{C(int)} [C^+] \exp(-F\psi_{C^+}/RT)\gamma_{C^+}$$

and neglecting the change of the activity coefficients γ_{H^+}, γ_{C^+} and γ_{A^-} with ionic strength. It may be preferable to describe the protolytic behavior of the surface hydroxyl groups by $K^S_{a1(I)}$ and $K^S_{a2(I)}$ values, rather than by intrinsic constants, but the work by Davis et al. [24] is, nonetheless, a very important contribution to the to the problem of surface protolysis. Table III collects some typical protolysis and values and suggests a correlation between the protolytic behavior of the metal ions in solution and the surface properties of the corresponding oxides. The acidity of the metal ions increases in the order $Al^{3+}_{aq} < Fe^{3+}_{aq} < Th^{4+}_{aq} < Ti^{4+}_{aq} < Si^{4+}_{aq}$. The same order is observed for the acidity of the surface hydroxyl groups.

Table III. Log $Ka^S_{(I)}$ Values of Surface Hydroxyl Groups of 25°

Group	Solid	Ionic medium	$pKa^S_{1(I)}$	$pKa^S_{2(I)}$	Ref.
Si-OH	Amorphous silica	0.1 M NaClO$_4$	(-3)[a]	6.8	21
		1.0 M NaClO$_4$		6.71	34
		1.0 M LiClO$_4$		6.57	34
		1.0 M CsCl		5.97	34
Ti-OH	Anatase	3.0 M NaClO$_4$	4.98	7.80	22
	Rutile	1.0 M NaClO$_4$	4.46	7.75	35
		10^{-3} M LiNO$_3$	2.75	9.1	24[b]
		10^{-2} M LiNO$_3$	3.25	8.9	24[b]
		0.1 M LiNO$_3$	3.6	8.4	24[b]
		1.0 M LiNO$_3$		7.2	24[b]
Th-OH	ThO$_2$	1.0 M NaClO$_4$	5.15	7.90	3
Al-OH	γ-Al$_2$O$_3$	0.1 M NaNO$_3$	6.51	8.43	1
Fe-OH	α-FeOOH	10^{-2} M NaCl	5.4		24[c]
		0.1 M NaCl	5.6		24[c]
		1.0 M NaCl	6.1		24[c]

[a] Assuming $pH_{zpc} \simeq 2$.
[b] Calculated from Yates [23].
[c] Calculated from Hingston et al. [6].

Ligand properties of surface hydroxyl groups

As indicated by Figure 2b, the specific adsorption of metal ions will be discussed in terms of complex formation with deprotonated surface hydroxyl groups. The pertinent equilibria are, therefore:

$$XOH + M^{z+} \rightleftharpoons XOM^{(z-1)+} + H^+$$

$$2\, XOH + M^{z+} \rightleftharpoons (XO)_2 M^{(z-2)+} + 2\, H^+$$

$$n\, XOH + M^{z+} \rightleftharpoons (XO)_n M^{(z-n)+} + n\, H^+$$

The corresponding intrinsic stability constants are defined by the following equations:

$$*K^S_{1(\text{int})} = \frac{[H^+]\{XOM^{(z-1)+}\}}{[M^{z+}]\{XOH\}} \times \frac{\gamma_{H^+}\,\gamma_1}{\gamma_{M^{z+}}\gamma_0} \exp\left[\frac{F}{RT}(z\psi_1 - \psi_{H^+})\right] \quad (11)$$

$$*\beta^S_{2(\text{int})} = \frac{[H^+]^2\{(XO)_2 M^{(z-2)+}\}}{[M^{z+}]\{XOH\}^2} \times \frac{\gamma^2_{H^+}\,\gamma_2}{\gamma_{M^{z+}}\gamma^2_0} \exp\left[\frac{F}{RT}(z\psi_2 - 2\psi_{H^+})\right] \quad (12)$$

$$*\beta^S_{n(\text{int})} = \frac{[H^+]^n\{(XO)_n M^{(z-n)+}\}}{[M^{z+}]\{XOH\}^n} \times \frac{\gamma^n_{H^+}\,\gamma_n}{\gamma_{M^{z+}}\gamma^n_0} \exp\left[\frac{F}{RT}(z\psi_n - n\psi_{H^+})\right] \quad (13)$$

where $\gamma_1, \gamma_2 \ldots \gamma_n$ = activity coefficients of $XOM^{(z-1)+}$, $(XO)_2 M^{(z-2)+}$, $\ldots (XO)_n M^{(z-n)+}$, and

$\psi_1, \psi_2 \ldots \psi_n$ = potentials of M^{z+} in the complexes $XOM^{(z-1)+}$, $(XO)_2 M^{(z-2)+}, \ldots (XO)_n M^{(z-n)+}$.

Occasionally the abbreviations

$$\Gamma_n = \frac{\gamma^n_{H^+}\,\gamma_n}{\gamma_{M^{z+}}\gamma^n_0} \;;\; \exp[n] = \exp\frac{F}{RT}(z\psi_n - n\psi_{H^+}) \quad (14)$$

will be used to simplify the nomenclature.

1. *Stoichiometry of the surface complexes*: Several authors have investigated the overall stoichiometry of the surface complex formation by measuring the number of hydrogen ions released per adsorbed metal ion. The pertinent quantity is

$$Z = \frac{\text{moles of } H^+ \text{ released}}{\text{moles of } M^Z \text{ bound}} \qquad (15)$$

In evaluating Z, the contribution of the previously discussed reactions

$$XOH_2^+ \rightleftharpoons XOH + H^+ \quad \text{and} \quad XOH_2^+A^- \rightleftharpoons XOH + H^+ + A^-$$

$$XOH \rightleftharpoons XO^- + H^+ \quad \text{and} \quad XOH + C^+ \rightleftharpoons XO^-C^+ + H^+$$

must be taken into account. This fact has obviously been overlooked by several investigators, especially Z values derived from Kurbatov plots, which often suffer from this neglect. Representative results are listed in Table IV and indicate that (1) the range of Z is given by $1 \leqslant Z \leqslant 2$; and (2) Z is a function of $[H^+]$; for low values $[H^+]$ Z approaches a value 2.

On the basis of our model Z is given by the equation

$$Z = \frac{\sum\limits^n n\left\{(XO)_n M^{(z-n)+}\right\}}{\sum\limits^n \left\{(XO)_n M^{(z-n)+}\right\}} = \frac{\sum\limits^n n\, {^*\beta^s_{n(\text{int})}}[H^+]^{-n}\{XOH\}^n\, \Gamma^{-1}(\exp[n])^{-1}}{\sum\limits^n {^*\beta^s_{n(\text{int})}}[H^+]^{-n}\{XOH\}^n\, \Gamma_n^{-1} \exp[n]^{-1}} \qquad (16)$$

Hence it is concluded that only two complexes (n = 1,2) are formed.

2. *Evaluation of the stability constants*: Studies on metal ion adsorption generally reveal the following experimental parameters:

(a) M , is the total amount of adsorbed metal. Since we have already noted that but two surface complexes are formed, $\{M\}$ is composed of

$$M = \left\{XOM^{(z-1)+}\right\} + \left\{(XO)_2 M^{(z-2)+}\right\} \qquad (17)$$

(b) XOM_{tot}, which is the total amount of modified and unmodified hydroxyl groups:

Table IV. Overall Stoichiometry of the Surface Reaction

System (M^{z+}, solid oxide)	Z	$-\log[H^+]$	Ref.
Mn^{2+}, δ-MnO_2	1.0	4.5	37
	1.4	7.5	37
	1.7	8.2	37
Co^{2+}, δ-MnO_2	2.1±0.05	4	38
Zn^{2+}, δ-MnO_2	2.1±0.26	4	38
Pb^{2+}, δ-MnO_2	1.4	6	39
Cd^{2+}, δ-MnO_2	1.3	6	39
Zn^{2+}, δ-MnO_2	1.1	6	39
Pb^{2+}, "FeooH"	1.18	5	40
	1.59	6	40
Pb^{2+}, γ-Al_2O_3	1.5	4-7	41
UO_2^{2+}, SiO_2, amorphous	1.3-1.8	~3	42
Zn^{2+}, SiO_2, amorphous	1.19	6	43
Co^{2+}, SiO_2, amorphous	1.18	6.3	43
	1.34	7.6	43
Ni^{2+}, SiO_2, amorphous	1.36	6.3-6.8	43
Fe^{3+}, SiO_2, amorphous	1.1-1.6	2-3	44

$$\{XOH\}_{tot} = \{XOH\} + \{XO^-\} + \{XO^-C^+\} + \{XOH_2^+\} + \{XOH_2^+A^-\}$$
$$+ \{XOM^{(z-1)+}\} + 2\{(XO)_2 M^{(z-2+)}\} \qquad (18)$$

It is favorable to conduct adsorption studies in such a way that the condition $M \ll XOH_{tot}$ is met. In this case, Equation 18 simplifies to

$$\{XOH\}_{tot} = \{XOH\} + \{XO^-\} + \{XO^-C^+\} + \{XOH_2^+\}$$
$$+ \{XOH_2^+A^-\} \qquad (19)$$

XOH is then obtained from acid-base titrations as follows:
For $pH < pH_{PZC}$:

$$(\{XO^-\} + \{XO^-C^+\}) = 0$$

$$\{XOH\} = \{XOH\}_{tot}/(1 + (Q^s_{a1(I)})^{-1} [H^+])$$

For pH > pH$_{PZC}$:

$$(\{XOH_2^+\} + \{XOH_2^+ A^-\}) = 0$$

$$\{XOH\} = \{XOH\}_{tot}/(1 + Q^s_{a2(I)} [H^+]^{-1})$$

(c) [M^{z+}] can be obtained directly in cases where ion-selective electrodes are available. Other analytical methods such as atomic absorption give [M] the total concentration of dissolved metal. [M] and [M^{z+}] are related by

$$[M] = [M^{z+}] + \sum_{j}\sum_{n}\sum_{m} m\, [M_m(L_j)_n]$$

where L_j stands for the ligands present in the aqueous solution. The evaluation of [M^{z+}] from [M] requires the knowledge of the pertinent stability constants.

(d) [H^+] is easily obtained from electromotive force (emf) measurements in solutions of constant ionic strength.

No direct information is available on the magnitude of the exponential terms, so it seems thus extremely difficult, if not impossible to evaluate *$K^s_{1(int)}$ and *$\beta^s_{2(int)}$ from the available experimental quantities. Thus, it is necessary to introduce some simplifications. In a medium of constant ionic strength, γ_{H^+} and $\gamma_{M^{z+}}$ are constants. Moreover, making the approximation

$$\frac{\gamma_n}{\gamma_0^n} = 1 \qquad (20)$$

one can, at constant ionic strength, define the stability constants in the following way:

$$*K^S_{1(\text{int},I)} = \frac{[H^+]\{XOM^{(z-1)+}\}}{[M^{z+}]\{XOH\}} \exp[1] = *Q^S_{1(I)} \exp[1] \quad (21)$$

$$*\beta^S_{2(\text{int},I)} = \frac{[H^+]^2\{(XO)_2 M^{(z-2)+}\}}{[M^{z+}]\{XOH\ _2\}} \exp[2] = *Q^S_{2(I)} \exp[2] \quad (22)$$

At first glance it seems that $*K^S_{1(\text{int},I)}$ and $*\beta^S_{2(\text{int},I)}$ can be evaluated from $*Q^S_{1(I)}$ and $*Q^S_{2(I)}$ by extrapolation to zero charge conditions. Such a procedure requires, however, that $*Q^S_{1(I)}$ and $*Q^S_{2(I)}$ can be evaluated separately. Actually it can be shown that a separate and independent evaluation of both $*Q^S_1$ and $*Q^S_2$ is possible from a set of data including M, $[M^{z+}]$, $[H^+]$, XOH and Z. Unfortunately, the published values for Z are by far too imprecise for this purpose, so it is not yet possible to evaluate directly the intrinsic constants $*K^S_{1(\text{int},I)}$ and $*\beta^S_{2(\text{int},I)}$. Alternatively, an indirect method can be used. Equations 17, 21 and 22 yield the following:

$$M = \frac{[M^{z+}]\{XOH\}}{[H^+]}(*K^S_{1(\text{int},I)}(\exp[1])^{-1})$$

$$+ *\beta^S_{2(\text{int},I)} \exp[2]^{-1}\frac{\{XOH\}}{[H^+]}) \quad (23)$$

Assuming that exp [1] is constant:

$$\exp[1] = \text{constant} = a \quad (24)$$

Equation 23 is then rearranged to obtain the following:

$$v = \frac{1}{a} *K^S_{1(\text{int},I)} + *\beta^S_{2(\text{int},I)} \exp[2]^{-1} u \quad (25)$$

where $v = \dfrac{M\ [H^+]}{[M^{z+}]\{XOH\}}$, $u = \dfrac{\{XOH\}}{[H^+]}$.

Next compute

$$\frac{dv}{du} \left(\simeq \frac{\Delta v}{\Delta u}\right) = *\beta^S_{2(\text{int},I)} \exp[2]^{-1} \tag{25a}$$

and extrapolate the obtained values to zero charge conditions. Alternatively, assume that exp [2] is constant:

$$\exp[2] = \text{constant} = b \tag{26}$$

Equation 23 is then rearranged to obtain

$$y = \frac{1}{b} *\beta^S_{2(\text{int},I)} + *K^S_{1(\text{int},I)} (\exp[1])^{-1} x \tag{27}$$

where $y = \dfrac{\{M\}[H^+]^2}{[M^{z+}]\{XOH\}^2}$, $x = \dfrac{[H^+]}{\{XOH\}}$

Again compute

$$\frac{dy}{dx} \left(\simeq \frac{\Delta y}{\Delta x}\right) = *K^S_{1(\text{int},I)} \exp[1]^{-1} \tag{27a}$$

and extrapolate the obtained values to zero charge conditions. The outlined procedure will now be illustrated taking experimental data from Fürst [35] (systems Cd^{2+} - SiO_2 and Cu^{2+} - TiO_2 both at 25° and I = 1 M $NaClO_4$). Figures 3a and 3b show plots of $\Delta v/\Delta u$ and $\Delta y/\Delta x$ versus log $[H^+]$, i.e., the logarithm of the predominantly potential determining ion. Surprisingly, both assumptions (Equations 24 and 26) are simultaneously met over a limited pH range. Hence the apparent constants can be calculated.

$$\log *K^S_{1(I)} = \frac{1}{a} \log *K^S_{1(\text{int},I)} \tag{28a}$$

$$\log *\beta^S_{2(I)} = \frac{1}{b} \log *\beta^S_{2(\text{int},I)} \tag{28b}$$

with the aid of Equations 25 and 27 (Figures 4a and 4b), whereas the intrinsic constants are not available. At this point it must be emphasized that al-

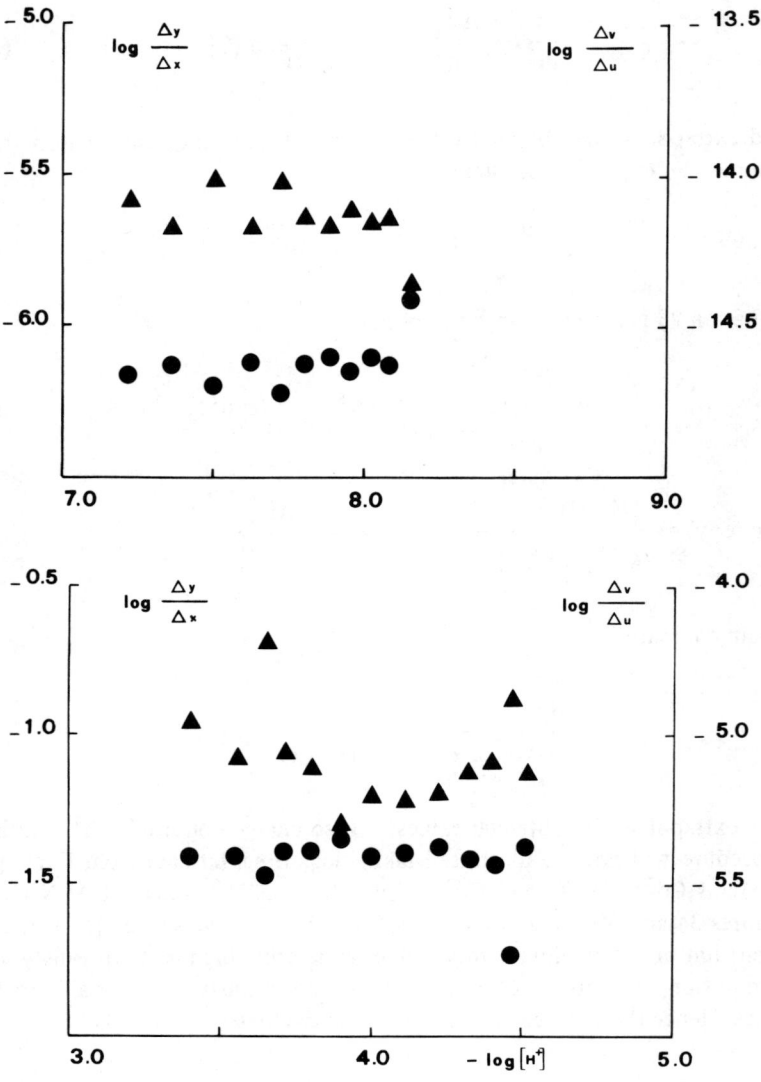

Figure 3. Tentative evaluation of $(\Delta u/\Delta v) = {}^*\beta^s_{2(\text{int},I)}(\exp[2])^{-1}$ (▲) and $(\Delta y/\Delta x) = {}^*K^s_{1(\text{int},I)}\exp[1]^{-1}$ (●).
(a) System Cd(II) - SiO_2.
(b) System Cu(II) - TiO_2.
The obtained values show no systematic change with $\log[H^+]$. The observed fluctuations are symmetrical, thus indicating the mutual dependence of the two functions.

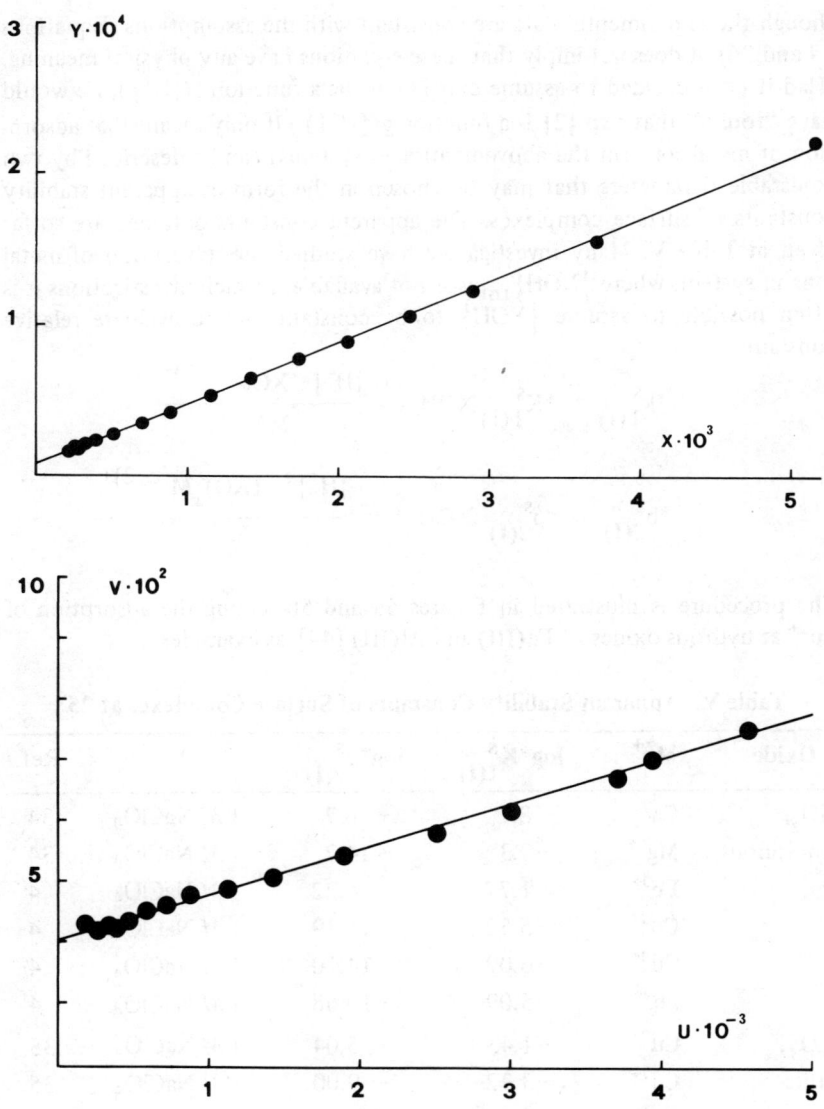

Figure 4. System Cu(II) - TiO$_2$ (1 M NaClO$_4$, 25°) [35]. Evaluation of the apparent stability constants:
(a) *$K^s_{1(I)}$ from Equation 27; (b) *$\beta^s_{2(I)}$ from Equation 25. The straight lines were calculated with the constants given in Table V.

though the experimental data are consistent with the assumptions (Equations 24 and 26), it does not imply that these equations have any physical meaning. (Had it been decided to assume exp [1] to be a function $f([H^+])$, we would have "found" that exp [2] is a function $g([H^+])$.) It only means that adsorption of metal ions (in the abovementioned systems) can be described by two adjustable parameters that may be chosen in the form of apparent stability constants of surface complexes. The apparent constants obtained are so far given in Table V. Many investigators have studied the adsorption of metal ions in systems where $\{XOH\}_{tot}$ was not available. In such investigations it is often possible to assume $\{XOH\}$ to be constant and to evaluate relative constants

$$*k^S_{1(I)} = *K^S_{1(I)} \, XOH = \frac{[H^+] \, XOM^{(z-1)+}}{M^{2+}}$$

$$*b^S_{2(I)} = *\beta^S_{2(I)} \, XOH^{\,2} = \frac{[H^+]^2 \, (XO)_2 M^{(z-2)+}}{M^{2+}}$$

The procedure is illustrated in Figures 5a and 5b, taking the adsorption of Zn^{2+} at hydrous oxides of Fe(III) and Al(III) [44] as examples.

Table V. Apparent Stability Constants of Surface Complexes at 25°

Oxide	M^{z+}	$\log{*K^S_{1(I)}}$	$\log{*\beta^S_{2(I)}}$	I	Ref.
SiO_2, amorphous	Ca^{2+}	−8.1	−16.7	1 M $NaClO_4$	34
	Mg^{2+}	−7.3	−14.7	1 M $NaClO_4$	34
	Fe^{3+}	−1.77	−4.22	3 M $NaClO_4$	4
	Cu^{2+}	−5.52	−11.19	1 M $NaClO_4$	4
	Cd^{2+}	−6.09	−14.20	1 M $NaClO_4$	4
	Pb^{2+}	−5.09	−10.68	1 M $NaClO_4$	4
TiO_2, rutile	Cu^{2+}	−1.43	−5.04	1 M $NaClO_4$	35
	Cd^{2+}	−3.32	−9.00	1 M $NaClO_4$	35
	Pb^{2+}	0.44	−1.95	1 M $NaClO_4$	35
δMnO_2	Ca^{2+}	−5.5	−	0.1 M $NaNO_3$	1
$\gamma\text{-}Al_2O_3$	Ca^{2+}	−6.1	−	0.1 M $NaNO_3$	2
	Mg^{2+}	−5.4	−	0.1 M $NaNO_3$	2
	Ba^{2+}	−6.6	−	0.1 M $NaNO_3$	2
	Pb^{2+}	−2.2	−8.1	0.1 M $NaClO_4$	41
	Cu^{2+}	−2.1	−7.0	0.1 M $NaClO_4$	41

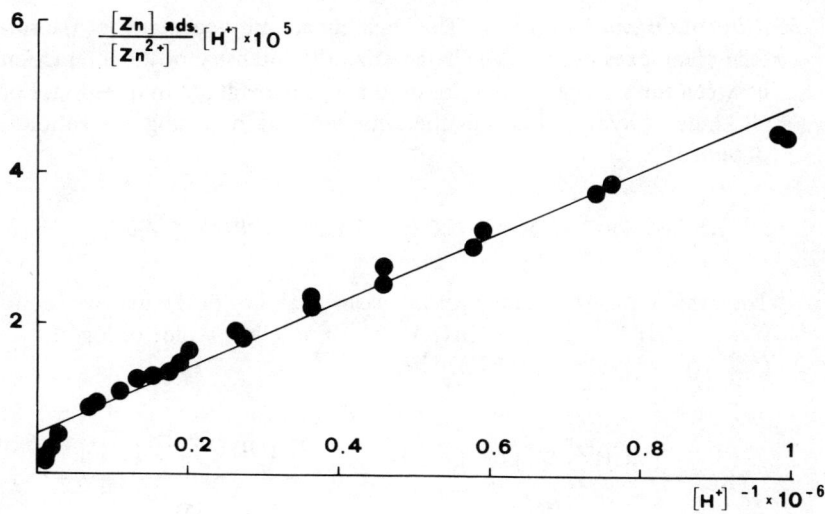

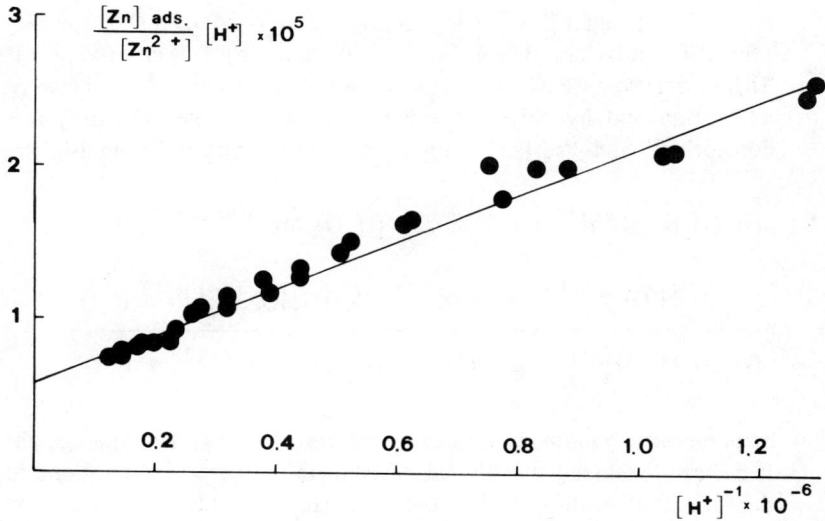

Figure 5. Adsorption of Zn(II) at iron(III) and aluminum hydrous oxide (1 M NaNO$_3$, room temperature) [44]: evaluation of the relative stability constants. The original work gives y, the fraction of adsorbed Zn(II), as a function of pH. Plotting $[H^+]y/(1-y) = [H^+][Zn]_{ads}/[Zn^{2+}]$ vs $[H^+]^{-1}$ results in approximately straight lines. From intercept and slope one obtains the relative stability constants $*k^S_{1(I)} = *K^S_{1(I)}\{XOH\}$ (A/V), $*\beta^S_{2(I)}\{XOH\}^2$(A/V).
(a) System Zn(II) - hydrous iron(III) oxide: $*k^S_{1(I)} = 5.40 \times 10^{-6}$, $*b^S_{2(I)} = 4.36 \times 10^{-11}$.
(b) System Zn(II) - hydrous aluminum oxide: $*k^S_{1(I)} = 5.83 \times 10^{-6}$, $*b^S_{2(I)} = 1.58 \times 10^{-11}$.

3. *Adsorption and hydrolysis*: The apparent stability constants of the surface complexes (Table IV) characterize the intensity of the interaction between the surface hydroxyl groups and the metal ion in question. For all kinds of hydroxyl groups the same series of increasing interaction is observed:

$$Ba^{2+} < Ca^{2+} < Mg^{2+} < Cd^{2+} < Cu^{2+} < Pb^{2+} < Fe^{3+}$$

The same series is obtained for increasing stability of the hydroxo complexes $MOH^{(z-1)+}$ and $M(OH)_2^{(z-2)+}$. For silica, a plot of log $*K_1^S{}_{(I)}$ ($*\beta_{2(I)}^S$) versus log $*K_1$ ($*\beta_2$), where

$$*K_1 = \frac{[MOH^{(z-1)+}][H^+]}{[M^{z+}]} \quad , \quad *\beta_2 = \frac{[M(OH)_2^{(z-2)+}][H^+]^2}{[M^{z+}]}$$

results in a straight line [4] log $*K_{1(I)}^S$ ($*\beta_{2(I)}^S$) = -0.09 + 0.62 log $*K_1$ ($*\beta_2$). A similar correlation has been observed by Dugger et al. [45]. This correlation seems to explain the well-known coincidence between adsorption and hydrolysis. Actually the two schemes "hydrolysis + adsorption" and "surface complex formation" are indistinguishable:

$$(H_2O)_5M(OH_2)^{z+} + H_2O = (H_2O)_5MOH^{(z-1)+} + H_3O^+$$

$$\underline{(H_2O)_5M(OH)^{(z-1)+} + HOX = (H_2O)_5MOX^{(z-1)+} + H_2O}$$

$$(H_2O)_5M(OH_2)^{z+} + HOX = (H_2O)_5MOX^{(z-1)+} + H_3O^+$$

4. *Structure and bonding in surface complexes*: In the previous paragraphs it has been assumed that the adsorbed metal ions are directly bound to deprotonated surface OH-groups. Actually, equilibrium analysis in aqueous solution does not permit differentation between "inner sphere" (I) and "outer sphere" (II) complexes:

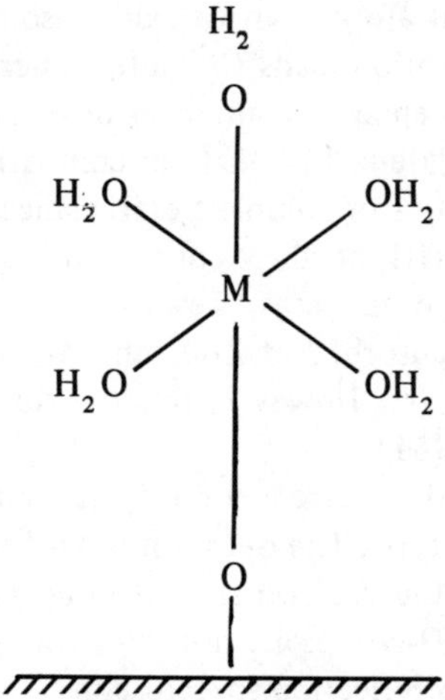

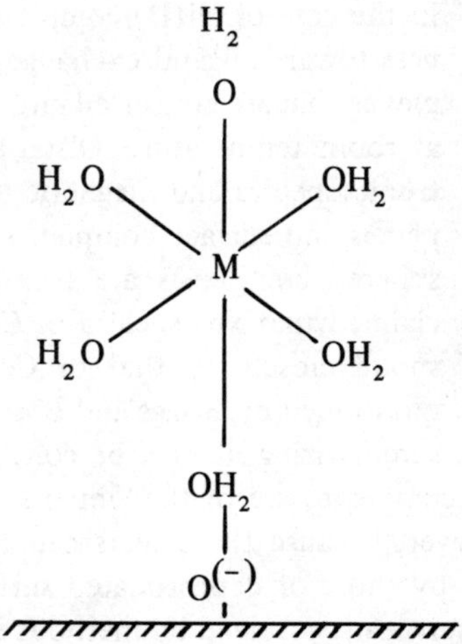

SURFACE COMPLEXES AT OXIDE-WATER INTERFACES

I   II

The problem of the coordinative environment of metal ions at oxide surfaces has been attacked by ultraviolet (UV), electron spin resonance (ESR) and X-ray photoelectron spectroscopy (XPS) techniques. Hathaway and Lewis [46-48] examined the electronic spectra of Ni(II), Co(II) and Cu(II) species such as $M(H_2O)_6^{2+}$, $M(en)(H_2O)_4^{2+}$, $M(en)_2(H_2O)_2^{2+}$ and $M(en)_3^{2+}$ adsorbed at silica gel. In all cases, the systems could be described in terms of ligand competition for the coordination sites of the adsorbed metal ion, the ligands being water, deprotonated surface OH groups and N-compounds. In some cases, the spectra seemed to indicate that the formation of surface complexes is connected to a change in the coordination number. The adsorption of $Co(H_2O)_6^{2+}$ at silica gel leads to a tetrahedral $[CoO'O_3'']$ chromophore that may be identified either by (III) or by (IV):

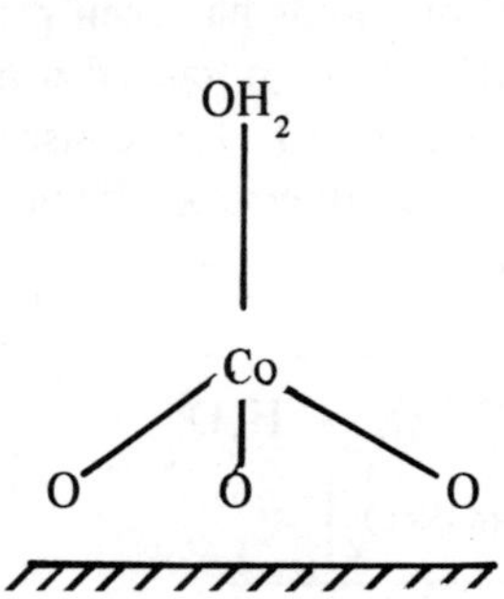

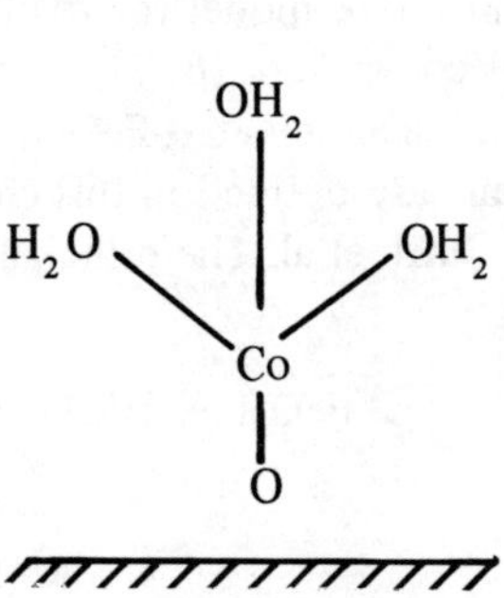

III   IV

In the case of Ni(II) complexes that are known to exhibit some inertness towards ligand exchange, adsorption leads to "outer sphere" complexes that are converted into "inner sphere" complexes on dehydration at room temperature. Olson and O'Melia [27, 49], in comparing electronic spectra and magnetic properties of soluble Fe(III) silicato complexes and surface complexes of Fe(III) at silica, concluded that "inner sphere" complexes are formed in both cases. Tewari and Lee [50] claimed that xps spectra of Co(II) adsorbed at ZrO_2 and Al_2O_3 correspond closely to that of $Co(OH)_2(s)$. However, this conclusion was questioned by Briggs and Bosworth [51].

In summarizing, it may be concluded that there is some evidence that surface complexes are of the "inner sphere" type. The question is hard to settle, however, because the effects caused by the replacement of O ligands from H_2O by those of deprotonated surface OH-groups are not large enough to be unequivocally resolved by methods such as reflection spectroscopy.

Replacement of surface hydroxyl groups by specifically adsorbed anions

Since the metal ions in compounds such as SiO_2, TiO_2, AlOOH and FeOOH are hard Lewis acids, it can be expected that their surface hydroxyl groups may be replaced by fluoride and by oxyanions, as shown by Figure 2c. Numerous studies have confirmed that fluoride, phosphate, arsenate and other oxyanions are specifically adsorbed. The quantitative treatment of the observed data in terms of an equilibrium model seems very difficult, however. The difficulties may arise from two sources. First, it seems that adsorption of anions is often only partially reversible. Second, most of the studies on adsorption of anions were carried out under conditions in which the surface densities of the adsorbed anions are high. Hence, the interaction between the surface and the dissolved anions may be superimposed by lateral interactions between the adsorbed species.

A quantitative model for the adsorption of anions has been proposed by Quirk and co-workers [6, 7]. Their approach will be discussed using the adsorption of selenite at α-FeOOH as an example [6]. For consistency, the symbols already defined in this chapter are used rather than the notation preferred by Quirk et al. The principal reaction is

$$>\text{FeOH} + \text{HSeO}_3^- \rightleftharpoons \text{FeOSeO}_2^- + H_2O$$

$$K^S = \frac{\{\text{FeOSeO}_2^-\}}{\{\text{FeOH}\}[\text{HSeO}_3^-]} \tag{29}$$

Introducing

$$K_a = \frac{[H^+][SeO_3^{2-}]}{[HSeO_3^-]}$$

one obtains

$$\frac{[HSeO_3^-]}{[HSeO_3^-] + [SeO_3^{2-}]} = \frac{[HSeO_3^-]}{[Se]_{tot}} = \frac{[H^+]}{[H^+] + K_a}$$

For constant pH one can define the conditional constant as follows:

$$K = \frac{\{FeOSeO_2^-\}}{\{FeOH\}[Se]_{tot}} = K^s \frac{[H^+]}{[H^+] + K_a} \qquad (30)$$

Now $\{FeOH\}$ is given by

$$\{FeOH\} = \{FeOH\}_{tot} - \{FeOSeO_2^-\} - \{FeOH_2^+\}_e - \{FeO^-\}_e \qquad (31)$$

Quirk et al. assume that

$$\{FeOH_2^+\}_e + \{FeO^-\}_e = 0 \qquad (32)$$

With this simplification they obtain from Equations 30 and 31

$$\{FeOSeO_2^-\} = K[Se]_{tot}\{FeOH\}_{tot} - K[Se]_{tot}\{FeOSeO_2^-\}$$

or

$$\{FeOSeO_2^-\} = \{FeOH\}_{tot} \frac{K[Se]_{tot}}{1 + K[Se]_{tot}} \qquad (33)$$

Partially from the simplification (Equation 32), but also for other reasons, the quantity $\{FeOH\}_{tot}$ as used in Equation 33 is not a constant but is dependent on pH. The curve that describes the change of $FeOH_{tot}$ with pH is called the adsorption envelope. Once this envelope is established, values for K are obtained from a Langmuir plot at constant pH. Within a limited pH range the conditional constants K show the pH dependence given by Equation 30.

The unpredictability of the adsorption envelope obviously limits the predictive capabilities of this model. Gupta [52, 53] tried a somewhat different approach for modeling the adsorption of phosphate at γ-FeOOH. Within this model, the pertinent reactions and the related apparent equilibrium constants are as follows:

$$>FeOH_2^+ \rightleftharpoons >FeOH + H^+ \qquad Q_{a1}^s = \frac{[H^+]\{FeOH\}}{\{FeOH_2^+\}_e}$$

$$>FeOH \rightleftharpoons >FeO^- + H^+ \qquad Q_{a2(I)}^s = \frac{[H^+]\{FeO^-\}_e}{\{FeOH\}}$$

$$>FeOH + H_3PO_4 \rightleftharpoons >FeOPO_3H_2 + H_2O \qquad K_{1(I)}^s = \frac{\{FeOPO_3H_2\}}{\{FeOH\}[H_3PO_4]}$$

$$>FeOPO_3H_2 \rightleftharpoons >FeOPO_3H^- + H^+ \qquad K_{a3(I)}^s = \frac{[H^+]\{FeOPO_3H^-\}}{\{FeOPO_3H_2\}}$$

$$>FeOPO_3H^- \rightleftharpoons >FeOPO_3^{2-} + H^+ \qquad K_{a4(I)}^s = \frac{[H^+]\{FeOPO_3^{2-}\}}{\{FeOPO_3H^-\}}$$

$$2 >FeOH + H_3PO_4 \rightleftharpoons (>FeO)_2PO_2H + 2H_2O \qquad \beta_{2(I)}^s = \frac{\{(FeO)_2PO_2H\}}{\{Fe-OH\}^2[H_3PO_4]}$$

$$(>FeO)_2PO_2H \rightleftharpoons (>FeO)_2PO_2^- + H^+ \qquad K_{a5(I)}^s = \frac{[H^+]\,(FeO)_2PO_2^-)}{\{(FeO)_2PO_2H\}}$$

Note that $Q_{a1(I)}^s$, $Q_{a2(I)}^s$, $K_{a3(I)}^s$, $K_{a4(I)}^s$ and $K_{a5(I)}^s$ are conditional constants depending on log $[H^+]$.

The total concentration of surface groups and adsorbed phosphate species are given by the following equations:

$$FeOH_{tot} = C[1 + (Q^S_{a1(I)})^{-1}[H^+] + Q^S_{a2(I)}[H^+]^{-1} + \frac{[P]}{S} a]$$

$$+ 2C^2 \frac{[P]}{S} b \quad (34)$$

$$P = \frac{[P]}{S}(Ca + C^2 b) \quad (35)$$

where $C = \{FeOH\}$,
[P] = total concentration of dissolved phosphate,
$$a = K^S_1 (1 + K^S_{a3}[H^+]^{-1} + K^S_{a3}K^S_{a4}[H^+]^{-2}), \quad (36)$$
$$b = \beta^S_2 (1 + K^S_{a5}[H^+]^{-1}), \text{ and} \quad (37)$$
$s = (1 + K_{a1}[H^+]^{-1} + K_{a1}K_{a2}[H^+]^{-2} + K_{a1}K_{a2}K_{a3}[H^+]^{-3})$,
K_{a1}, K_{a2}, K_{a3}: acidity constants of H_3PO_4.

A useful function is

$$F = \frac{\{FeOH\}_{tot} - \{P\}}{\{FeOH\}_{tot} - 2\{P\}} \quad (38)$$

Combining Equations 34, 35 and 38 one obtains

$$\frac{(F-1)(1 + (Q^S_{a1(I)})^{-1}[H^+] + Q^S_{a2(I)}[H^+]^{-1})S}{[P] F} = K^S_1 \cdot a + \frac{C}{F}\beta^S \cdot b \quad (39)$$

K^S_{a1}, K^S_{a2} and the acidity constants of H_3PO_4 were obtained from separate titrations. From a series of adsorption measurements at constant log [H⁺], the quantities a and b were evaluated by iteration using Equations 34 or 35 and 39. From a set of data (a, b) for different log [H⁺] values, the individual constants are obtained from Equations 36 and 37 and are found tabulated in Table VI. Figure 6 compares calculated and experimental values of the distribution coefficient

Table VI. Modeling of Phosphate Adsorption at γ-FeOOH—Numerical Values for the Pertinent Equilibrium Constants (0.5 M NaClO$_4$, 25°) [52,53]

---

a) Acidity constants of the surface hydroxyl groups

$\log [H^+] = -3.0$, $\log Qa^S_{1(I)} = -2.39$

$\log [H^+] = -5.0$, $\log Qa^S_{1(I)} = -4.11$

$\log [H^+] = -7.0$, $\log Qa^S_{1(I)} = -5.33$

$\log [H^+] = -9.0$, $\log Qa^S_{2(I)} = -9.94$

b) Equilibrium constants of phosphate adsorption

$\log K^S_{1(I)} = 2.3$, $\log K^S_{2(I)} = 4.0$

c) Acidity constants of adsorbed phosphate species

$\log K^S_{a3(I)} = -2.0$, $\log K^2_{a4(I)} = -6.7$, $\log K^S_{a5(I)} = -12.3$

d) Acidity constants of H_3PO_4

$\log K_{a1} = -1.73$, $\log K_{a2} = -6.43$, $\log K_{a3} = -10.62$

$$D = \frac{\{P\}}{[P]} \qquad (40)$$

A closer inspection shows that $(FeO)_2 PO_2^-$ is the dominating species in the range $-3 > \log [H^+] > -7$, whereas $FeOPO_3^{2-}$ becomes more important at $\log [H^+] < -7$. These results are partially supported by the work of Parfitt et al. [10] who found spectroscopic evidence for $(FeO)_2 PO_2^-$. They explain Boehm's [14] observation of the different stoichiometry of the reactions with D_2O and H_3PO_4 since at low pH each H_3PO_4 replaces two hydroxyl groups.

Ternary Surface Complexes

The adsorption of metal ions by complex formation with deprotonated surface hydroxyl groups has been discussed. Within the scope of this model one or two places in the coordination shell of an adsorbed metal ion are occupied by the surface ligands. In the systems discussed so far, it has been assumed

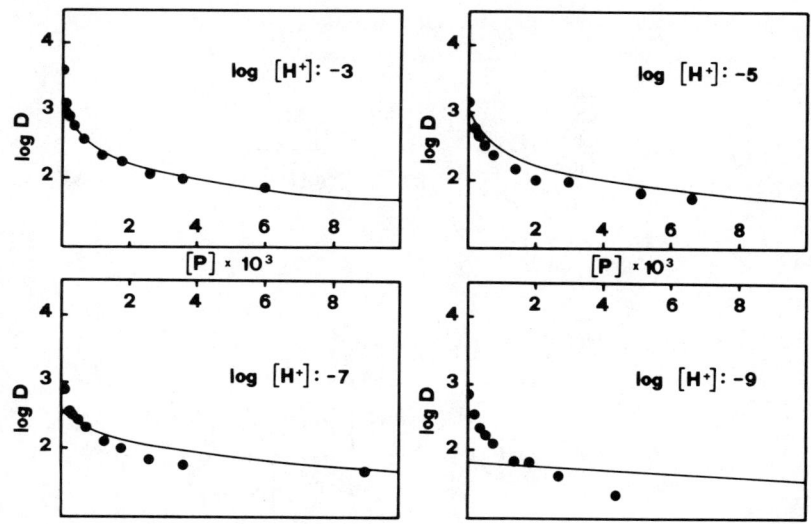

Figure 6. Adsorption of phosphate at γ-FeOOH (0.5 M NaClO$_4$, 25°) [52]. The curves are calculated with the equilibrium constants given in Table VI. Although the model gives a reasonable fit of the experimental data at log $[H^+]$ = -3,-5,-7, it clearly fails at log $[H^+]$ 9. This breakdown of the model is believed to be a consequence of lateral interactions of adsorbed phosphate species. Lateral interaction becomes dominiating at log $[H^+]$ values, where the adsorbed phosphate species are fully deprotonated.

that the remaining places are occupied by water molecules. It is conceivable, however, that these water molecules are at least partially replaced by other ligands (Figure 2d). The inverse situation is depicted in Figure 2e, where the polydentate ligand L bridges the adsorbed metal ion M and the metal ion of the oxide lattice. The species

$$X - O - M - L \quad \text{and} \quad X - L - M$$
$$(I) \qquad\qquad\qquad (II)$$

are designated by the term "ternary surface complexes." Roman letters (I, II) are used to distinguish between the different structures.

Knowledge on the formation of the ternary surface complexes is based mainly on studies of the effect of dissolved ligands on the adsorption of metal ions. Our discussion, therefore, shall begin by considering the stability of the ternary complexes with respect to the stability of the corresponding binary complexes X-O-M and X-L. In comparing the equilibria

$$n\ XOH + M^{z+} \rightleftharpoons (XO)_n M^{(z-n)+} + n\ H^+$$

$$*\beta^s_{n(I)}(M) = \frac{[H^+]^n \{(XO)_n M^{(z-n)+}\}}{[M^{z+}] \{XOH\}^n}$$

and

$$n\ XOH + M(L)_\ell^{z+} \rightleftharpoons (XO)_n M(L)_\ell^{(z-n)+} + n\ H^+$$

$$*\beta^s_{n(I)}(ML_\ell) = \frac{[H^+]^n \{(XO)_n M(L)_\ell^{(z-n)+}\}}{[M(L)_\ell^{z+}] \{XOH\}^n}$$

a ratio can be defined as follows:

$$R_I = \frac{*\beta^s_{n(I)}(ML_\ell)}{*\beta^s_{n(I)}(M)} \qquad (41)$$

Further, a comparison may be made between

$$n\ XOH + L \rightleftharpoons (X)_n L^{n+} + n\ OH^-$$

$$\beta^s_{n(I)}(L) = \frac{[OH^-]^n \{(X)_n L^{n+}\}}{[L] \{XOH\}^n}$$

and

$$n\ XOH + LM^{z+} \rightleftharpoons (X)_n LM^{(n-z)+} + n\ OH^-$$

$$\beta^s_{n(I)}(LM) = \frac{[OH^-]^n \{(X)_n LM^{(n-z)+}\}}{[LM^{z+}] \{XOH\}^n}$$

by defining the ratio

$$R_{II} = \frac{\beta^s_{n(I)}(LM)}{*\beta^s_{n(I)}(L)} \qquad (42)$$

In a very general sense, both R_I and R_{II} can be understood as a product of different factors that may individually increase or decrease the stability of the ternary complex. Tentatively, and with reference to Sigel [54], one may write

$$R = R_{stat} \cdot R_{steric} \cdot R_{charge} \cdot R_{disp} \cdot R_{el}$$

A statistical factor R_{stat} can be evaluated by the method of Bjerrum [55]. Alternatively, statistical factors may be estimated from stability constants of dissolved complexes [54]. The range is $R_{stat} \leq 1$. The effect of steric hindrance, described by R_{steric}, also covers the range $R_{steric} \leq 1$. Electrostatic interaction between the surface and L (in type I complexes) or M (in type II complexes) is represented by R_{charge}. Depending on the signs, R_{charge} is smaller or larger than unity. Weak interactions (such as van der Waals interaction and H-bonding) between L and the surface in type I complexes are included in R_{disp} (Range $R_{disp} \geq 1$). Other electronic interactions R_{el} may occur mainly in type I complexes. As an example for such (possibly electronic) interactions reference should be made to the surprisingly high stability of the complex (bipy) Cu(oxalate)^0 [54]. The range will thus be $R_{el} \geq 1$. After this general consideration our knowledge on type I and type II complexes shall be summarized.

1. *Type I complexes*: Bourg and Schindler [11,56] have investigated the adsorption of Cu(II) at silica in the presence of ethylenediamine (en). They found that addition of en does not prevent adsorption of Cu(II), although the percentage of adsorbed Cu(II) at a given log [H$^+$] value decreases with increasing concentration of en. They presumed that Cu^{2+}-en complexes might participate in the surface reactions. The considerations by Sigel suggest the following statistical factors:

$$n = 1, \ell = 1: \quad R_{I(1,1)stat} = \frac{1}{3} \text{ to } \frac{2}{3}$$

$$n = 2, \ell = 1: \quad R_{I(2,1)stat} = \frac{1}{12} \text{ to } \frac{5}{12}$$

$$n = 1, \ell = 2 : R_{I(1,2)stat} \ll 1$$

$$n = 2, \ell = 2 : R_{I(1,2)stat} \ll 1$$

The experimental data could be explained by these equilibria:

$$SiOH + Cu^{2+} \rightleftharpoons SiOCu^+ + H^+$$

$$SiOH + Cu(en)^{2+} \rightleftharpoons SiOCu(en)^+ + H^+$$

$$2\,SiOH + Cu^{2+} \rightleftharpoons (SiO)_2 Cu^o + 2\,H^+$$

$$2\,SiOH + Cu(en)^{2+} \rightleftharpoons (SiO)_2 Cu(en)^o + 2\,H^+$$

$Cu(en)_2^{2+}$ does not participate in the surface reaction, although a weak nonspecific adsorption was observed. The observed ratios $R_{I(1,1)} = 2$, $R_{I(2,1)} = 4.2 \times 10^{-2}$ ($R_{I(1,2)} = R_{I(2,2)} = 0$) are thus governed by statistical factors. Charge effects and H-bonding are probably responsible for the high stabilities of ternary complexes composed of clay minerals, Fe^{3+} (Al^{3+}), and humic acids. Charge effects may, on the other hand, prevent the formation of type I complexes [57]. An unusually high electronic effect was found in the system silica-Cu(II)-bipy, where the addition of bipy dramatically enhanced the adsorption of Cu(II) [56].

2. *Type II complexes*: Evidence for type II complexes was recently presented by Davis and Leckie [12] in a study on the simultaneous adsorption of metal ions and ligands at amorphous iron hydroxides. Silver, itself not adsorbed at low pH, becomes adsorbed after addition of thiosulfate. The pertinent reaction is presumably

$$FeOH + AgS_2O_3^- \rightleftharpoons FeOSO_2 SAg^o + OH^-.$$

In this case, where the hard Lewis acid Fe(III) coordinates with oxygen whereas the soft Lewis acid Ag^+ coordinates with sulfur, the statistical factor is close to unity. Actually, Davis and Leckie [12] found that $AgS_2O_3^-$ is adsorbed with approximately the same energy as $S_2O_3^{2-}$. Formation of type II complexes was also found in the systems iron hydroxide-glutamic acid-Cu(II) and iron hydroxide - 2,3-pyrazinedicarboxylic acid - Cu(II) [12]. It may further explain the increased adsorption of Ag^+ at montmorillonite after addition of thiourea [58].

PREDICTIVE CAPABILITIES OF THE COORDINATION CHEMISTRY MODEL

The following sections deal with the predictive capabilities of the coordination chemistry model. Although the model seems capable of explaining the adsorption of cations and anions, the quantitative aspects of anion adsorption are so far not well understood. Therefore, the discussion is restricted to phenomena related to the adsorption of metal ions. Moreover, since our knowledge on ternary surface complexes is poor, we shall not attempt to deal with possible formation of these species.

Effect of System Parameters Upon the Adsorption of Metal Ions

The extent of adsorption of a given metal ion from an aqueous solution at the surface of a given solid oxide depends on many variables, most of them being related to the composition of the aqueous phase. Many scientists have followed the effect of pH on adsorption. Most of the pertinent articles present adsorption isotherms at constant pH; i.e., the effect of the concentration of the dissolved metal on adsorption. Some authors have reported on effects of dissolved ligands on the extent of adsorption. Some of the functions used to describe the distribution of a metal ion between the solid oxide and the aqueous solution are strongly dependent on the ratio

$$\frac{A}{B} = \frac{\text{amount of suspended oxide}}{\text{volume of the aqueous solution}}$$

A closed model system is constructed (Figure 7) consisting of an adsorbing solid oxide and an aqueous solution. The pertinent system parameters are summarized in Table VII.

Interest should be focused on a given metal ion i and its adsorption at a given solid oxide. The important quantity is

$$M_{i(ads)} = \{M_i\} A/V = \{M_i\} * AS/V \qquad (43)$$

Some authors prefer to describe the extent of adsorption by the quantity

$$\% \text{ ads} = \frac{100\, M_{i(ads)}}{M_i} = \frac{100\, (A/V)\{M_i\}}{[M_i] + (A/V)\{M_i\}} \qquad (44)$$

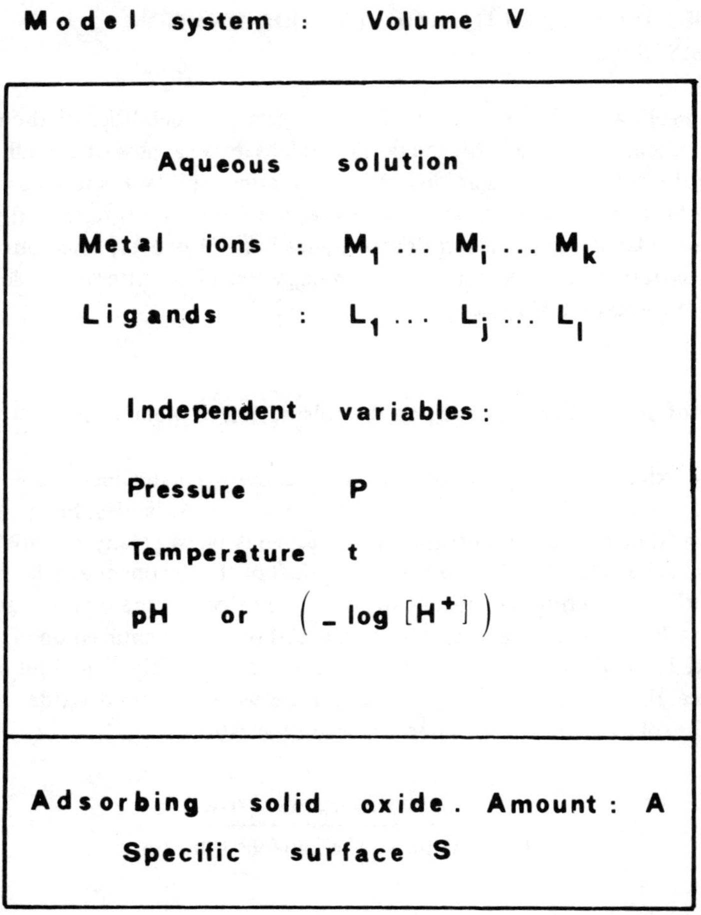

```
Model system :    Volume V

┌─────────────────────────────────────┐
│                                     │
│      Aqueous    solution            │
│                                     │
│   Metal  ions  :  M₁ ... Mᵢ ... Mₖ  │
│                                     │
│   Ligands      :  L₁ ... Lⱼ ... L_l │
│                                     │
│   Independent  variables :          │
│                                     │
│   Pressure       P                  │
│                                     │
│   Temperature    t                  │
│                                     │
│   pH    or    ( _ log [H⁺] )        │
│                                     │
├─────────────────────────────────────┤
│                                     │
│ Adsorbing  solid  oxide. Amount : A │
│      Specific   surface  S          │
└─────────────────────────────────────┘
```

Figure 7. Model system for the adsorption of metal ions from aqueous solution at the oxide-water interface.

or by the distribution coefficient

$$D = \{M_i\}/[M_i] \qquad (45)$$

Combining Equations 43, 44 and 45 with Equations 23, 28a and 28b, we obtain

$$M_{i(ads)} = (A/V)[M_i^{z+}](*K_{1(I)}^{s,i}\{XOH\}[H^+]^{-1} + *\beta_{2(I)}^{s,i}\{XOH\}^2[H^+]^{-2}) \qquad (46)$$

SURFACE COMPLEXES AT OXIDE-WATER INTERFACES

Table VII. Model Parameters

Parameter	Symbol	Unit
Volume of the system	V	dm^3
Total concentration of metal i	M_i	mol/dm^3
Total concentration of dissolved metal i	$[M_i]$	mol/dm^3
Concentration of free metal ion i	$[M_i^{z+}]$	mol/dm^3
Amount of adsorbed metal i		
Related to the system	$M_{i(ads)}$	mol/dm^3
Related to the solid oxide	$\{M_i\}$	mol/kg^1
or	$\{M_i^*\}$	mol/m^2
Amount of solid oxide	A	kg
Specific surface of the solid oxide	S	m^2/kg^1
Total concentration of surface ligand	$\{XOH\}_{tot}$	mol/kg^1
Concentration of surface ligand	$\{XOH\}$	mol/kg^1
Concentration of free ligand j	$[L_j]$	mol/dm^3
Pressure	P	atm
Temperature	t	°C
pH (NBS scale) or ($-\log [H^+]$)		

$$\% \text{ ads} = \frac{100(A/V)[M_i^{z+}](*K_{1(I)}^{S,i}\{XOH\}[H^+]^{-1} + *\beta_{2(I)}^{S,i}\{XOH\}^2[H^+]^{-2})}{[M_i] + (A/V)[M_i^{z+}](*K_{1(I)}^{S,i}\{XOH\}[H^+]^{-1} + *\beta_{2(I)}^{S,i}\{XOH\}^2[H^+]^{-2})} \quad (47)$$

$$D = \frac{[M_i^{z+}](*K_{1(I)}^{S,i}\{XOH\}[H^+]^{-1} + *\beta_{2(I)}^{S,i}\{XOH\}^2[H^+]^{-2})}{[M_i]} \quad (48)$$

$[M_i]$ and $[M_i^{z+}]$ are related by

$$[M_i] = [M_i^{z+}] \overset{j}{\Sigma} \overset{n}{\Sigma} \overset{m}{\Sigma} m[M_m(L_j)_n] \quad (49)$$

Since adsorption studies are mostly carried out at low concentrations of metal ion i, the occurrence of polynuclear complexes is often negligible. Equation 49 simplifies then to

$$[M_i] = [M_i^{z+}](1 + \overset{j}{\Sigma}\overset{n}{\Sigma}\beta_{j(n)}[L_j]^n) \qquad (49a)$$

where $\beta_{j(n)} = [M_i(L_j)_n]/[M_i^{z+}][L_j]^n$.

The total concentration of the surface ligand is given by

$$\begin{aligned}\{XOH\}_{tot} &= \{XOH\} + \{XOH_2^+\} + \{XOH_2^+A^-\} + \{XO^-\} + \{XO^-C^+\} \\ &\quad + \overset{k}{\Sigma}\{XOM_i^{(z-1)+}\} + 2\overset{k}{\Sigma}\{(XO)_2 M_i^{(z-2)+}\} \\ &= \{XOH\}(1 + (Ka_{1(I)}^S)^{-1}[H^+] + Ka_{2(I)}^S[H^+]^{-1} \\ &\quad + \overset{k}{\Sigma}[M_i^{z+}]*K_{1(I)}^{S,i}[H^+]^{-1} + 2\{XOH\}\overset{k}{\Sigma}[M_i^{z+}]*\beta_{2(I)}^{S,i}[H^+]^{-2}) \quad (50)\end{aligned}$$

Equation 50 implies that the concentration of the surface ligands is not altered by adsorption of dissolved ligands.

The effect of system parameters on the adsorption of metal ion i can be calculated from Equations 43 to 50 if the pertinent equilibrium constants are available. Analytical solutions are only obtained for simple cases where Equation 50 can be simplified to ($\{XOH\} = \{XOH\}_{tot}$). In most of the practical cases, the numerical evaluation requires iteration procedures.

Discussed in the subsequent paragraphs is the effect of the individual system parameters, assuming all the other variables to remain constant. The predictions of the model are then compared with experimental results. This comparison is sometimes impeded as some authors prefer to evade the standard rule of the experimental scientist and to change two or three variables during the same experiment.

Effect of pH on Adsorption

The pH dependence of adsorption was first recognized in 1951 by Kurbatov et al. [36]. Since then it has become progressively recognized that pH is the master variable that governs the extent of adsorption [59]. Starting at low pH with negligible adsorption, an increase of pH leads to a narrow region of 1-2 pH units, where adsorption increases from almost 0 to 100%. A further increase in pH often results in progressive desorption [60-62]. Bearing this

Table VIII. Effect of pH on Adsorption

Model System	$Cu(II)$ - silica - H_2O
Model Parameters	$V = 1\ dm^3$, $A = 0.05\ kg$, $Cu(II) = 10^{-7}\ mol/dm^3$ $t = 25°$, $p = 1\ atm$, Ionic strength: $1\ M\ (NaClO_4)$
Pertinent Equations	$\%\ ads = \dfrac{100 \times (A/V)\ (*K_{1(I)}^{S}\{SiOH\}[H^+]^{-1} + *\beta_{2(I)}^{S}\{SiOH\}^2[H^+]^{-2})}{1 + \Sigma(A/V)\ (*K_{1(I)}^{S}\{SiOH\}[H^+]^{-1} + *\beta_{2(I)}^{S}\{SiOH\}^2[H^+]^{-2})}$ $\Sigma = *K_1[H^+]^{-1} + *\beta_2[H^+]^{-2} + *\beta_3[H^+]^{-3} + *\beta_4[H^+]^{-4}$
Stability Constants	Surface complexes: $\log *K_{1(I)}^{S} = -5.52$, $\log *\beta_{2(I)}^{S} = -11.19$ Hydroxo complexes [63]: $*\beta_n = [Cu(OH)_n^{(n-2)-}][H^+]^n[Cu^{2+}]^{-1}$ $\log *K_1 = -8.26$, $\log *\beta_2 = -17.55$, $\log *\beta_3 = -27.8$, $\log *\beta_4 = -40.78$
Properties of Silica	$SiOH_{tot} = 1.45\ mol/kg^1$ $SiOH$ was calculated from empirical equations [35] $-\log[H^+] < 7.5:\ \{SiOH\} = 1.45 + (\dfrac{\log[H^+]}{10.7})$ [5,26] $-\log[H^+] \geqslant 7.5:\ \{SiOH\} = 1.45 + \dfrac{\log[H^+] + 5.9}{9.9}$

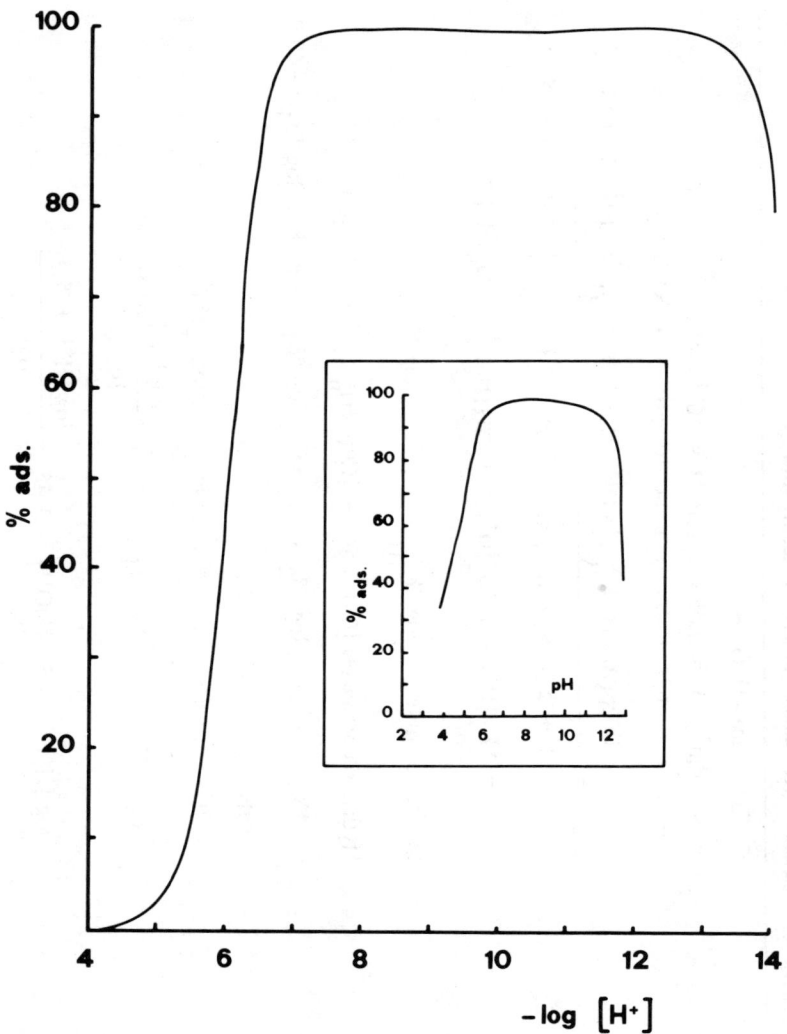

Figure 8. Adsorption of Cu(II) at silica as a function of log [H$^+$] (1 M NaClO$_4$, 25°). The curve was calculated with the equilibrium constants given in Table VIII. Inset: adsorption of Cu(II) from artificial river water at illite as a function of pH [61].

general feature in mind, the predictions of the model shall be evaluated, taking the system Cu(II)-silica as an example. The pertinent equations and the numerical parameters are given in Table VIII. The results of the model calculations (Figure 8) show a nice similarity with the experimental data of O'Connor and Kester [61] (system Cu(II)-illite). For a more detailed analysis, the interesting pH range is separated into two parts: (1) the pH range of increasing adsorption; and (2) the pH range of increasing desorption.

Region a is obviously governed by the term

$$(A/V)(*K^S_{1(I)}\{XOH\}[H^+]^{-1} + *\beta^S_{2(I)}\{XOH\}^2 [H^+]^{-2}.$$

The pH of beginning adsorption depends on the interaction between the adsorbed metal ion and the surface hydroxyl groups (as expressed by $*K^S_{1(I)}$ and $*\beta^S_{2(I)}$), the concentration of the surface hydroxyl group and the system parameter (A/V). Region b originates from two kinds of processes. As already mentioned by O'Connor and Kester [61] desorption of Cu(II) is based on the formation of anionic hydroxo complexes:

$$XOCu^+ + 3\,OH^- = XO^- + Cu(OH)_3^-$$

$$XOCu^+ + 4\,OH^- = XO^- + Cu(OH)_4^{2-}$$

$$(XO)_2 Cu^0 + 3\,OH^- = 2\,XO^- + Cu(OH)_3^-$$

$$(XO)_2 Cu^0 + 4\,OH^- = 2\,XO^- + Cu(OH)_4^{2-}$$

From the stability constants of anionic hydroxo complexes [63] one expects similar desorption for Sn(II) and Zn(II), but not for Hg(II). The progressive desorption of Hg(II) from quartz between pH 7 and 10 originates from a decrease of $\overline{XOH}$ by protolysis of the surface hydroxyl groups [62].

Adsorption Isotherms

As already mentioned, many authors present their results by adsorption isotherms at constant pH. Both $\{M_i\}$ vs $[M_i]$ and log $\{M_i\}$ vs log $[M_i]$ plots are frequently used. Some authors attempt to fit their data by a linearized Langmuir isotherm. Small deviations from linearity are generally attributed to experimental errors. In simulating adsorption isotherms (Table IX, Figure 9), one observes that at low pH and, thus, at low coverage, the $\{M_i\}$ vs $[M_i]$ curve approaches a straight line. At high pH values, the curve rises steeply to reach a plateau. Although the coordination chemistry model differs from the Langmuir model, the Langmuir plot of the simulated data (Figure 9c) differs but slightly from linearity. This demonstrates the low sensitivity of the Langmuir plot.

Table IX. Adsorption Isotherms

Model System	$Cd(II)$ - TiO_2 - H_2O
Model Parameters	$t = 25°$, $p = 1$ atm, ionic strength: $1\,M$ $(NaClO_4)$
Pertinent Equations	$\{Cd\} = [Cd^{2+}]\,(*K^S_{1(I)}\{TiOH\}[H^+]^{-1} + *\beta^S_{2(I)}\{TiOH\}^2[H^+]^{-2})$ (a)
	$\{TiOH\}_{tot} = \{TiOH\}(1 + (Ka^S_1)^{-1}[H^+] + Ka^S_2[H^+]^{-1} + *K^S_{1(I)}[Cd^{2+}][H^+]^{-1})$
	$\quad\quad\quad\quad\quad + \{TiOH\}^2 \, *\beta^S_{2(I)}[Cd^{2+}][H^+]^{-2}$ (b)
	At low pH where adsorption of Cd^{2+} is small, Equation (b) reduces to
	$\{TiOH\}_{tot} = \{TiOH\}(1 + (Ka^S_1)^{-1}[H^+] + Ka^S_2[H^+]^{-1})$ (c)
	Combining Equations (a) and (c) for constant values of H^+ gives
	$\overline{Cd} = [Cd^{2+}] \times \text{constant}$ (d)
Stability Constants of Surface Complexes:	$\log *K^S_{1(I)} = -3.32$, $\log *\beta^S_{2(I)} = -9.00$
Properties of TiO_2 [35]	$\{TiOH\}_{tot} = 0.15$ mol/kg^1
	$\log Ka^S_1 = -3.77$ (pH = 5)
	$\log Ka^S_2 = -7.75$ (pH 6), -8.28 (pH 7), -8.80 (pH 8)

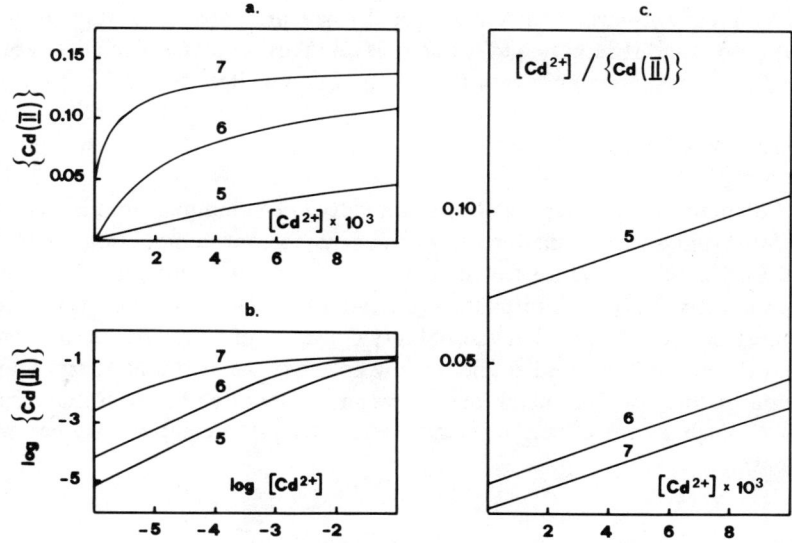

Figure 9. Adsorption of Cd(II) at TiO_2 (1 M $NaClO_4$, 25°)[35]: simulated adsorption isotherms for $-\log[H^+]$ = 5, 6, 7. Note that the linearized isotherm (c) gives an almost correct value for XOH_{tot}.

Effect of Dissolved Ligands

In the early stage of the coordination chemistry model, it was assumed that only the free metal ion M^{z+} participates in the surface reaction [4,64]. With this assumption, the effect of dissolved ligands can be calculated from Equations 43-50. The discovery of the formation of ternary surface complexes certainly questions the validity of such calculations. Thus, the discussion must be mainly qualitative.

The effect of dissolved ligands can be discussed, in part, in terms of competition with the surface ligands. If the dissolved ligands succeed in occupying all of the sites in the coordination shell of the metal, ion-specific adsorption is obviously prevented. Inhibition of adsorption is thus observed after equimolar addition of EDTA [56,57] and for an excess of monodentate ligands such as ammonia [65]. If the dissolved ligands occupy only part of the sites of the coordination shell, the adsorption may proceed by formation of type I ternary surface complexes. The results presented by Farrah and Pickering [57,66, 67] seem to indicate formation of type I complexes with en, glycine and other bidentate ligands. On the other hand, Kinniburgh and Jackson [68] have shown that coordination of one Cl^- prevents Hg(II) from reaction with the surface ligands.

As already discussed, dissolved ligands may promote adsorption by formation of type II ternary surface complexes. This mechanism requires polydentate ligands, preferably with three or four donor sites.

Competition by Other Metal Ions

Competition by other metal ions can indeed be computed from Equations 43-50. From such calculations it was concluded that competition by Mg^{2+} and Ca^{2+} is of importance for the extent of trace metal adsorption in marine environments [64]. The ultimate significance of such calculations is limited. It must be remembered that the stability constants of metal ion surface complexes were derived from studies at low coverage, i.e., without lateral interaction of the adsorbed metal ions. However, competition becomes important only at high coverage, i.e., in a region where lateral interaction may become important.

Effect of Adsorption on the Surface Charge

Adsorption phenomena are responsible for surface charge and colloidal stability. Within the scope of our model, adsorption affects the surface charge by the following reactions:

$$XOH_2^+ \rightleftharpoons XOH + H^+ \quad \text{and} \quad XOH_2^+ A^- \rightleftharpoons XOH + H^+ + A^- \quad \text{(A)}$$

$$XOH \rightleftharpoons XO^- + H^+ \quad \text{and} \quad XOH + C^+ \rightleftharpoons XO^-C^+ + H^+ \quad \text{(B)}$$

$$M^{z+} + n\,XOH \rightleftharpoons (XO)_n M^{(z-n)+} + n\,H^+ \quad (n \neq z) \quad \text{(C)}$$

In the absence of metal ions the surface charge is controlled by the equilibria A and B and the isoelectric point (iep) is given by

$$\log [H^+]_{iep} = \frac{1}{2}(\log K_{a1(I)}^s + \log Ka_{2(I)})$$

$$pH_{iep} = \frac{1}{2}(pK_{a1(I)}^s + pK_{a2(I)}^s)$$

pH_{iep} is the pH of a charge reversal. No simple equation is available for case in which the equilibria A and B are superimposed with C. Measurements of electrophoretic mobility show that adsorption of metal ions may cause additional charge reversals [69]. James and Healy [69] observed three charge reversals when systems consisting of a colloid oxide and metal ions were observed in an extended pH range. The coordination chemistry model at its present stage certainly does not permit the calculation of electrokinetic parameters. The model does permit, however, the computation of the surface charge and the occurrence of charge reversal. Figures 10, 11 and 12 present the calculated effect of adsorbed Cd(II) on the surface charge of silica. This system was chosen to compare the results with experimental data from James and Healy on the system Co(II)-SiO_2. The pertinent parameters are given in Table X.

Figure 10 shows the surface charge of silica as a function of pH in the presence of varying amounts of Cd(II). Although not strictly comparable, the calculated curves are very similar to the experimental curves. Figure 11 shows the effect of the important system parameter A/V. Its dominating influence explains the sometimes contradictory reports on charge reversals studied at the same systems by different investigators. Again we note the similarity of calculated and observed curves. Figure 12 shows the distribution of the different surface species. It is seen that charge reversals in the system Cd(II) silica results from a competition between $SiOCd^+$ and (SiO^- + SiO^- Na^+).

CONCLUSIONS

It is both possible and useful to describe the adsorption of Lewis acids (H^+ and metal ions) and Lewis bases (charged and uncharged ligands) from aqueous solution by formation of binary and ternary surface complexes at the oxide-water interface. These surface complexes can be characterized by apparent stability constants, depending on temperature, pressure, ionic strength, and chemical nature of the adsorbed species and the adsorbing oxide, and on the amount of surface hydroxyl groups per unit weight of the adsorbing oxide. These stability constants form an arithmetic basis for reliable predictions. As in the field of solution chemistry, one might proceed by experimental evaluation of a broad set of pertinent equilibrium constants.

It must be remembered that the stability constants of surface complexes contain contributions from electrostatic interactions between the adsorbed species and the charged interface. In addition, lateral interactions of adsorbed species are of importance. The future development of the coordination chemistry model must tend to include these effects. Theoretical efforts to improve double layer models might be desirable, but it is felt that extended and careful experimental work is even more urgent.

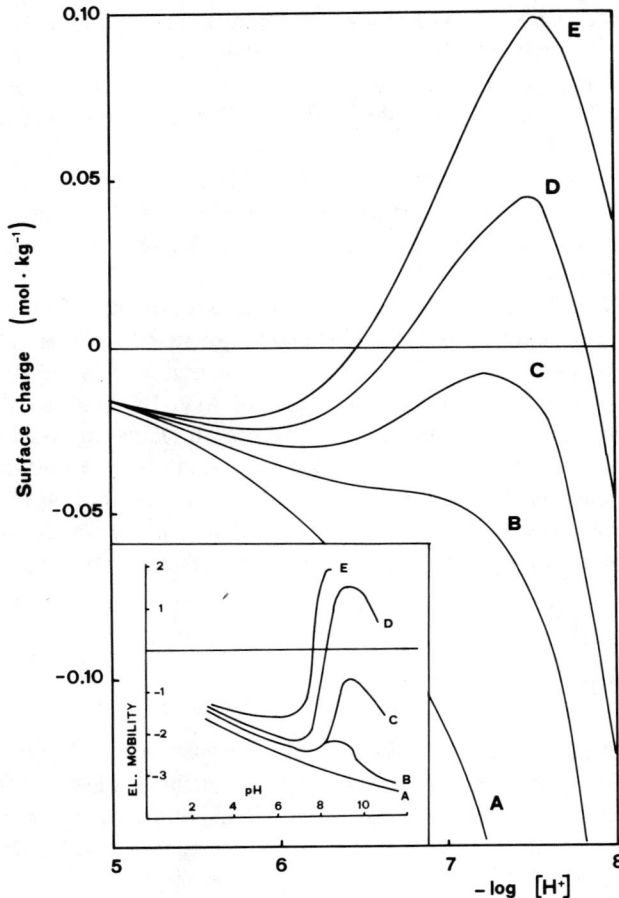

Figure 10. Simulated charge reversals in the system Cd(II)-silica (1 M NaClO$_4$, 25°) [35]. (A/V) = 0.05 kg/dm^3.
(A) Cd(II) = 0.
(B) Cd(II) = 0.01 mol/dm^3.
(C) Cd(II) = 0.015 mol/dm^3.
(D) Cd(II) = 0.02 mol/dm^3.
(E) Cd(II) = 0.025 mol/dm^3.
Inset: observed charge reversals in the system Co(II) − SiO$_2$ (I 10^{-3}, 25°). (A/V) = 10^{-4} kg/dm^3.
(A) Co(II) = 0.
(B) Co(II) = 10^{-5} mol/dm^3.
(C) Co(II) = 2 x 10^{-5} mol/dm^3.
(D) **Co(II)** = 10^{-4} mol/dm^3.
(E) Co(II) = 10^{-3} mol/dm^3 [69].

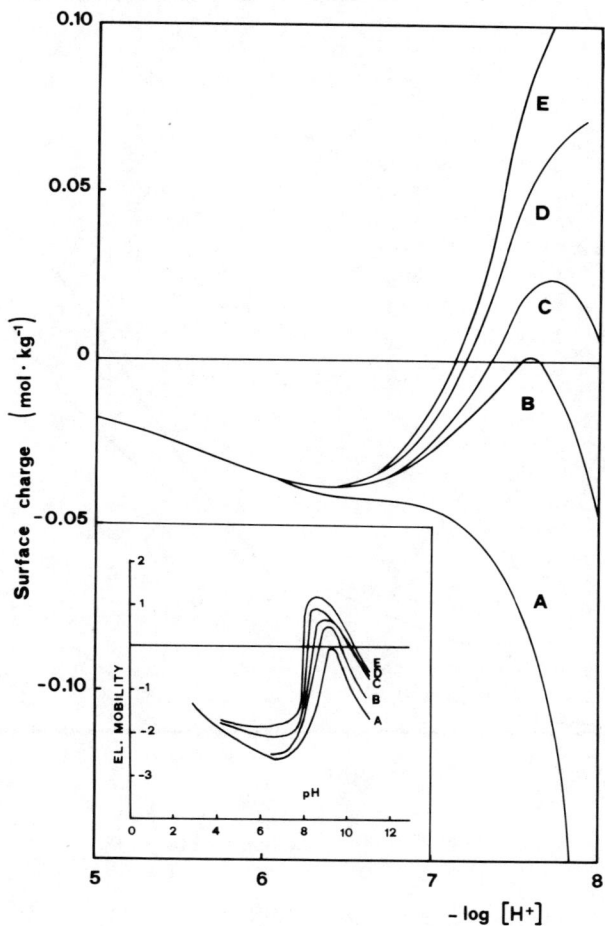

Figure 11. Simulated charge reversals in the system Cd(II)-silica (1 M NaClO$_4$, 25°) [35]. Cd(II) = 0.01 mol/dm^3.
(A) (A/V) = 0.05 kg/dm^3.
(B) (A/V) = 0.02 kg/dm^3.
(C) (A/V) = 0.015 kg/dm^3.
(D) (A/V) = 0.01 kg/dm^3.
(E) (A/V) = 0.005 kg/dm^3.
Inset: observed charge reversals in the system Co(II) - SiO$_2$ (I 10^{-3}, 25°). Co(II) = 10^{-4} mol/dm^3.
(A) (A/V) = 5 x 10^{-4} kg/dm^3.
(B) (A/V) = 2 x 10^{-4} kg/dm^3.
(C) (A/V) = 10^{-4} kg/dm^3.
(D) (A/V) = 5 x 10^{-5} kg/dm^3.
(E) (A/V) = 2 x 10^{-5} kg/dm^3 [69].

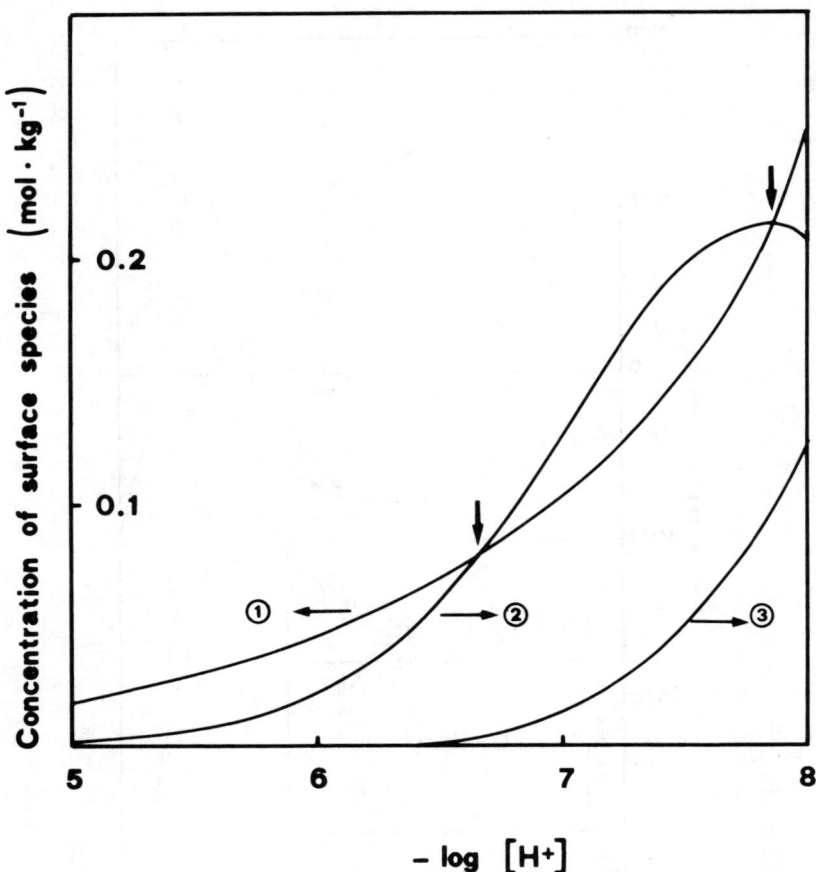

Figure 12. Calculated distribution of surface species in the system Cd(II)-silica (1 M NaClO$_4$, 25°) [35] as a function of log [H$^+$] : Cd(II) = 0.025 mol/dm^3, (A/V) = 0.05 kg/dm^3.
(1) SiO$^-$ + SiO$^-$C$^+$
(2) SiOCd$^+$.
(3) (SiO)$_2$Cd .
Charge reversals occur at the intersections of curves 1 and 2.

Table X. Effect of Adsorption on the Surface Charge

Model System	Cd(II) - silica - H_2O
Model Parameters	V/A: variable, Cd(II): variable t = 25°, p = 1 atm, I = 1 M (NaClO$_4$)
Pertinent Equation	Surface charge (mol/kg^1) = {SiOCd$^+$} - ({SiO$^-$} + {SiO$^-$Na$^+$})
Stability Constants of Surface Complexes:	$\log {}^*K^s_{1(I)}$ = -6.09, $\log {}^*\beta^s_{2(I)}$ = -14.20
Properties of Silica [35]	{SiOH}$_{tot}$ = 1.45 mol/kg^1
	{SiOH} = 1.45 + $\left(\dfrac{\log [H^+]}{10.7}\right)$ 5.26 (-log [H$^+$] < 7.5)
	{SiOH} = 1.45 + $\dfrac{\log [H^+] + 5.9}{9.9}$ (-log [H$^+$] > 7.5)
	$\log Ka^2_{(I)}$ = -5.7 + 8.9 (1.45 - {SiOH}) - 10.1 (1.45 - {SiOH})$^{1/2}$

ACKNOWLEDGMENTS

The author wishes to express his gratitude to Drs. Werner Stumm, Herbert Hohl and Alain Bourg for stimulating discussions. The work was supported financially by the Swiss National Foundation (Project No. 2.901-0.77).

REFERENCES

1. Stumm, W., C. P. Huang and S. R. Jenkins. "Specific Chemical· Interaction Affecting the Stability of Dispersed Systems," *Croat. Chem. Acta* 42: 223-1245 (1970).
2. Huang, C. P., and W. Stumm. "Specific Adsorption of Cations on Hydrous $\gamma\text{-}Al_2O_3$," *J. Colloid Interface Sci.* 43: 409-420 (1973).
3. Schindler, P. W., E. Wälti and B. Fürst. "The Role of Surface-Hydroxyl Groups in the Surface Chemistry of Metal Oxides," *Chimia* 30: 107-109 (1976).
4. Schindler, P. W., B. Fürst, B. Dick and P. U. Wolf. "Ligand Properties of Surface Silanol Groups. I. Surface Complex Formation with Fe^{3+}, Cu^{2+}, Cd^{2+}, and Pb^{2+}," *J. Colloid Interface Sci.* 55: 469-475 (1976).
5. Stumm, W., H. Hohl and F. Dalang. "Interaction of Metal Ions with Hydrous Oxide Surfaces," *Croat. Chem. Acta* 48: 491-504 (1976).
6. Hingston, F. J., A. M. Posner and J. P. Quirk. "Adsorption of Selenite by Goethite," in *Adsorption From Aqueous Solution*, R. F. Gould, Ed., **Advances in Chemistry Series, No. 79, Washington, DC** (1968), pp. 82-90.
7. Hingston, F. J., A. M. Posner and J. P. Quirk. "Anion Adsorption by Goethite and Gibbsite. I. The Role of the Proton in Determining Adsorption Envelopes," *J. Soil Sci.* 23: 177-192 (1972).
8. Raghavan, S., and D. W. Fuerstenau. "The Adsorption of Aqueous Octylhydroxamate on Ferric Oxide," *J. Colloid Interface Sci.* 50: 319-330 (1975).
9. Yates, D. E., and T. W. Healy. "Mechanism of Anion Adsorption at the Ferric and Chromic Oxide/Water Interfaces," *J. Colloid Interface Sci.* 52: 222-228 (1975).
10. Parfitt, R. L., R. J. Atkinson and R. St. C. Smart. "The Mechanism of Phosphate Fixation by Iron Oxides," *Soil Sci. Soc. Am. Proc.* 39: 837-841 (1975).
11. Bourg, A. C. M., and P. W. Schindler. "Ternary Surface Complexes I. Complex Formation in the System Silica-Cu(II)-Ethylenediamine," *Chimia* 32: 166-168 (1978).
12. Davis, J. A., and J. O. Leckie. "Effect of Adsorbed Complexing Ligands on Trace Metal Uptake by Hydrous Oxides," *Environ. Sci. Technol.* 2: 1309-1315 (1978).
13. Davydov, V. Ya., A. V. Kiselev and L. T. Zhuravlev. "Surface and Bulk Hydroxyl Groups of Silica by Infrared Spectra and D_2O-Exchange," *Trans. Faraday Soc.* 60: 2254-2284 (1964).
14. Boehm, H. P. "Acidic and Basic Properties of Hydroxylated Metal Oxide Surfaces," *Disc. Faraday Soc.* 52: 264-275 (1971).

15. Morimoto, T., K. Shiomi and H. Tanaka. "Heat of Immersion of Al_2O_3 in Water," *Bull. Soc. Chem. Japan* 37: 392-395 (1964).
16. McCafferty, E., and A. C. Zettlemoyer. "Adsorption of Water Vapour on α-Fe_2O_3," *Disc. Faraday Soc.* 52: 239-254 (1971).
17. Nagao, M. "Physisorption of Water on Zinc Oxide Surface," *J. Phys. Chem.* 75: 3822-3828 (1971).
18. Zettlemoyer, A. C., and E. McCafferty. "Water on Oxide Surfaces," *Croat. Chem. Acta* 45: 173-187 (1973).
19. Peri, J. B. "Infrared and Gravimetric Study of the Surface Hydration of γ-Alumina," *J. Phys. Chem.* 69: 211-219 (1965).
20. Peri, J. B. "A Model for the Surface of γ-Alumina," *J. Phys. Chem.* 69: 220-230 (1965).
21. Schindler, P. W., and H. R. Kamber. "Die Acidität von Silanolgruppen," *Helv. Chim. Acta* 51: 1781-1786 (1968).
22. Schindler, P. W., and H. Gamsjäger. "Acid-Base-Reactions of the TiO_2 (Anatase)-Water Interface and the Point of Zero Charge of TiO_2 Suspensions," *Kolloid-Z. u. Z. Polymere* 250: 759-763 (1972).
23. Yates, D. E. "The Structure of the Oxide Aqueous Electrolyte Interface," Ph.D. Thesis, University of Melbourne, Melbourne, Australia (1975).
24. Davis, J. A., R. O. James and J. O. Leckie. "Surface Ionisation and Complexation at the Oxide-Water Interface Computation of Electrical Double Layer Properties in Simple Electrolytes," *J. Colloid Interface Sci.* 63: 480-499 (1978).
25. Lagerström, G. "Equilibrium Studies of Polyanions, III. Silicate Ions in $NaClO_4$ Medium," *Acta Chem. Scand.* 13: 722-736 (1959).
26. Weber, W. J., Jr., and W. Stumm. "Formation of a Silicato-Iron(III) Complex in Dilute Aqueous Solution," *J. Inorg. Nucl. Chem.* 27: 237-239 (1965).
27. Olson, L. L., and C. R. O'Melia. "The Interactions of Fe(III) with $Si(OH)_4^4$," *J. Inorg. Nucl. Chem.* 35: 1977-1985 (1973).
28. Santschi, P. H., and P. W. Schindler. "Complex Formation in the Ternary Systems Ca^{II} - H_4SiO_4 - H_2O and Mg^{II} - H_4SiO_4 - H_2O," *J. Chem. Soc. Dalton* 2: 181-184 (1974).
29. Riesen, W., H. Gamsjäger and P. W. Schindler. "Complex Formation in the Ternary System Mg(II) - CO_2 - H_2O," *Geochim. Cosmochim. Acta* 41: 1193-1200 (1977).
30. Fahrni, H. P., H. Rohrer and P. W. Schindler. "Interactions between H_4SiO_4 and Fluoride," (to be published).
31. Schwarzenbach, G. "Koordinationsselektivität und Thermodynamik der Komplexbildung in Lösung," *Chimia* 27: 1-15 (1973).
32. Chan, D., J. W. Perram, L. R. White and T. W. Healy. "Regulation of Surface Potential at Amphoteric Surfaces During Particle-Particle Interaction," *J. Chem. Soc. Faraday Trans.* 1: 1046-1057 (1975).
33. Rossotti, F. J. C., and H. Rossotti. *The Determination of Stability Constants* (New York: McGraw-Hill Book Co., 1961), p. 23.
34. Sigg, L., and P. W. Schindler. Unpublished results.
35. Fürst, B. "Das koordinationschemische Adsorptionsmodell: Oberflächenkomplexbildung von Cu(II), Cd(II) und Pb(II) an SiO_2 (Aerosil) und TiO_2 (Rutil)," Ph.D. Thesis, University of Bern (1976).

36. Kurbatov, M. H., G. B. Wood and J. D. Kurbatov. "Isothermal Adsorption of Cobalt from Dilute Solutions," *J. Phys. Colloid Chem.* 55: 1170-1182 (1951).
37. Morgan, J. J., and W. Stumm. "Colloid-Chemical Properties of Manganese Dioxide," *J. Colloid Sci.* 19: 347-359 (1964).
38. Loganathan, P., and R. G. Burau. "Sorption of Heavy Metal Ions by A Hydrous Manganese Oxide," *Geochim. Cosmochim. Acta* 37: 1277-1293 (1973).
39. Gadde, R. R., and H. A. Laitinen. "Studies of Heavy Metal Adsorption by Hydrous Iron and Manganese Oxides," *Anal. Chem.* 46: 2022-2026 (1974).
40. Gadde, R. R., and H. A. Laitinen. "Study of the Sorption of Lead by Hydrous Ferric Oxide," *Environ. Lett.* 5: 223-235 (1973).
41. Hohl, H., and W. Stumm. "Interaction of Pb^{2+} with Hydrous $\gamma\text{-}Al_2O_3$," *J. Colloid Interface Sci.* 55: 281-288 (1976).
42. Ahrland, S., I. Greuthe and B. Norén. "The Ion Exchange Properties of Silica Gel I. The Sorption of Na^+, Ca^{2+}, Ba^{2+}, UO_2^{2+}, Gd^{3+}, $Zr(IV)$ + Nb, U(IV) and Pu(IV)," *Acta Chem. Scand.* 14: 1059-1076 (1960).
43. Vydra, F., and J. Galba. "Sorption von Metallkomplexen an Silikagel V. Sorption von Hydrolyseprodukten des Co^{2+}, Mn^{2+}, Cu^{2+}, Ni^{2+}, Zn^{2+} an Silicagel," *Coll. Czech. Chem.* 34: 3471-3478 (1969).
44. Kinniburgh, D. G., K. Sridhar and M. L. Jackson. "Specific Adsorption of Zinc and Cadmium by Iron and Aluminum Hydrous Oxides," *Proc. Fifteenth Hanford Life Sciences Symp. on Biological Implications of Metals in the Environment,* USERDA Natl. Techn. Information Service, Springfield, VA (1977), pp. 231-239.
45. Dugger, D. L., J. H. Stanton, B. N. Irby, B. L. McConell, W. W. Cumings and R. W. Maatman. "The Exchange of Twenty Metal Ions with the Weakly Acidic Silanol Group of Silica Gel," *J. Phys. Chem.* 68: 757-780 (1964).
46. Hathaway, B. J., and C. E. Lewis. "Electronic Properties of Transition-metal Complex Ions Adsorbed on Silica Gel. Part I. Nickel(II) Complexes," *J. Chem. Soc. (A)* 1176-1182 (1969).
47. Hathaway, B. J., and C. E. Lewis. "Electronic Properties of Transition-metal Complex Ions Adsorbed on Silica Gel. Part II. Cobalt(II) and Cobalt(III)," *J. Chem. Soc. (A)* 1183-1188 (1969).
48. Hathaway, B. J., and C. E. Lewis. "Electronic Properties of Transition-metal Complex Ions Adsorbed on Silica Gel. Part III. Copper(II)," *J. Chem. Soc. (A)* 2295-2299 (1969).
49. Olson, L. L., and C. R. O'Melia. Personal communication.
50. Tewari, P. H., and W. Lee. "Adsorption of Co(II) at the Oxide-Water Interface," *J. Colloid Interface Sci.* 52: 77-88 (1975).
51. Briggs, D., and Y. M. Bosworth. "Comment on Adsorption of Co(II) at the Oxide-Water Interface," *J. Colloid Interface Sci.* 59: 194 (1977).
52. Gupta, S. K. "Phosphate-Removal in Systems H_3PO_4 - γ-FeOOH and H_3PO_4 - $FeCl_3$ and Characteristics of Sludge-Phosphate," Ph.D. Thesis, University of Bern (1976).
53. Gupta, S. K., and P. W. Schindler. "Adsorption of Phosphate at the γ-FeOOH Water Interface," (In preparation).
54. Sigel, H. "Stabilität, Struktur und Reaktivität von ternären Cu^{2+} - Komplexen," *Angew. Chem.* 87: 391-400 (1975).

55. Bjerrum, J. *Metal Amine Formation in Aqueous Solution* (Copenhagen: P. Haase and Son, 1941).
56. Bourg, A. C. M., and P. W. Schindler. (To be published).
57. Farrah, H., and W. F. Pickering. "The Sorption of Copper Species by Clays. I. Kaolinite," *Aust. J. Chem.* 29: 1167-1176 (1976).
58. Pleysier, J., and A. Cremers. "Stability of Silver-Thiourea Complexes in Montmorillonite Clay," *J. Chem. Soc. Faraday I* 256-264 (1975).
59. Stryker, L. J., and E. Matijević. "Adsorption of Hydrolyzed Hafnium Ions on Glass," in *Adsorption From Aqueous Solution*, R. F. Gould, Ed., Advances in Chemistry Series No. 79, Washington, DC (1968), pp. 44-61.
60. James, R. O., and T. W. Healy. "Adsorption of Hydrolyzable Metal Ions at the Oxide-Water Interface I. Co(II) Adsorption on SiO_2 and TiO_2 as Model Systems," *J. Colloid Interface Sci.* 40: 42-52 (1972).
61. O'Connor, T. P., and D. R. Kester. "Adsorption of Copper and Cobalt from Fresh and Marine Systems," *Geochim. Cosmochim. Acta* 39: 1531-1543 (1975).
62. MacNaughton, M. G., and R. O. James. "Adsorption of Aqueous Mercury(II) complexes at the Oxide/Water Interface," *J. Colloid Interface Sci.* 47: 431-440 (1974).
63. Baes, C. F., Jr., and R. Mesmer. *The Hydrolysis of Cations* (New York: John Wiley & Sons, Inc., 1976).
64. Schindler, P. W. "Removal of Trace Metals from the Oceans: A Zero Order Model," *Thal. Jugoslav.* 11: 101-111 (1975).
65. Simon, J., W. Schulze and M. Völtz. "Sorptionseffekte an Metall(III)-Hydroxide Fällungen II. Mitfällung von Cu^{2+}-Ionen mit Aluminiumhydroxide in Gogenwart von Ammoniak," *Z. Anal. Chem.* 257: 184-187 (1971).
66. Farrah, H., and W. F. Pickering. "The Sorption of Copper Species by Clays. II. Illite and Montmorillonite," *Aust. J. Chem.* 29: 1177-1184 (1976).
67. Farrah, H., and W. F. Pickering. "The Sorption of Zinc Species by Clay Minerals," *Aust. J. Chem.* 29: 1649-1656 (1976).
68. Kinniburgh, D. G., and M. L. Jackson. "Adsorption of Mercury(II) by Iron Hydrous Oxide Gel," *Soil Sci. Soc. Am. J.* 42: 45-47 (1978).
69. James, R. O., and T. W. Healy. "Adsorption of Hydrolyzable Metal Ions at the Oxide-Water Interface II. Charge Reversal of SiO_2 and TiO_2 Colloids by Adsorbed Co(II), La(III) and Th(IV) as Model Systems," *J. Colloid Interface Sci.* 40: 53-64 (1972).

CHAPTER 2

A REVIEW OF ANION ADSORPTION

Frank J. Hingston
Division of Land Resources Management
Commonwealth Scientific and Industrial Research Organization
Perth, Australia

INTRODUCTION

In the last decade, experimental and theoretical studies of adsorption of anions at aqueous solution–mineral interfaces have been intensified in an attempt to understand the diverse natural phenomena in which adsorption plays a key role. Soil scientists interested in assessing the availability of biologically important elements such as P, B, Mo, Se and S, which occur in anionic form, have been responsible for many applied investigations of anion adsorption [1]. Mechanisms of adsorption are also relevant in pollution studies [2,3], geochemistry of waters, lacustrine and sea floor sediments [4], flotation [5,6] and water treatment.

Studies utilizing natural materials as experimental media are subject to many problems because the solid phase is a heterogeneous mixture of mineral and organic adsorbent surfaces and the aqueous phase contains a similarly complex mixture of soluble organic and inorganic species. Adsorption is readily estimated in such mixtures; however, because of difficulties of separating the reactions, it is difficult to interpret mechanisms from the data.

More success in interpreting mechanisms of adsorption has been achieved using as experimental systems suspensions of crystalline oxides, such as geothite and gibbsite, in aqueous electrolyte solutions containing adsorbate. In these systems, adsorbent characteristics, surface area and charge can be estimated, and the composition of the aqueous phase can easily be controlled. Ionic interactions in the interfacial region between aqueous solution and oxide lattices can be interpreted from the detailed mathematical treatments available from electrochemical theory. It must be remembered, however, that these experimental systems are only convenient models. Even where the same oxides occur as components of natural substrates, their behavior will probably be modified by interactions with other components.

This chapter examines some of the basic concepts and hypotheses relating to adsorption of anions and applies these hypotheses to interpret surface reactions in natural systems. Although the emphasis here is on specific adsorption of anions, many aspects of the approach may be equally applicable to nonspecific adsorption and adsorption of cations.

CONCEPTS AND TERMS

Understanding the nature of the interfacial region is basic to modeling reactions of ions at mineral surfaces. Early ideas of the structure, electrical potential and charge relationships arose from studying the polarizable mercury surface [7]. The Gouy-Chapman theory of distribution of ions in a diffuse double layer and Stern's treatment of ion size and specific adsorption [8] form the basis of adsorption theory for ions. In Stern's model of the mercury surface, the number of sites for adsorption is limited by surface area and ion size. The standard molal free energy of adsorption is considered to consist of electrostatic and chemical terms:

$$\overline{\Delta G}^o_{ads} = \overline{\Delta G}^o_{chem} + zF\psi_s \qquad (1)$$

The chemical component of the free energy is constant and accounts for specific adsorption of some ionic species, regardless of the sign of the surface charge. Inner and outer Helmholtz planes were introduced to double layer theory to explain the variation in charge and capacity of the interfacial region in relation to changes in surface potential [7]. The inner Helmholtz plane is defined as the locus of electrical centers of adsorbed ions bound by covalent or van der Waals forces, or both. The outer Helmholtz plane is the locus of centers of charge of hydrated or solvated ions in contact with the mercury surface. Since hydration water is more readily removed from anions than cations, many anions tend to be adsorbed at the inner Helmholtz plane. The term "specific adsorption" is used for all adsorption than cannot be accounted for solely by electrostatic forces [7,10].

The surface charge on "nonpolarizable" mineral surfaces such as silica, iron oxides and aluminum oxides is due to unequal adsorption of potential-determining ions (H_3O^+ and OH^-) from aqueous solution [11,12]. Potential-determining ions are components of both the solvent phase and the solid, and occupy the inner Helmholtz plane with respect to the metal ions of the oxide lattice. Ions that are not constituents of the solid lattice but are adsorbed into the inner Helmholtz plane have been termed "specifically adsorbed" or "ligand exchanged" ions [13,14].

Recently, a number of workers [12,15,17] have drawn attention to the higher surface charge density and differential capacity of oxide surfaces compared with classical Hg and AgI surfaces. Stumm et al. [18] suggest that the difference is due primarily to a strongly structured, extensively hydrogen-bonded and chemisorbed water layer adjacent to oxide surfaces. Blok and de Bruyn [17], Atkinson et al. [12] and Bowden et al. [19] proposed that potential-determining ions protrude from the surface of oxides so that counterions can lie almost in the same plane as charged sites, thus minimizing electrostatic repulsion between like charges. This allows higher-charge densities to be accommodated on oxides rather than on Hg surfaces.

Yates et al. [20] reproduced the properties of oxide-aqueous electrolyte solution interfaces in a discrete acid-base site model. A form of ion-pairing through localized hydrolysis [21] is suggested as a mechanism in charging of oxide surfaces. However, these authors were not able to disprove the hypothesis that gel or porous layers at oxide surfaces may be responsible for their high surface charge density and differential capacity [22].

Consideration has been given in many studies and reviews [23] to the effects of hydration of adsorbate species and interactions with oriented water dipoles at discrete charged sites on adsorbent surfaces. James and Healy [24] stress the importance of solvation energy in their model of adsorption of hydrolyzable metal ions, in which they divide the free energy of adsorption (ΔG_{ads}) into three terms: coulombic (ΔG_{coul}), solvation (ΔG_{solv}) and chemical (ΔG_{chem}):

$$\Delta G_{ads} = \Delta G_{coul} + \Delta G_{solv} + \Delta G_{chem} \qquad (2)$$

An abrupt change in the balance between coulombic and chemical energy on the one hand, and solvation energy on the other, was put forward to explain the relationship between pH and adsorption of hydrolyzable metal ions. Although anions are considered to be less hydrated than cations, the variation of adsorption of anionic species with pH may relate to analogous changes in the solvation component of the free energy of adsorption for anions.

Težak [25] has described a model of the interfacial region ("methorical region") at solid-liquid surfaces with features approaching "real world" behavior of oxides. In this model the interfacial region commences at the ordered crystal lattice and has sublayers consisting of a crystalline condensed phase with a distorted lattice structure; potential-determining complexes or complexoids; counterions; co-ions; and is terminated at the bulk solution. The electrokinetic plane of shear is located in the layer of counterions close to the surface of the sublayer containing potential determining complexes. One feature of the disordered crystalline phase model is that it may provide an explanation for slow reactions with adsorbed anions and "aging" to produce a more ordered crystalline surface. A second important feature is the layer of potential-determining complexes. Težak stresses the formation of Bjerrum ion pairs between potential-determining ions and counterions in the interfacial region is a mechanism for coagulation of colloids, and specific adsorption is a mechanism for peptization. On the basis of Bjerrum's theory, association of oppositely charged ions into ion pairs should not be significant until the distance of closest approach is less than $3.57|z_1 z_2|$ Å where z_1, and z_2 are the valences of the two ions. The degree of association increases with decreasing distance of closest approach [26].

Relationship Between pH and Anion Adsorption

Adsorption of anions on geothite is described by isotherms approximating a Langmuir shape, which appear to reach maximum values with increasing concentration of anions in solution. Adsorption isotherms for gibbsite appear to conform to a Langmuir shape at low concentrations of anions in solution but commonly do not reach maxima; adsorption increases appoximately linearly with adsorbate concentration (Figure 1). In studies of anion adsorption at a range of pH values, a relationship was obtained between the "apparent" Langmuir maxima and pH. This was termed an adsorption envelope [14] and attempts were made to relate the characteristics of the envelope to properties of the adsorbent and the adsorbate [27]. Apparent maxima in the envelope were found at the pK_a for anions with monoprotic conjugate acids and breaks of slope were found at pK_a values for anions of polyprotic conjugate acids. A good correlation was found between points of inflection in adsorption envelopes and pK_a values for conjugate acids (Figure 2).

The correlation between the tendency for anions to react at oxide surfaces through ligand exchange, the most common type of specific adsorption, and the pK_a of their conjugate acids is analogous with the correlation between the logarithms of stability constants for complex formation with metal ions and pK_a values for acids corresponding with various ligands [28]. Indeed, chemical bonding for anions specifically adsorbed on oxides would be expected to be similar to the bonding in complexes and crystalline compounds. The char-

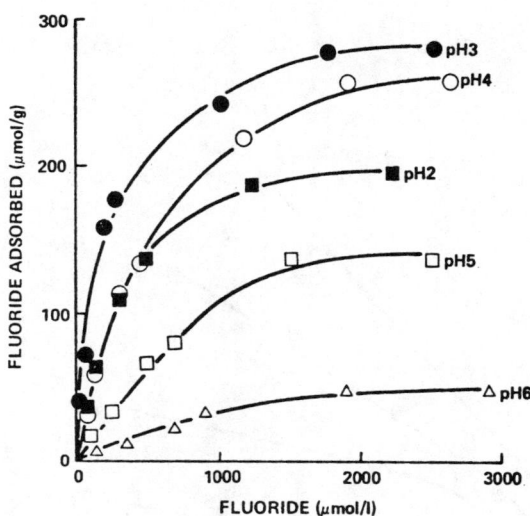

Figure 1a. Isotherms for adsorption of fluoride on goethite (surface area, 28 m²/g). Temperature, 23°C, pH values as shown, supporting electrolyte 0.1 M NaCl.

acteristics used here to distinguish between adsorption and compound formation are that adsorption occurs only at the interfaces. It generally involves very low concentrations of adsorbate, the composition of the surface phase is nonstoichiometric; and surface charge is one of the factors determining the amount of anion adsorption. The behavior of anions in natural substrates and in experimental oxide-solution systems has been explained by reference to the solubility products of known crystalline compounds. However, even in simple model experiments using oxides and low concentrations of anions, solubility product relationships for pure compounds are not reproduced.

In attempts to explain the variation in the amount of adsorption of anions with changes in pH, Hingston et al. [27] postulated that both the anion and its conjugate acid could be adsorbed from solution and that the acid was dissociated to form coordinate complexes at oxide surfaces. It was later realized that the hypothesis used to explain the maximum adsorption given by adsorption envelopes [27] was inadequate because in the range of experimental solution concentrations used experimentally there is clearly no maximum for anion adsorption of gibbsite. The existence of a maximum is difficult to prove, even for goethite. The hypothesis was proposed to explain the observation that anions such as silicate were adsorbed at a low pH, where essentially only the undissociated conjugate acids of these anions are present in solution and adsorption continued into the pH range where dissociation in

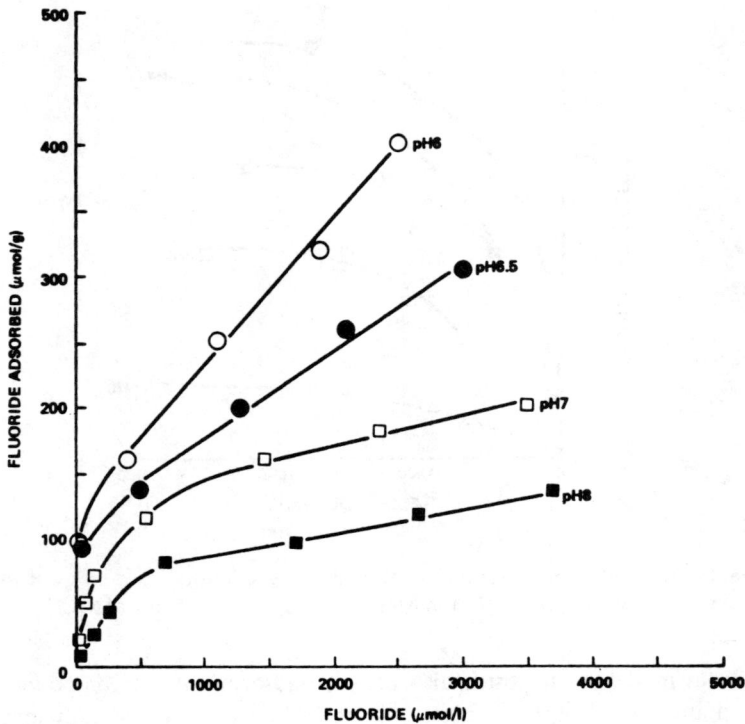

Figure 1b. Isotherms for adsorption of fluoride on gibbsite (surface area, 47 m^2/g). Temperature, 23°C, pH values as shown, supporting electrolyte 0.1 M NaCl.

solution predominated. The lowering of pH of oxide suspensions when silicic acid was adsorbed suggested that either the acid was dissociated at the oxide surface or adsorption caused a change in the charging characteristics of the oxide surface.

If both conjugate acid and anions are adsorbed, the Gibbs-Duhem relation states that the sum of the chemical potentials for acid and anions in solution is equal to the chemical potential of the adsorbate in the interface phase. This approach was used by Cabrera et al. [29] to explain adsorption envelopes against pH at constant concentrations of adsorbate (i.e., conjugate acids plus anions) in solution, which approximately reproduce adsorption envelope shapes. However, since adsorption is dependent on the concentration of adsorbate in solution, maxima are not explained.

A model based on adsorption of only the anionic form of the adsorbate [30] is consistent with an increase in adsorption as the pH is increased until the dissociation is essentially complete. The decrease in adsorption as the pH

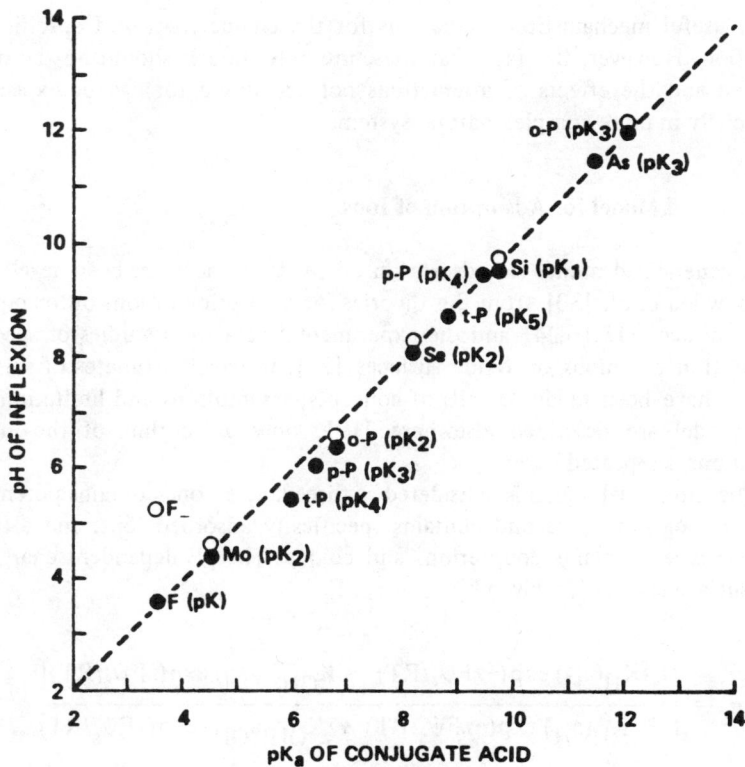

Figure 2. Relationship between pK_a and pH at the change in slope of adsorption envelopes. Adsorbents: ●, goethite; ○, gibbsite. Adsorbates: F, *fluoride;* Mo, *molybdate*; t-P, *tripolyphosphate*; p-P, *pyrophosphate*; o-P, *orthophosphate*; Se, *selenite*; Si, *silicate*; As, *arsenate*.

is increased past the pK_a is attributed to the increase in negative charge of the surface. The only maximum allowed for in this treatment is if all the possible sites on the surface become filled. The number of possible sites at the surface is constant (not dependent on pH), but normally only a relatively small proportion of these sites is occupied when adsorption occurs in the range of solution concentrations used in adsorption studies.

Even for goethite it is uncertain whether the apparent maxima for isotherms at constant pH are real or whether there is a slow increase with increasing solution concentration. If the latter is true, the fitting of Langmuir-type equations is inappropriate. Thus, models that (1) do not require consideration of maxima at constant pH, (2) incorporate adsorption of particular anionic species and (3) estimate the effects of surface charge [19,30,33] provide the most useful mechanistic explanations for the characteristics of specific adsorption. However, the fact that these are only models should not be overlooked and the effects of interactions not accounted for can be expected, especially in more complex natural systems.

Generalized Model for Adsorption of Ions

A generalized model for adsorption on oxide surfaces has been developed by Bowden et al. [30] from the theories for adsorption of ions onto amphoteric surfaces [15,16,24] and the experimental data from studies of specific adsorption of anions on oxide surfaces [27], in which estimates of surface charge have been made. Details of concepts, assumptions and limitations of the model are described elsewhere [19]; only an outline of the major equations is repeated here.

The interfacial region is considered as three layers: one contains potential-determining ions; a second contains specifically adsorbed ions; and a third diffuse layer contains counterions and coions. The pH-dependent charge in the surface layer, σ_s, is given by

$$\sigma_s = \frac{N_s[K_H(a_H)\exp(-zF\psi_s/RT) - K_{OH}(a_{OH})\exp(zF\psi_s/RT)]}{1 + K_H(a_H)\exp(-zF\psi_s/RT) + K_{OH}(a_{OH})\exp(zF\psi_s/RT)} \quad (3)$$

where σ_s = the surface charge (mol/cm^2),
N_s = the maximum site density for the surface (mol/cm^2),
ψ_s = the potential at the surface (mV),
K_H and K_{OH} = the binding constants for H_3O^+ and OH^- at activities, (a_H) and (a_{OH}), respectively, and
F, R and T have their usual physical chemical meaning.

The charge, σ_i, in the second layer containing specifically adsorbed species is given by a similar expression:

$$\sigma_i = \frac{N_t \sum K_i C_i \exp(-z_i F\psi_d/RT)}{1 + \sum K_i C_i \exp(-z_i F\psi_d/RT)} \quad (4)$$

where σ_i = the charge in the second layer (mol/cm^2),
N_t = the maximum possible sites in the second layer (mol/cm^2),
ψ_d = the potential in the diffuse layer (mV),

A REVIEW OF ANION ADSORPTION

z_i = the valence of the ionic species i,
K_i = the binding constant for species i, and
C_i = the mole fraction of species i in solution, approximately $0.018\, C_{io}$, where C_{io} is the molar concentration of species i.

The charge, σ_d, developed in the diffuse layer as a result of adsorption of potential determining ions and specifically adsorbed anions is given by the Gouy-Chapman theory as

$$\sigma_d = -1.22 \times 10^{-10}\, C^{\frac{1}{2}} \sinh(0.0195|z|\psi_d) \qquad (5)$$

where σ_d = the diffuse layer charge (mol/cm^2),
 C = the total ionic concentration (mol/l), and
 $|z|$ = the modulus of the valance of the supporting electrolyte.

Gauss's law is used to relate the change in potential from the surface to the diffuse layer:

$$(\psi_s - \psi_d) = (4\pi d/\epsilon_d)\sigma_s = \sigma_s/G \qquad (6)$$

where $G = \epsilon_d/4\pi d$ is the electrical capacitance of the inner and outer Helmholtz layers and ϵ_d is the permittivity of the region.

Combining Equations 3, 4, 5 and 6 with the condition for electrical neutrality,

$$\sigma_s + \sigma_i + \sigma_d = 0 \qquad (7)$$

Bowden et al. [30] were able to evaluate $\sigma_d, \sigma_i, \sigma_s$ and ψ_s values and modeled charging and adsorption data.

THE MODEL OF BOWDEN, POSNER AND QUIRK

Effect of pK$_a$ of Conjugate Acids

Using the model of Bowden et al. [30], constant values for K_i and N_t are assigned for the whole range of pH values. This contrasts with simple models of adsorption data where the coefficients were evaluated by fitting Langmuir equations without regard to charge or the species in the equilibrium solution.

In these simple analyses, the binding coefficients and adsorption maxima obtained varied with pH [34]. Variations in the coefficients can be attributed to the exponential term, $\exp(-z_i F\psi_d/RT)$, in Equation 4 and the variation in conjugate acid and anionic species with pH. At constant concentration of adsorbate (conjugate acid + anionic species), the concentration of the anionic form increases from very low levels about two pH units below the pK_a until essentially only the anionic form is present about two pH units above it. The term, $\exp(-z_i F\psi_d/RT)$, decreases continuously as pH increases. Since only the anionic species are considered to be adsorbed, the model accounts for an increase in adsorption with pH from a value several units lower than the pK_a because of the corresponding increase in proportion of anionic species with dissociation of the conjugate acid. Adsorption then decreases with increasing pH because of the decrease in the exponential term, $\exp(-z_i F\psi_d/RT)$. Inflections at pK_a values for polyprotic acids are explained by the changes in adsorbate speciation and diffuse layer potential (ψ_d) with pH.

Bowden et al. [30] are able to achieve reasonable fits to adsorption isotherms and the relationships between pH and adsorption for a range of anions with goethite as the absorbent. They report that adjustments and modifications to the model are necessary to fit individual situations. To fit the data for F^- adsorption, for example, it was necessary to adjust the model to allow the small F^- ion to enter the same plane as coordinated OH^- and H_2O. Similarly, the poor fit of silicate adsorption versus pH curves at higher pH was attributed to dimeric and polymeric forms of silica in solution under alkaline conditions.

Adsorption of Charged Plus Uncharged Species

The wide range of pH over which adsorption of anions such as silicate occurs suggests that consideration should be given to the possibility that both the anion and its conjugate acid are adsorbed. Bowden et al. [30] considered adsorption of anions and uncharged species, but the neutral species were not adsorbed to a significant extent; therefore, only adsorption of anions is allowed for in their model. However, Rendall and Smith [32] treat adsorption of both neutral and positive species of α-picoline (4-methylpyridine) at silica-water interfaces. Their data showed a maximum in the adsorption versus pH curve at $pH \approx pK_b$ ($pK_b = 6.02$) and the adsorption process should be comparable with hydrolyzable anions.

Langmuir-Stern isotherms were fitted:

$$n_{s^+} = N_s X_+ \exp[-(ze_0\psi_\beta + \Phi_+)/kT] \tag{8}$$

$$n_{sn} = N_s X_n \exp(-\Phi_n/kT) \qquad (9)$$

where n_{s+} and n_{sn} = the amounts of positively charged and neutral species adsorbed,
N_s = the density of adsorption sites,
X_+ and X_n = the mole fractions of positive and neutral species,
ψ_β = the potential at the plane of adsorption,
Φ_+ and Φ_n = the specific adsorption potentials of positive and neutral species, and
z, e_o, K and T = the valence of the adsorbed ion, electronic charge, Boltzmann constant and temperature, respectively.

The total adsorption at low coverage is given by

$$n_s = A (1 + a_H K)^{-1} [a_H K \exp(\psi_r) \exp(\Phi_{r+}) + \exp(\Phi_{rn})] \qquad (10)$$

where A = $N_s \, c/55.51$ with the total concentration c in mol/dm^3,
a_H = the hydrogen ion activity,
K = the dissociation constant for the base,
ψ_r = $-e_o \psi_\beta/kT$, and
Φ_{rn} = $-\Phi_n/kT$ and $\Phi_{r+} = -\Phi_+/kT$.

Equation 10 gives the relationship between pH and adsorption at a particular concentration of adsorbate in the aqueous phase. The adsorption passes through a maximum (i.e., $dn_s/d(pH) = 0$) when

$$\exp(\psi_r + \Delta\Phi_r) = [1 - p(1 + a_H K)]^{-1} \qquad (11)$$

where $\Delta\Phi_r = \Phi_{r+} - \Phi_{rn}$ and $p = (e_o/2.303 \, kT) [d\psi_\beta/d(pH)]$.

Relationship of Various Isotherm Types

According to the model [30], the equation of the isotherm describing adsorption of anions at charged oxide surfaces is

$$a_{is} = a_i K_i \exp(-z_i F\psi_s/RT) \qquad (12)$$

where a_{is} = the activity of an anionic species "i" at the surface,

a_i = the activity of the species in solution, and
K_i = the binding constant for that species and surface.

Binding constants (K_i) for various adsorbate species, surface charge (σ_s), diffuse layer potential (ψ_d, which varies as more anions are adsorbed) and the number of surface sites (N_t) determine isotherm shapes. Therefore, approximate fits to equations such as Langmuir or Freundlich can have little meaning in relation to reaction mechanism. Bowden et al. [35], commenting on the practice of dividing isotherms into arbitrary regions, maintain that their model can be made to account for all isotherm shapes, including apparent composites fitted by other workers [36].

The model [19] is based on anions being adsorbed to fill a maximum number of sites on the oxide surface; however, these are rarely all taken up. The effect of surface charge and the charge added through adsorption make it increasingly difficult to fill these sites as the negative charge on the surface is increased. Crystal structure studies and identification of particular types of surface OH groups using infrared (IR) spectroscopy in combination with deuteration [37] and measurements of surface area provide a means for estimating the maximum number of surface sites. Mean areas of the oxide surface occupied by anions at high concentrations of adsorbate in solution and near optimum pH values (Table I) show that although anions are not closely packed, coverage is appreciable. When steric factors involving the anions and the replaceable hydroxyls on crystal surfaces are taken into account, the areas obtained must be close to the maximum possible for a monolayer.

In applied studies, adsorption data are often fitted to Langmuir equations:

$$(A)_s = X_m K_A (A) / [1 + K_A (A)] \qquad (13)$$

where $(A)_s$ and (A) = are concentrations of adsorbate at the surface and in solution, respectively,
X_m = the apparent maximum of adsorption curve, and
K_A = a coefficient that varies with pH.

This variation with pH is due to the change in the proportions of anionic species and their conjugate acids, as well as to the variation in surface charge given by the term $\exp(-z_i F \psi_s / RT)$ in the model [30].

Table I. Mean Areas Occupied by Anions at High Solution Concentrations and Near Optimum pH

Anion	pH	Absorbate	No. of Samples	Area/Anion ($Å^2$)
Fluoride	3.5	Goethite	1	20
	5.5	Gibbsite	1	18
Molybdate	4.0	Goethite	3	31,32,38
		Gibbsite	1	39
Selenite	4.0	Goethite	3	50,53,58
		Gibbsite	2	96,206
Selanate	4.0	Goethite	1	91
Sulfate	3.0	Goethite	1	90
		Gibbsite	1	120
Arsenate	3.0	Goethite	1	61
Phosphate	4.0	Goethite	3	61,63,68
		Gibbsite	2	46,60
Pyrophosphate	3.0	Goethite	1	61
Tripolyphosphate	3.0	Goethite	1	91
Silicate	9.0	Goethite	3	60,61,64

Effect of Ionic Strength

Ionic strength affects the activity coefficients for OH^-, H_3O^+ and specifically adsorbable ions and thus, indirectly, ψ_s and ψ_d, as represented in Equations 3, 4 and 5. According to the model [19], increasing the ionic strength of the solution increases adsorption at pH values below the point of zero charge (pzc) and decreases adsorption at pH values above the pzc. The reason for this is that increasing the ionic strength increases the positive charge of the surface below the pzc, resulting in greater attraction of anions, and increases the negative charge of the surface above the pzc, resulting in greater repulsion of anions.

Experimental studies show that varying supporting electrolyte concentrations from 0.01 M to 1.0 M has a relatively small effect on adsorption of selenite on goethite [34]. No effect was observed for phosphate adsorption on gibbsite when the concentration of supporting electrolytes NaCl, KCl and $MgCl_2$ was varied from 0.002 to 0.02 M [38].

Watson et al. [39] found a decrease in adsorption of 2,4-D (2,4-dichlorophenoxyacetic acid) by goethite on increasing the concentration of supporting electrolyte from 0.01 to 1 M NaCl. They attributed this to contraction of the diffuse double layer and consequent competition for surface sites by chloride. There was a reversal of this behavior at higher concentrations of 2,4-D in the aqueous phase, apparently because higher concentrations of NaCl caused salting out, or precipitation, of 2,4-D on the oxide surface. Later studies by Kavanagh et al. [40] showed that phenoxyacetic acids were only adsorbed on goethite at pH values below the pzc. Their studies indicated that ligand exchange was probably not the mechanism for specific adsorption in this case and attributed the specificity to van der Waals interactions at the oxide surface.

The change in surface charge per mole of anion adsorbed on goethite and gibbsite increases with ionic strength and varies with pH [27] but is reasonably constant for the range of surface coverages examined. However, Ryden and Syers [41], working with soils and ferric oxide gel, found significant variation in the charge per anion adsorbed with surface coverage by phosphate.

Adsorption of silicate on goethite and gibbsite increases with ionic strength for concentrations of about 1 mM Si in the aqueous phase [42]. At this concentration, the solution is about half saturated with silica for much of the pH range. Polymerization of silica at oxide surfaces may be responsible for the increase in adsorption with the concentration of supporting electrolyte.

In view of the importance of surface charge in models for specific adsorption and the dependence of surface charge on ionic strength of the aqueous phase, there seems to be a need for further experimentation to clarify some aspects of adsorption models. For example, as ionic strength affects surface charge, how can the stoichiometry of adsorption reactions involving H_3O^+ and OH^- ions be interpreted? If redistribution of charge is possible through electronic inductive effects of anions acting along coordinate bonds to oxygen ions forming part of the oxide lattice, what effect would this have on interpretations of mechanisms? As most experiments have been conducted in the presence of supporting electrolyte, what significant changes, if any, would be observed for specific adsorption at minimum electrolyte concentration? Additional effects of ionic strength on adsorption reactions are also referred to in Chapter 6 of this text.

Effects of Cations on Adsorption

Studies of phosphate adsorption on gibbsite [38] at pH 5.5 showed that among the chlorides of common cations only calcium increased adsorption in the concentration range to 20 mM. This suggests formation of calcium phosphate complexes at the gibbsite surfaces. Formation of surface complexes of phosphate with strontium [43], cadmium and zinc [44] is also indicated by the adsorption data.

Calcium-phosphate surface complexes are particularly relevant for interpretation of phosphate adsorption in soils where congruent dissolution of Ca^{2+} and phosphate could explain the observation that the calcium phosphate potential defined as, ½ pCa^{2+} + $pH_2PO_4^-$, is constant for many soils [45]. This has been a point for argument [46] about the relevance of surface charge to competition between anions adsorbed and changes in pH. Recent studies by Jensen [47] and Helyar et al. [43] found that phosphate potentials tended to be constant when phosphate was adsorbed on soils and on gibbsite; when the Ca^{2+} ion concentration in the equilibrium solution was between 1 and 10 mM; and for a wide range of coverage about the mean for surface adsorbed phosphate. However, when the phosphate concentration was high or low, the phosphate potential diverged from the relatively constant value obtained for the middle range.

Reversibility of Anion Adsorption on Oxides

In many practical applications of anion adsorption, the apparent irreversibility of the surface complexes formed is an important characteristic. Adsorption is reversible with respect to changes in pH and temperature [48,49], but desorption by washing the adsorption complex at constant pH and ionic strength varies between complete reversibility and almost complete irreversibility. A simple measure of reversibility is "desorbability," defined as the amount of desorption calculated from the concentration of the anion in the desorbing solution and the adsorption isotherm, the measure being expressed as a percentage [50]. Desorbability estimated for anions adsorbed on goethite and gibbsite is dependent on the nature of the adsorbent, the anion, pH, ionic strength, surface coverage and temperature.

Desorbability of some anions has been studied under different conditions of pH and ionic strength, but this aspect does not appear to have been examined extensively enough to produce generalizations. Desorbability of phosphate increases with pH and concentration of supporting electrolyte (NaCl); it is lower for goethite than gibbsite and increases with surface coverage. Selenite is reversibly adsorbed on gibbsite at pH 6-7 and in 0.1 M NaCl [48] and almost irreversibly adsorbed on goethite [50]. Fluoride is reversibly adsorbed on goethite at pH 4.5 and in 0.01 M NaCl [50].

Desorption of phosphate from surface complexes on goethite has been studied using a flow procedure that allows renewal of the desorbing solution 3 x 10^6 times per hour [51]. The concentration of phosphate in desorbing solutions can be reduced almost to zero, and desorption was conducted for long periods (8 days in these experiments). Thus, all reversibly adsorbed phosphate in the surface complex would be expected to be desorbed. This technique confirmed the low desorbability of surface phosphate on goethite, which was suggested in repeated washing and equilibration experiments. The desorbability of phosphate was shown to be significantly decreased by "aging" phosphated oxide in a moist condition for 15 days at 23°C. A feature of the results was the much increased desorption when the experiment was conducted at 80°C rather than 23°C, indicating that the apparent failure to desorb all the phosphate from oxides may be due to the slowness of reactions at sites requiring high activation energies for desorption.

Although the reasons for the low desorbability are not clearly understood, experiments suggest that the desorption reaction is probably complex, perhaps relating to redistribution of charge in the surface [50], the effects of heterogeneous distribution activation energies for desorption among the various surface sites [51] and the bonding structure for surface complexes [49, 52].

The slow rates of desorption and isotopic exchange reactions have led workers to suggest multidentate structures for phosphate and selenite adsorbed on goethite [50,51] and phosphate adsorbed on gibbsite [53]. Infrared and deuteration techniques applied in studies of the nature of hydroxyls on goethite surfaces [37,54] show the presence of three types of hydroxyl: (1) singly coordinate "A-type" hydroxyls, which are involved in reactions with anions, (2) doubly coordinated "B-types," and (3) triply coordinated "C-types," which are unreactive. Binuclear bridging complexes formed by anions replacing adjacent A-type hydroxyls are suggested for adsorption of phosphate, selenite, sulfate, oxalate and carbonate on goethite [55-60]. Monodentate complexes are suggested for benzoate, nitrate and halide ions adsorbed on goethite, where vacuum-dry films have been examined using infrared spectroscopy [57,59]. Air-dry films examined in infrared studies [59] also appear to indicate that NO_3^-, Cl^-, Br^- and I^- react with goethite to replace OH^- and, therefore, are bound as ligands at the surface rather than by purely coulombic forces. This contrasts with evidence from reactions of anions with goethite in aqueous suspensions, which indicate that in suspensions it is unlikely that NO_3^- and Cl^- are bound by covalent forces.

KINETICS OF SURFACE REACTIONS

Adsorption and Desorption

Kinetic studies are valuable for interpreting mechanisms of homogeneous reactions, but interpretation of the kinetics of heterogeneous reactions has been less successful because their analysis is particularly difficult. In attempts to analyze the data for adsorption and desorption of anions at solution-solid surfaces, the reactions have been treated as pseudofirst-order [31,42] and diffusion-controlled reactions. The data is often split into an arbitrary series of steps to fit the hypothetical model for reaction. In some cases there is independent evidence to suggest that the reaction may be broken up in this way, but generally the validity of such a procedure is questionable.

In studies of the kinetics of chemisorption at gas-solid interfaces, this same problem has been approached by an alternative procedure using the Elovich equation [61], the simplest form of which is

$$\frac{dq}{dt} = a \exp(-bq) \qquad (14)$$

where q = the adsorbate taken up in time, t,
       a = a constant relating to the initial velocity of reaction, and
       b = a constant relating to activation energy for adsorption.

The Elovich equation has been derived theoretically for uniform and nonuniform surfaces [61]. An essential feature for uniform surfaces is that the energy of activation increases linearly with surface coverage. The derivation for nonuniform surfaces involves dividing the surface into a number of uniform elements and assuming that the activation energy for adsorption increases linearly as each element is filled. The equation may also be derived if the total number of sites is a function of both the amount of adsorption and the temperature. Essentially, however, the equation is empirical. Its application to chemisorption has been treated in detail by Allen and Scaife [62] and some of the problems in its use have been discussed by Aharoni and Ungarish [63,64].

Integrating Equation 14 with the boundary condition q = 0 at t = 0 gives

$$q = b^{-1}\{\ln\ (t+k)/k\} \qquad (15)$$

where $k = (ab)^{-1}$. However, in analyses of gas adsorption kinetics, a pre-Elovichian stage is postulated, leading to the boundary conditions $q = q_o \neq o$ at t = 0, giving the integrated form

$$q = b^{-1} \ln\{(t+k')/k'\} + q_o \qquad (16)$$

where $k' = (ab)^{-1} \exp(bq_o)$.

In a further elaboration of the theory, Allen and Scaife [62] suggest a generalized equation to account for interaction between adsorbed species:

$$\frac{dq}{dt} = a_\gamma \exp\{-b\,(q-\gamma q_o)\} \qquad (17)$$

where γ is a constant to account for interaction. The integrated form is

$$q = b^{-1} \ln\{(t+k'')/k''\} + q_o \qquad (18)$$

where $k'' = (a_\gamma b)^{-1} \exp -(\gamma-1)bq_o$.

Treatment of kinetic data for chemisorption using Elovich plots based on Equations 15, 16 and 17 can give discontinuities if more than one type of reaction occurs, but many sets of data yield linear plots of $\ln\{(t+k)/k\}$ against q for the whole reaction period. The attractive feature of this type of model is that it is continuous, generally requiring no arbitrary breakup of the data.

Despite its common use for chemisorption of gases, the Elovich equation has not been applied to adsorption at oxide-aqueous solution interfaces. However, it offers the possible advantage that it may be applicable to adsorption in soils and sediments where there is wide variation in activation energies because the mixture of adsorption surfaces is so complex. The model yields activation energies for the initial part of adsorption and desorption reactions, from the constant "a," and a measure of the range of activation energies, from the constant "b."

Recent theoretical studies of the Elovich model, its applicability and method of fitting kinetic data [62-64] show that accurate data will be required, particularly at short reaction times, to fit results from adsorption and desorption in oxide-aqueous solution systems. As an example of the possible application of Elovich-type kinetics, consider the reaction betweeen silicate and crystalline aluminum hydroxide. The adsorption reaction at pH 9.2 was analyzed as a series of pseudofirst-order reactions [42]. Later, in a similar study, Huang [31] determined the effects of pH, surface area and silicate concentration on specific rate constants. Using the data of Hingston and Raupach [42], Elovich equations can be fitted as shown in Figures 3(a) and 3(b).

The parameters of the Elovich equation fitted to adsorption data varied as follows: (1) When the initial concentration of silicate in solution was increased, the value for "q_o" increased, the value for "a" decreased and "b" was not affected; and (2) increasing the temperature from 10°C and 35°C increased the values for "a," but did not affect "q_o" or "b." The activation energy for silicate adsorption, where the initial solution was about 1 mM Si, was estimated to be 23 kcal/mol (15-24 kcal/mol in the previous analysis of the data [42]). A possible interpretation of this analysis is that "a," the initial rate of adsorption, was decreased at higher initial concentrations of silicate in solution because the surface was covered to a greater extent (i.e.,

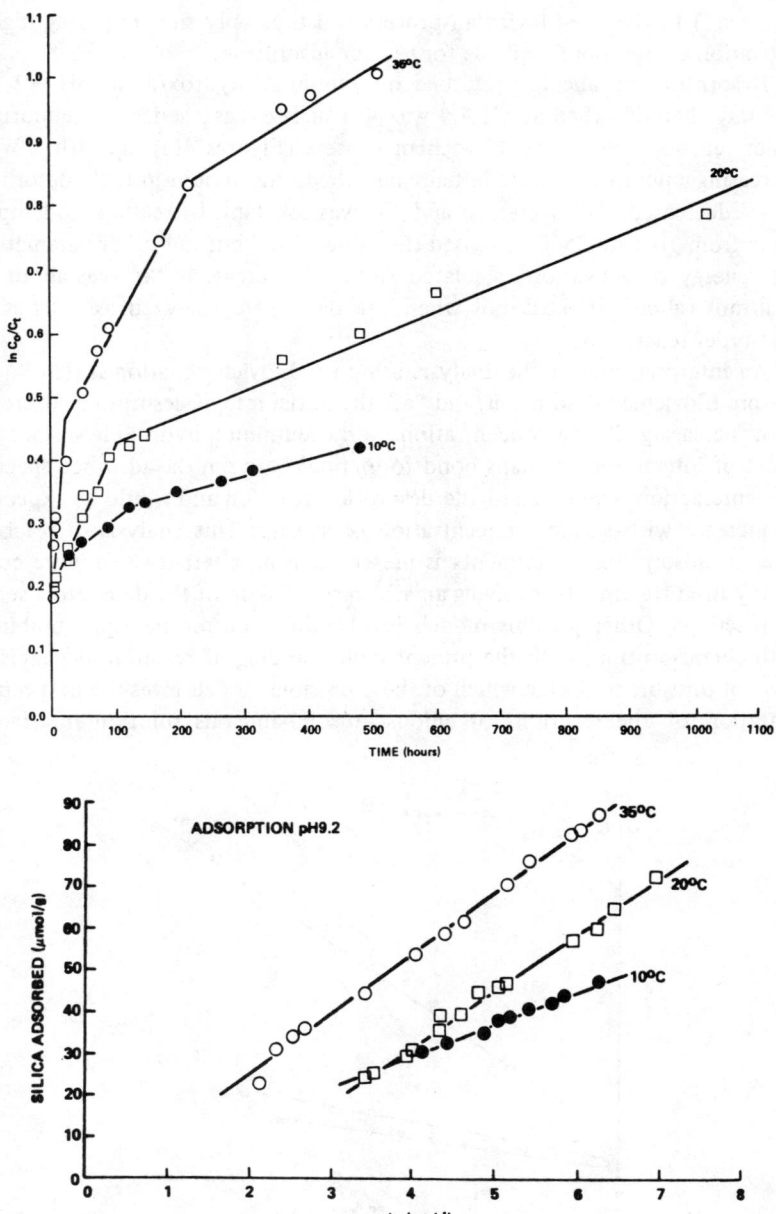

Figure 3. Kinetic data for adsorption of silicate on aluminum hydroxide at pH 9.2: (a) Data plotted as a series of pseudofirst-order reactions; (b) Elovich plots: $10°C$, $K' = 35$, initial silicate concentration, $1.39 \times 10^{-3} M$; $20°C$, $K' = 26$, initial silicate concentration $1.34 \times 10^{-3} M$; $35°C$, $K' = 7.3$, initial silicate concentration $1.35 \times 10^{-3} M$; and $K' = (ab)^{-1} \exp(bq_o)$ (see Equation 16).

larger q_o) in the pre-Elovichian process and thus only sites requiring higher activation energy were available for further adsorption.

Desorption of silicate presorbed on aluminum hydroxide at pH 9.2 for one day then desorbed at pH 4.4 was also analyzed as a series of pseudofirst-order reactions and as an Elovichian process (Figures 4(a) and 4(b)). With increasing amount of silicate initially adsorbed, the amount initially desorbed, "q_o," decreased, "a" decreased and "b" was constant. Increasing the temperature from 10°C to 35°C increased the value of "a" but not other parameters. The energy of activation calculated from the increase in "a" was about 12 kcal/mol (about 14 kcal/mol when the data were analyzed as a series of first-order reactions).

An interpretation of the analysis using the Elovich equation is that "q_o," the pre-Elovichian desorption, and "a," the initial rate of desorption, decrease with increasing silicate concentration on the aluminum hydroxide surface because of interaction (perhaps bond formation) between the adsorbed species. The interaction would retard the desorption reaction and would be expected to increase with surface concentration of silicate. This analysis of reaction rates in adsorption experiments is presented as an alternative to more commonly used treatments involving an arbitrary breakup of the data into a series of reactions. Other possible models involve diffusion mechanisms combined with chemisorption. With the present understanding of reaction mechanisms it is not possible to decide which of these possible models gives the best representation of the reactions of anions at solid-aqueous solution interfaces.

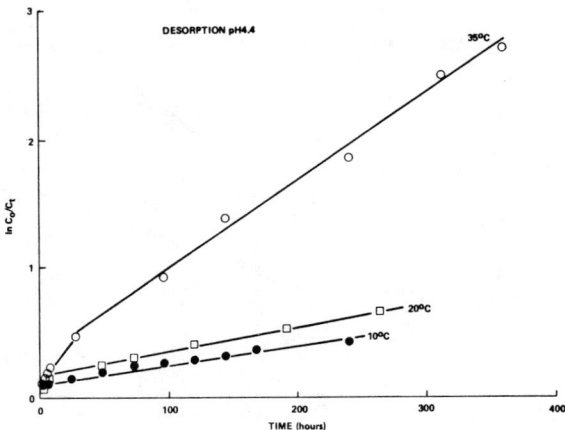

Figure 4a. Kinetic data for desorption of silicate from aluminum hydroxide (silicate presorbed for 24 hr at pH 9.2). Desorption at pH 4.4. Data plotted as a series of pseudofirst-order reactions: C_o = initial concentration of silicate on the surface; C_t = concentration of silicate on the surface at time t.

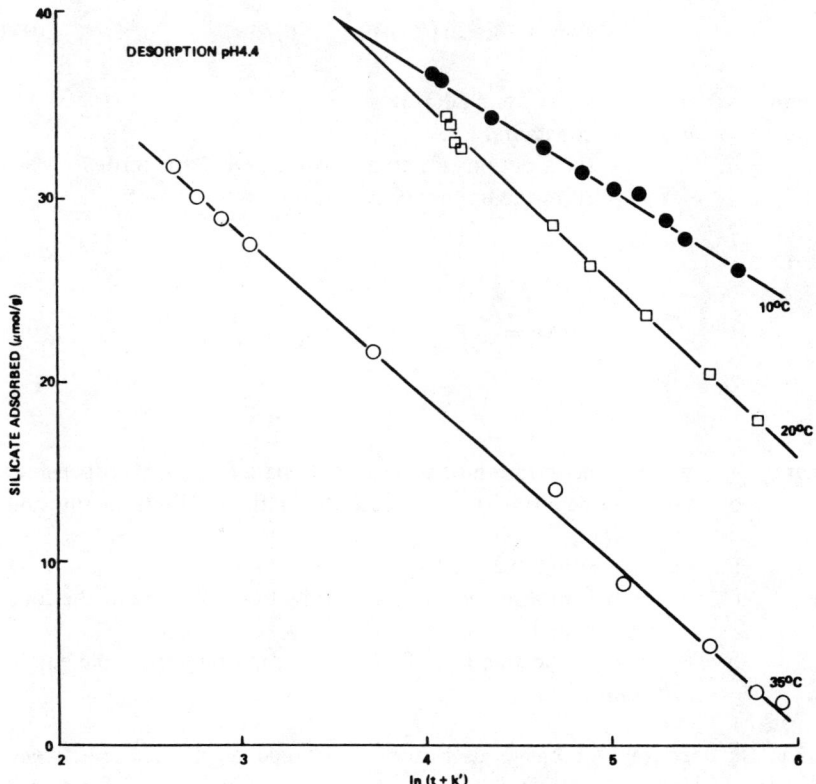

Figure 4b. Kinetic data for desorption of silicate from aluminum hydroxide (silicate presorbed for 24 hr at pH 9.2). Desorption at pH 4.4. Elovich plots: 10°C, $K' = 54$, initial adsorbed silicate 40 μmol/g; 20°C, $K' = 60$, initial adsorbed silicate 30 μmol/g; 35°C, $K' = 12.9$, initial adsorbed silicate 35 μmol/g; and $K' = (ab)^{-1} \exp(bq_o)$ (see Equation 16).

Isotopic Exchange

Isotopic exchange between anions in aqueous solution and adsorbed on oxides has been studied to gain insight into reaction mechanisms and the nature of binding of ligands at mineral surfaces. Using soils as adsorbents, the kinetic data have been analyzed using the McKay equation with arbitrary splitting of the data into linear regions [65,66]. However, the Elovich equation was considered to be more appropriate for studies of isotopic exchange on goethite and gibbsite [52,53,67].

The integrated form of the equation corresponding with Equation 16 was used:

$$\theta = B^{-1} \ln[(t+k)/k] + \theta_o \qquad (19a)$$

where k = $(AB)^{-1} \exp(BQ_o)$,
A and B = constants, and
θ = the fraction of surface coverage and the term θ_o arises from the boundary condition $\theta = \theta_o$ at t = 0.

Also,

$$\theta = bF/[b + a(1-F)] \qquad (19b)$$

$$= f\infty \, F/f^*$$

where a = the concentration of surface phosphate (mol/1 of suspension),
b = the concentration of phosphate in the equilibrium solution (mol/1),
F = the fraction exchanged,
f* = the radioisotope in solution at time t as a fraction of the total isotope, and
f∞ = b/(a+b), the fraction of isotope in solution at isotopic equilibrium.

Kinetic data for heterogeneous isotopic exchange of phosphate between aqueous solution and Fe(III) phosphate complexes on goethite surfaces have been fitted to an Elovich equation model [52]. The rate of exchange is first order with respect to the concentration of Fe(III) phosphate complexes and subject to acid-base catalysis with a minimum at about pH 9. The relatively inert nature of the Fe(III) complexes is attributed to formation of a bridging ligand structure formed by the phosphate anion between two Fe atoms in the geothite surface. Isotopic exchange of selenite in solution with Fe(III) selenite complexes at geothite surfaces follows a similar pattern. It can be fitted by a single Elovich equation and is acid-base catalyzed with a minimum about pH 7 [68].

Isotopic exchange of phosphate adsorbed on gibbsite also fits an Elovich model and is acid-base catalyzed with a minimum rate at pH 5.5 [53]. Rates of isotopic exchange are greater for Al(III) phosphate complexes at gibbsite surfaces than for Fe(III) phosphate, or for Fe(III) selenite complexes at goethite surfaces. Kyle et al. [53] consider that the orders of lability of surface complexes indicated by these isotopic exchange rates correlate well with the ease of hydrolysis of anions absorbed on goethite and gibbsite.

EXCHANGE INVOLVING SPECIFICALLY ADSORBABLE ANIONS

By contrast to nonspecific adsorption, where anions such as NO_3^- and Cl^- exchange on an equivalent basis, the reactions between anions that form coordinate surface complexes cannot be explained as simple ion exchange. In specific adsorption the surface coverage varies with anion species, pH and the presence of other specifically adsorbable anions.

In an early attempt to explain some of the results from experiments to investigate competitive adsorption, Hingston et al. [69] postulated three kinds of sites, i.e., common sites and those where only either one or the other of the anions could be adsorbed. However, this may be unnecessary in the general model [30], where the amounts of anions adsorbed are determined by the surface potential, ψ_d, the surface charge, σ_i, and the relative values for adsorption constants for anions, K_i. Measurements of both surface charge and adsorption are required to evaluate the various parameters in the model from Equations 3, 4, 5 and 6.

Competition between specifically adsorbed anions can be evaluated from the model [19] from Equation 12 describing the isotherms. The form of the relationship is

$$a_{1s}/a_{2s} = (a_1/a_2)(K_1/K_2) \qquad (20a)$$

where a_{1s} and a_{2s} = the activities of anion 1 and 2 on the surface,
a_1 and a_2 = their activities in solution, and
K_1 and K_2 = the binding constants for the anions.

For applied studies, however, where each adsorbate may consist of a number of anionic species and charge measurements may not have been made, the following alternative procedure may be useful:

The concentrations of two competing adsorbates such as phosphate, and selenite in solution and surface phases can be related by the equation

$$(P)_s/(Se)_s = K_p(P)/K_{Se}(Se)$$

$$= K_{Se}^p[(P)/(Se)] \qquad (20b)$$

where K_p and K_{Se} = Langmuir coefficients (Equation 13),
(P) and (Se) = the concentrations of phosphate and selenite in solution, and
$(P)_s$ and $(Se)_s$ = their concentrations in the surface phase.

The term K^P_{Se} is a selectivity coefficient [50,69], which is characteristic of the adsorbates and the surface (Table II) and varies with pH (Figure 5). K^P_{Se} also decreases slightly with surface coverage; however, the reasons for this have not been investigated using the model [19]. A hypothesis put forward earlier to explain variation of K^P_{Se} with surface coverage was that the adsorbates do not compete at all sites [69].

Variation of K^P_{Se} with pH is due partly to the changes in proportions of anionic species and conjugate acids for both adsorbates, and partly to the variation of surface charge. K_i for each anionic species (Equation 12) is incorporated in the Langmuir constants K_p and K_{Se} and does not vary with pH. The surface charge and the variation of surface charge with pH differ among adsorbent surfaces. Consequently, K_i also varies between adsorbents such as gibbsite and goethite. If the values for selectivity coefficients are similar and vary similarly with pH for all aluminum oxide-type surfaces on the one hand and all iron oxide-type surfaces on the other, the coefficients may have application for understanding competitive effects between anions in more complex materials.

In studies of competition between phosphate and bicarbonate and between phosphate and citrate using kaolinite as the adsorbent, Nagarajah [70] found

Table 2. Selectivity Coefficients for Competitive Adsorption of Anions on Goethite and Gibbsite[a]

Absorbent	Competing Anions		Selectivity Coefficients[+] at Various pH Values		
			pH 5	pH 7	pH 9
Goethite	Phosphate-selenite	K^P_{Se}	0.5	1.0	1.6
	Phosphate-arsenate	K^P_{As}	0.7	0.9	1.0
	Phosphate-silicate	K^P_{Si}	27.0	9.0	2.2
Gibbsite	Phosphate-selenite	K^P_{Se}	7.0	11.0	1.6
	Phosphate-arsenate	K^P_{As}	5.0	5.0	3.0
	Phosphate-silicate	K^P_{Si}	84.0	12.0	0.9

[a+] $K^i_j = K_i/K_j$, where K_i and K_j are constants for adsorption of anions "i" and "j".

that where adsorbing anions were added in sequence rather than simultaneously, the relative amounts of each adsorbed at equilibrium (24 hours) depended on the order of adsorption of the anions. They attributed this effect to formation of bidentate complexes by whichever anion was adsorbed first and postulated that changes in surface charge may control the course of the adsorption reaction. However, it is also possible that the effect may be due to very slow approach to equilibrium and, if sufficient time is allowed, the same relative concentrations of anions on the surface may be achieved, regardless of the order of addition of adsorbates.

The total adsorption of anions from a mixture of two competitors is often approximately equal to the amount of adsorption of whichever is adsorbed in

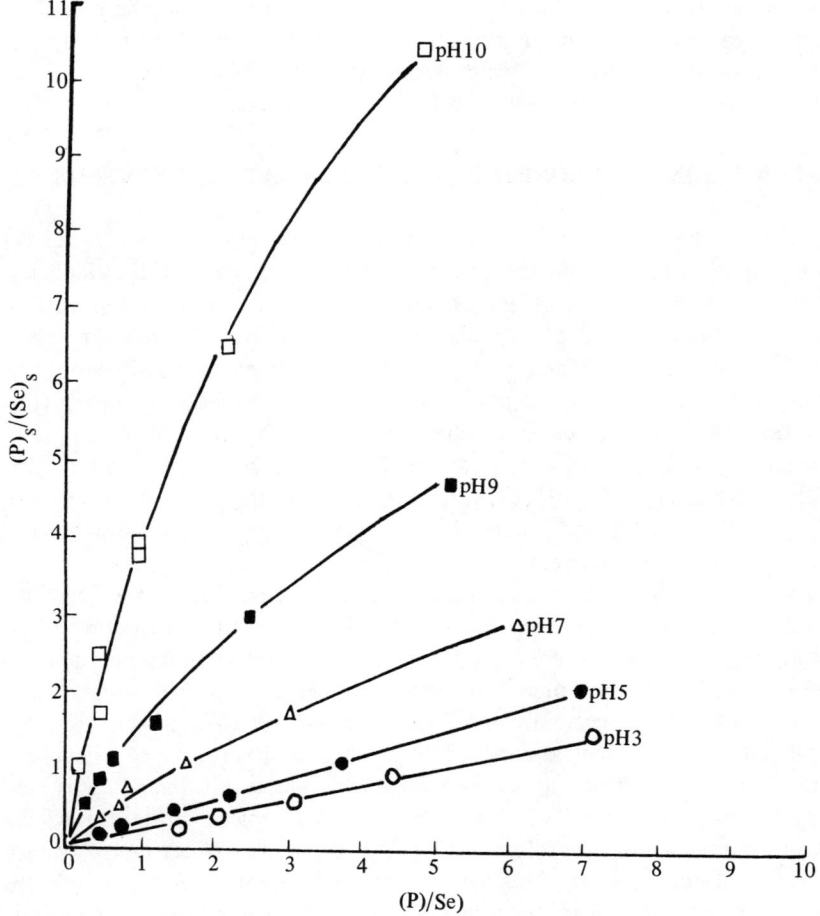

Figure 5. Competitive adsorption of phosphate and selenite on goethite (surface area 81 m^2/g). Supporting electrolyte 0.1 M NaCl, pH values as shown.

greatest amount when present alone [68,71]. However, in some instances, total adsorption is greater and adsorption of one component is enhanced when a mixture of anions is present [69]. For example, where phosphate and selenite were competitively adsorbed on goethite, the mean minimum area occupied by anions adsorbed from the mixture at pH 5 was about 37 Å^2, compared with 53 Å^2 for selenite and 69 Å^2 for phosphate when they were adsorbed separately from solutions of similar concentration. An explanation of this observation is suggested by Russell et al. [56] from infrared studies and consideration of steric effects involving the selenite anion and the "A-type" hydroxyls on the goethite surface. They propose that HPO_4^{2-} ion is bidentate, replacing two "A-type" hydroxyls, while adsorbed selenite must be present as the monodentate $HSeO_3^-$ ion if it is involved in charge transfer. The pyramidal structure of the $HSeO_3^-$ ion prevents it from replacing every "A-type" hydroxyl. However, when HPO_4^{2-} ions are also present, every "A-type" hydroxyl can be exchanged by alternating HPO_4^{2-} and $HSeO_3^-$ ions.

APPLICATION OF ADSORPTION STUDIES TO NATURAL SYSTEMS

Despite the complexity of natural systems, adsorption characteristics for anions in soils and sediments agree qualitatively with many of the characteristics found in reactions of anions with oxides. For example, BET surface areas of soils measured by N_2 adsorption are closely correlated with phosphate adsorption [72] and with adsorption of other anions. The positive charges developed in soils depend on pH and ionic strength of the aqueous solution [73,74] and determine the amount of NO_3^- and Cl^- that is non-specifically adsorbed. Observed effects of pH on adsorption of sulfate [75-77], molybdate [76,78,79], selenite [80,81], arsenate [81], phosphate [76,82,83], silicate [82] and borate [84-88] agree with the results from oxide/aqueous solution experiments.

Apparent differences between adsorption reactions with soils and with synthetic oxides can probably be explained by competition with other specifically adsorbed anions, complex formation with soil components, and the more amorphous nature of soil colloids [1]. Sulfate, silicate, bicarbonate and organic anions occur naturally in soils at sufficient concentrations to compete significantly with added anions. Adsorption and desorption of phosphate with kaolinite as adsorbent has been shown to involve silicate and hydroxide ions [89-91]. Formation of soluble complexes of anions [92] also affects adsorption in soils. Amorphous or poorly crystalline oxides and clays are probably responsible for the slow reactions with anions [93,94], which appear to be more significant for soils than for synthetic geothites and gibbsites.

Availability of Nutrients in Soils

Nutrient Uptake by Plant Roots

Models of uptake of nutrients from soils by plant roots incorporate adsorption of some ions by the soil and diffusion to the root surface [95-97].

Nye [97], in reviewing the mechanisms of plant uptake of nutrients, draws attention to the effects of adsorption on the diffusion coefficients for nutrient anions in soil. An approximate relationship between the flux of nutrient to any radial boundary around a root and soil properties is

$$I_r = 2\pi r D_1 \theta f_1 (dC_1/dr) + 2\pi r v C_1 \quad (21)$$

where I_r = the amount of nutrients crossing the radial boundary in unit time (mol/cm/sec),
r = the radius of the boundary (cm),
D_1 = the diffusion coefficient of the nutrient ion in free solution (cm^2/sec),
θ = the volumetric fraction,
f_1 = the diffusion impedance factor,
v = the flux of water towards the root (cm^3/cm^2/sec), and
C_1 = the concentration of the nutrient ion in solution (mol/cm^3).

Uptake of nutrient per unit length of root (I_a), is given by

$$I_a = 2\pi a \alpha C_{1r} \quad (22)$$

where α = the root absorbing power (cm/sec),
a = the root radius (cm), and
C_{1r} = the concentration at the root surface.

Uptake of anions such as phosphate and molybdate, which are specifically adsorbed in soils and present in low concentrations in solution, is regulated in the early stages of uptake by the spread of diffusion zone given by

$$(D_1 \theta f_1 t/b)^{1/2}$$

where b is the buffering power of the soil defined as dC_1/dC, where C is the concentration of nutrient in soil (mol/cm^3).

Thus, factors such as pH, surface area, surface charge, competing anions, reversibility and kinetics of desorption, which have been discussed for oxide-aqueous solution interfaces, will regulate the concentration of anions in solution (C_1) and the buffer capacity dC_1/dC in plant nutrition. Anions such as HCO_3^- and organic acids exuded by roots or formed in soil will increase C_1 and dC_1/dC. Tinker and Sanders [98] consider it unlikely that concentrations of exudates in the zone near the root would be high enough to increase uptake of phosphate. However, Nye [97,99] reports that in some experiments [100,101] uptake can be explained by diffusion and mass flow only if the solution concentration is higher in the vicinity of roots than in bulk soil.

Reactions of Fertilizers with Soils

Fertilizers are usually applied in concentrated form as granules or bands of granules. Thus, initially close to the point of application, the pH and concentrations of anions such as phosphate are extreme compared with adsorption experiments. Reactions and reaction products for phosphate fertilizers with calcareous and noncalcareous soils are described in a recent review by Mattingly [102]. Initially, crystalline calcium, aluminum and iron phosphates form in the zone closest to the point of application. Phosphate diffuses outward until equilibrium with the soil solution, determines plant uptake. In addition, reactions involving adsorbed phosphate also determine the long-term effectiveness, i.e., residual value, of phosphatic fertilizers.

To study the very slow reactions of anions such as phosphate and molybdate in soils, Barrow and Shaw [103] considered these anions in three hypothetical compartments:

$$A \rightleftharpoons B \rightleftharpoons C$$

where compartment "A" contains soluble phosphate, compartment "B" adsorbed phosphate and compartment "C" phosphate not in direct equilibrium with compartment "A." The empirical relationship for transfer of anions from compartment "B" to "C" is

$$d\alpha/dt = k(1-\alpha)^n \qquad (21)$$

where $(1-\alpha)$ is the proportion of anions in the form that equilibrates with the solution, k varies with temperature, and n is a coefficient. The form of the kinetic data for adsorption, desorption and isotopic exchange of phosphate obtained by Barrow and Shaw [103-106] suggest that the Elovich relationship (e.g., Equation 16) may also be applicable for the whole adsorption reaction. Barrow and Shaw [106] consider that the slow reaction into compartment "C" accounts for the observation that some adsorbed phosphate cannot be desorbed at constant pH. Munns and Fox [94] also consider that slow reactions are responsible for much of the apparent hysteresis in desorption of phosphate from tropical soils. White and Taylor [107] suggest that irreversibility of phosphate adsorption in soils is due to incomplete attainment of equilibrium during the adsorption process, a condition that is related to the unrealistically high initial concentrations of phosphate in the solution phase in experimental studies. However, in some experiments desorption has been measured for suspensions where the phosphate was adsorbed from solutions covering a wide range of concentrations, and hysteresis was observed even at the lowest adsorption. Some soils may be naturally at disequilibrium because of recent applications of fertilizer or severe depletion of adsorbed phosphate through plant uptake.

The effects of "aging" surface complexes of anions on the desorbability and rates of desorption have been investigated for soils [105,106], as well as for oxides. The pattern followed is qualitatively the same, i.e., with increasing "aging" time, anions become less desorbable and desorb at slower rates. The mechanism of aging is not clear. It possibly corresponds with the structure of surface phases becoming increasingly ordered and diffusion of anions to higher-energy, or less accessible, sites.

The kinetics of adsorption and desorption of anions in soils have been analyzed by fitting the data to empirical rate equations [103,108], to diffusion laws [109-111], and to first-order kinetics [112]. Enfield et al. [110], for example, compared five kinetic models for adsorption of phosphate on 25 mineral soils and found they all fitted reasonably well. The best fit was for diffusion-limited models incorporating either the Langmuir equation or the Freundlich equation to describe the isotherms.

Estimating Available Nutrients

Widely used methods for estimating the "availability" of nutrients to plants involve extracting soils with reagents capable of desorbing anions from the surface of soil minerals. Phosphate availability, for example, is estimated by extraction with bicarbonate [113] or fluoride [114]; sulfate is extracted with bicarbonate [115] or phosphate [116]; and molybdate is extracted with hydroxide or oxalate [117]. Extraction methods involve standard soil-extracting solution ratios and shaking times. Studies of competition between anions and rates of adsorption and desorption show that the amounts of nutrient anions desorbed depend on the amount of anions originally adsorbed on the soil, the adsorption characteristics of the soil, soil-extractant ratio, shaking time and the ratio of the concentration of competing anions in the extracting solution to desorbable anions. Detailed studies by Barrow and Shaw of extraction of phosphate by bicarbonate [118,119] and arsenate [120], of sulfate by phosphate [121] and of molybdate by phosphate and by hydroxide [122] illustrate the effects of conditions of extraction and soil properties on the experimental results.

Availability of phosphate has also been assessed using isotopic exchange to estimate the phosphate in soils that is adsorbed on surfaces [123]. However, the kinetics of isotopic exchange of phosphate on goethite and gibbsite are complex, so the significance of isotope exchange in soils may not be readily interpreted. Recent studies have shown that exchange is affected by pH and possibly by the presence of labeled phosphate complexes in solution [124] as well as by aging of phosphate added to soils [104,125].

Nutrient Cycling

In natural ecosystems, vegetative growth is sustained by nutrients accumulated slowly through contributions from the atmosphere and from weathering to the biogeochemical cycle [126]. Elements are retained within the system in biomass, accumulations of litter and in soil. An important retention mechanism for nutrient anions, phosphate, sulfate, molybdate and borate is adsorption on the surface of soil minerals.

Based on what is known about adsorption, some predictions can be made about the possible mobility of various ions in the system. Relevant characteristics are soil pH, surface area of soil minerals, the type of soil minerals present and the chemical properties of the ions. For example, nitrate and chloride will be adsorbed only at low pH, where soil colloids have a positive charge. These ions would also be easily leached since they are nonspecifically adsorbed and decreasing the ionic strength of the soil solution results in a lowering of the surface positive charge. Sulfate adsorption will decrease with increasing pH, with not much adsorption occurring above neutrality, but as this anion is specifically adsorbed it will be less easily leached than chloride and nitrate. Molybdate will be strongly adsorbed at pH values less than neutral and will be mobile under alkaline conditions. Borate and silicate are most strongly adsorbed at pH values greater than 7 and will be leached at low pH. Phosphate will be relatively immobile over most of the natural pH range, but adsorption will be less at high pH. Bicarbonate and carbonate will be adsorbed in the pH range above 4 and the adsorption will be greatest under alkaline conditions. Thus, there will tend to be separation of mobile and immobile anions during movement of water through soils.

Changes in soil pH, increases in soil organic matter and rates of organic matter decomposition to produce bicarbonate and organic acids are expected to change the mobility of other anions through effects on adsorption and competition reactions at mineral surfaces. A further effect is illustrated in recent studies, which indicate that bicarbonate and organic acids also play a key role as balancing ions in the leaching of cations through soils.

In natural systems, the importance of organic forms of elements must be stressed. The mobility of some substances may be completely altered by formation of organic compounds or organic complexes. In these instances, predictions about the cycling of the elements, which are based on the adsorption properties of simple anions and cations, are clearly inappropriate.

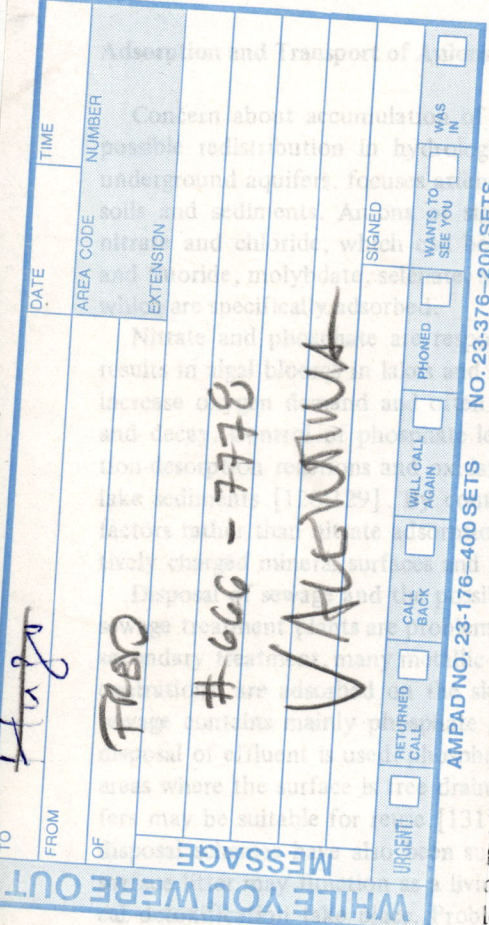

lutants in the environment and their ystems, such as surface streams and on natural processes of adsorption in cance as pollutants in water include nsidered as nonspecifically adsorbed, nite, phosphate, arsenate and arsenite,

ble for "nutritional" pollution, which eams [127]. Excessively heavy blooms oduce toxic substances when they die els in natural waters is through adsorp- n-reduction phenomena in stream and st, nitrate levels depend on biological , which only occurs at low pH on posi- not significant in most natural systems. le need to reuse renovated effluent from confronting modern large cities. During ations which could be toxic at high con- dge so that the effluent from domestic nd nitrate as anionic pollutants. If land te can be adsorbed in the soil [130]. In ing, water percolating to subsurface aqui-] if nitrate levels are not too high. Land ggested for forested areas where soil and ng filter and both adsorption and biologi- lems to be investigated in these schemes include the fate of viral pathogens, their survival, their adsorption on soil minerals and the effects of other substances such as phosphate on this adsorption.

In relation to disposal of solid and liquid wastes by landfill techniques, toxic pollutants such as molybdenum in sewage sludge [3] and arsenic and selenium in wastes [81] have been studied to determine potential hazards due to change in adsorption behavior with pH and other conditions. The phytotoxicity of arsenic accumulated in soil as a result of pesticide treatment has been studied in relation to chemical interaction with fertilizer phosphate [132,133]. Pollutants such as fluoride [134] are dispersed from industrial sources over considerable areas through the atmosphere, but adsorption by plant leaves seems to present more of a hazard than the amounts of these anions added to the soil and taken up by plant roots.

There seems to be little doubt that understanding adsorption mechanisms is relevant to finding solutions to problems associated with pollution and the mobility of pollutants; however, this should not draw attention away from equally important processes such as precipitation, coprecipitation, formation of organic and inorganic complexes, and the biological processes operative in complex natural systems.

CONCLUSIONS

Reactions involving anions at solid-aqueous solution interfaces have been studied extensively in recent years, and a much clearer picture of their mechanisms is emerging. This is due largely to careful studies of model oxide-aqueous solution systems, where adsorption and changes in surface charge have been estimated simultaneously. However, studies still appear to be necessary to clarify reaction mechanisms, particularly for desorption of anions and for competition between anions. Work on the kinetics of desorption at solution-solid interfaces may help to define these mechanisms, but the theoretical basis for analysis of kinetic data appears to be unsatisfactory.

Useful extensions of the work on model systems would include studies of mixtures of adsorbents, common clay minerals and soil features capable of being characterized through measurements of surface area and charge. The significance of organic matter and organic complexes in adsorption of anions also seems to have been studied relatively little and deserves more attention.

In natural soil-water systems, understanding and predicting the behavior of anions and cations still poses problems because of the mixture of reactions (including absorption, desorption, precipitation and soluble complex formation). Synthesis of information available from adsorption theory and experiments with simplified systems may give sufficient understanding for many purposes. A further refinement might be made through the development of a computer program similar to those that have been applied to evaluate complex formation between ions in natural waters [135,136]. Although it will not be possible to avoid the great amount of measurement needed to deal with environmental problems, it may help us to choose some of the right measurements to make.

REFERENCES

1. Jenne, E. A. "Trace Element Sorption by Sediments and Soils—Sites and Processes," in *Symposium on Molybdenum in the Environment, Volume 2, The Geochemistry, Cycling and Industrial Uses of Molybdenum*, W. R. Chappell and K. K. Peterson, Eds. (New York: Marcel Dekker Inc., 1977), p. 425.
2. Tofflemire, T. J., and M. Chen. "Phosphate Removal by Sands and Soils," in *Land as a Waste Management Alternative*, R. C. Loehr, Ed. (Ann Arbor, MI: Ann Arbor Science Publishers, Inc., 1977), p. 151.
3. Lahann, R. C. "Molybdenum Hazard in Land Disposal of Sewage Sludge," *Water Air Soil Poll.* 6: 38 (1976).
4. Stumm, W., and J. J. Morgan. *Aquatic Chemistry* (New York: John Wiley & Sons Inc., 1970), p. 445.
5. de Bruyn, P. L., and G. E. Agar. "Surface Chemistry of Flotation," in *Froth Flotation*, D. W. Fuerstenau (New York: American Institute of Mining, Metallurgical and Petroleum Engineers, 1962).
6. Fuerstenau, D. W. "Interfacial Processes in Mineral/Water Systems," *Int. Conf. of Pure and Applied Chemistry, Plenary Lectures,* 22: 135-154 (1969).
7. Grahame, D. C. "The Electrical Double Layer and the Theory of Electrocapillarity," *Chem. Rev.* 41: 441-501 (1947).
8. Stern, O. "On the Theory of the Electrical Double Layer," *Z. Electrochem.* 30: 508-516 (1924).
9. Bockris, J. O'M., M. A. V. Devanathan and K. Müller. "On the Structure of Charged Interfaces," *Proc. Roy. Soc. Ser.* A 274: 55-78 (1963).
10. Parsons, R. "Some Problems of the Electrical Double Layer," *Rev. Pure Appl. Chem.* 18: 91-108 (1968).
11. Parks, G. A. "Aqueous Surface Chemistry of Oxides and Complex Oxide Minerals," in *Equilibrium Concepts in Natural Water Systems*, Advances in Chemistry Series No. 67 (Washington, D.C.: American Chemical Society, 1967).
12. Atkinson, R. J., A. M. Posner and J. P. Quirk. "Adsorption of Potential Determining Ions at the Ferric Oxide-Aqueous Electrolyte Interface," *J. Phys. Chem.* 71: 550-558 (1967).
13. Bérubé, Y. G., and P. L. de Bruyn. "Adsorption at the Rutile-Solution Interface," *J. Colloid Interface Sci.* 27: 305-318 (1968).
14. Hingston, F. J., R. J. Atkinson, A. M. Posner and J. P. Quirk. "Specific Adsorption of Anions," *Nature* 215: 1459-1461 (1967).
15. Levine, S., and A. L. Smith. "Theory of the Differential Capacity of the Oxide/Aqueous Electrolyte Interface," *Disc. Faraday Soc.* 52: 290-301 (1971).

16. Wright, H. J. L., and R. J. Hunter. "Adsorption at Solid-Liquid Interfaces. I. Thermodynamics and the Adsorption Potential," *Aust. J. Chem.* 26: 1183-1189, 1191 (1973).
17. Blok, L., and P. L. de Bruyn. "The Ionic Double Layer at the ZnO/solution Interface," *J. Colloid Interface Sci.* 32: 518-526; 32: 527-532; 32: 533-538 (1970).
18. Stumm, W., C. P. Huang and S. R. Jenkins. "Specific Chemical Interaction Affecting the Stability of Dispersed Systems," in *The Chemistry of Solid/Liquid Interfaces,* B. Težak and V. Pravdić, Eds. (Zagreb: Croatica Chemica Acta, 1971).
19. Bowden, J. W., A. M. Posner and J. P. Quirk. "Ionic Adsorption on Variable Charge Mineral Surfaces. Theoretical-Charge Development and Titration Curves," *Aust. J. Soil Res.* 15: 121-136 (1977).
20. Yates, D. E., S. Levine and T. W. Healy. "Site-Binding Model of the Electrical Double Layer at the Oxide/Water Interface," *Trans. Faraday Soc.* 70: 1807-1818 (1974).
21. Robinson, R. A., and H. S. Harned. "Thermodynamics of Strong Electrolytes from Electomotive Force and Vapour Pressure Measurements," *Chem. Rev.* 28: 419-476 (1941).
22. Lyklema, J. "Electrical Double Layer on Oxides," *Croat. Chem. Acta* 43: 249-260 (1971).
23. Devanathan, M. A. V., and B. V. K. S. R. A. Tilak. "The Structure of the Electrical Double Layer at the Metal-Solution Interface," *Chem. Rev.* 65: 635-684 (1965).
24. James, R. O., and T. W. Healy. "Adsorption of Hydrolysable Metal Ions at the Oxide Water Interface," *J. Colloid Interface Sci.* 40: 65-81 (1972).
25. Težak, B. "Solid/Liquid Interfaces," in *Solid/Liquid Interfaces,* B. Težak and V. Pravdić, Eds. (Zagreb: Croatica Chemica Acta, 1971).
26. Robinson, R. A., and R. H. Stokes. *Electrolyte Solutions* (London: Butterworths, 1955), p. 384.
27. Hingston, F. J., A. M. Posner and J. P. Quirk. "Anion Adsorption by Goethite and Gibbsite. 1. The Role of the Proton in Determining Adsorption Envelopes," *J. Soil Sci.* 23: 177-192 (1972).
28. Martell, E. A., and M. Calvin. *The Chemistry of Metal Chelate Compounds* (Englewood Cliffs, NJ: Prentice-Hall, Inc., 1956), p. 153.
29. Cabrera, F., L. Madrid and P. De Ambarri. "Adsorption of Phosphate by Various Oxides. Theoretical Treatment of the Adsorption Envelope," *J. Soil Sci.* 28: 306-313 (1977).
30. Bowden, J. W., M. D. A. Bolland, A. M. Posner and J. P. Quirk. "Generalized Model for Anion and Cation Adsorption at Oxide Surfaces," *Nature* 245: 81-83 (1973).
31. Huang, C. P. "Adsorption of Phosphate at the Hydrous $\gamma - Al_2O_3$ - Electrolyte Interface," *J. Colloid Interface Sci.* 53: 178-186 (1975).
32. Huang, C. P. "The Removal of Aqueous Silica from Dilute Aqueous Solution," *Earth Planet. Sci. Lett.* 27: 265-274 (1975).
33. Rendall, H. M., and A. L. Smith. "Effects of Ionisation on Adsorption from Solution," *J. Chem. Soc. Faraday Trans.* 73: 101-110 (1977).
34. Hingston, F. J., A. M. Posner and J. P. Quirk. "Adsorption of Selenite on Goethite," in *Adsorption from Aqueous Solution,* Advances in Chemistry Series No. 79 (Washington, D.C.: American Chemical Society, 1968).

35. Bowden, J. W. Personal communication (1979).
36. Ryden, J. C., J. R. McLaughlin and J. K. Syers. "Mechanisms of Phosphate Sorption by Soils and Hydrous Ferric Oxide Gel," *J. Soil Sci.* 28: 73-92 (1977)
37. Parfitt, R. L., J. D. Russell and V. C. Farmer. "Confirmation of the Surface Structures of Goethite (α-FeOOH) and Phosphated Goethite by Infrared Spectroscopy," *J. Chem. Soc. Faraday Trans.* 72: 1082-1087 (1976).
38. Helyar. K. R., D. N. Munns and R. G. Burau. "Adsorption of Phosphate by Gibbsite. I. Effects of Neutral Chloride Salts of Calcium, Magnesium, Sodium and Potassium," *J. Soil Sci.* 27: 307-314 (1976).
39. Watson, J. R., A. M. Posner and J. P. Quirk. "Adsorption of the Herbicide 2,4-D on Goethite," *J. Soil Sci.* 24: 503-511 (1973).
40. Kavanagh, B. V., A. M. Posner and J. P. Quirk. "The Adsorption of Phenoxyacetic Acid Herbicides on Goethite," *J. Colloid Interface Sci.* 61: 545-553 (1977).
41. Ryden, J. C., and J. K. Syers. "Charge Relationships of Phosphate Sorption," *Nature* 255: 51-53 (1975).
42. Hingston, F. J., and M. Raupach. "The Reaction Between Monosilicic Acid and Aluminium Hydroxide," *Aust. J. Soil Res.* 5: 295-309 (1967).
43. Helyar, K. R., D. N. Munns and R. G. Burau. "Adsorption of Phosphate by Gibbsite 11. Formation of a Surface Complex Involving Divalent Cations," *J. Soil Sci.* 27: 315-323 (1976).
44. Bolland, M. D. A., A. M. Posner and J. P. Quirk. "Zinc Adsorption by Goethite in the Absence and Presence of Phosphate," *Aust. J. Soil Res.* 5: 279-286 (1977).
45. Schofield, R. K. "Can a Precise Meaning be Given to 'Available' Soil Phosphorus?" *Soils and Fertilizers* 18: 373-375 (1955).
46. White, R. E., and A. W. Taylor. "Effect of pH on Phosphate Adsorption and Isotopic Exchange in Acid Soils at Low and High Additions of Soluble Phosphate," *J. Soil Sci.* 28: 49-61 (1977).
47. Jensen, H. E. "Phosphate Potential and Phosphate Capacity of Soils," *Plant and Soil* 33: 17-29 (1970).
48. Muljadi, D., A. M. Posner and J. P. Quirk. "The Mechanism of Phosphate Adsorption by Kaolinite, Gibbsite and Pseudoboehmite 1. Isotherms and the Effect of pH on Adsorption," *J. Soil Sci.* 17: 212-228 (1966).
49. Kafkafi, U., A. M. Posner and J. P. Quirk. "Desorption of Phosphate from Kaolinite," *Proc. Soil Sci. Soc. Chem.* 31: 348-353 (1967).
50. Hingston, F. J., A. M. Posner and J. P. Quirk. "Anion Adsorption by Goethite and Gibbsite 11. Desorption of Anions from Hydrous Oxide Surfaces," *J. Soil Sci.* 25: 16-26 (1974).
51. Neoh, L. S. "Desorption of Phosphate from Goethite," Ph.D. Thesis, University of Western Australia (1975).
52. Atkinson, R. J., A. M. Posner and J. P. Quirk. "Kinetics of Heterogeneous Isotopic Exchange Reactions. Exchange of Phosphate at the alpha-FeOOH-Aqueous Solution Interface," *J. Inorg. Nucl. Chem.* 34: 2201-2211 (1972).
53. Kyle, J. H., A. M. Posner and J. P. Quirk. "Kinetics of Isotopic Exchange of Phosphate Adsorbed on Gibbsite," *J. Soil Sci.* 26: 32-42 (1975).
54. Russell, J. D., R. L. Parfitt, A. R. Fraser and V. C. Farmer. "Surface Structures of Gibbsite, Goethite and Phosphated Goethite," *Nature* 248: 220-221 (1974).

55. Atkinson, R. J., R. L. Parfitt and R. St. C. Swart. "Infra-red Study of Phosphate Adsorption on Goethite," *J. Chem. Soc. Faraday Trans.* 70: 1472-1479 (1974).
56. Russell, J. D., E. Paterson, A. R. Fraser and V. C. Farmer. "Adsorption of Carbon Dioxide on Goethite (α-FeOOH) Surfaces, and its Implications for Anion Adsorption," *J. Chem. Soc. Faraday Trans.* 71: 1623-1630 (1975).
57. Parfitt, R. L., V. C. Farmer and J. D. Russell. "Adsorption on Hydrous Oxides 1. Oxalate and Benzoate on Goethite," *J. Soil Sci.* 28: 29-39 (1977).
58. Parfitt, R. L., A. R. Fraser, J. D. Russell and V. C. Farmer. "Adsorption on Hydrous Oxides 11. Oxalate, Benzoate and Phosphate on Gibbsite," *J. Soil Sci.* 28: 40-47 (1977).
59. Parfitt, R. L., and J. D. Russell. "Adsorption on Hydrous Oxide IV. Mechanisms of Adsorption of Various Ions on Goethite," *J. Soil Sci.* 28: 297-305 (1977).
60. Parfitt, R. L., and R. St. C. Swart. "Infra-red Spectra from Binuclear Bridging Complexes of Sulphate Adsorbed on Goethite (α-Fe OOH)," *J. Chem. Soc. Faraday Trans.* 73: 796-802 (1977).
61. Hayward, D. O., and B. M. W. Trapnell. *Chemisorption* (London: Butterworth's, 1964), p. 93.
62. Allen, J. A., and P. H. Scaife. "The Elovich Equation and Chemisorption Kinetics," *Aust. J. Chem.* 15: 2023 (1966).
63. Aharoni, C., and M. Ungarish. "Kinetics of Activated Chemisorption Part 1—The Non-Elovichian Part of the Isotherm," *J. Chem. Soc. Faraday Trans.* 72: 400-408 (1976).
64. Aharoni, C., and M. Ungarish. "Kinetics of Activated Chemisorption Part 2.—Theoretical Models," *J. Chem. Soc. Faraday Trans.* 73: 456-464 (1977).
65. Ulrich, B., H. Lin and H. Katapurka. "Kinetics of Isotopic Exchange Between Soil Phosphates, Soil Solution and Plant," in *Radioisotopes in Soil-Plant Nutrition Studies* (International Atomic Energy Agency, Vienna, 1962).
66. Mattingly, G. E. G., and O. Talibudeen. *Topics in Phosphorus Chemistry,* Vol. 4, M. Grayson and E. J. Griffith, Eds. (New York: Wiley-Interscience, 1967).
67. Atkinson, R. J., F. J. Hingston, A. M. Posner and J. P. Quirk. "Elovich Equation for the Kinetics of Isotope Exchange Reactions at Solid-Liquid Interfaces," *Nature* 226: 148-149 (1970).
68. Hingston, F. J. "Specific Adsorption of Anions on Goethite and Gibbsite," Ph.D. Thesis, University of Western Australia (1970).
69. Hingston, F. J., A. M. Posner and J. P. Quirk. "Competitive Adsorption of Negatively Charged Ligands on Oxide Surfaces," *Disc. Faraday Soc.* 52: 334-342 (1971).
70. Nagarajah, S., A. M. Posner and J. P. Quirk. "Desorption of Phosphate from Kaolinite by Citrate and Bicarbonate," *Soil Sci. Soc. Am. Proc.* 32: 507-510 (1968).
71. Hingston, F. J., R. J. Atkinson, A. M. Posner and J. P. Quirk. "Specific Adsorption of Anions on Goethite," *9th Int. Cong. Soil Sci. Trans.* 1: 669-678 (1968).

72. Juo, A. S. R., and R. L. Fox. "Phosphate Sorption Characteristics of Some Benchmark Soils of West Africa," *Soil Sci.* 124: 370-376 (1977).
73. El-Swaify, S. A., and A. H. Sayegh. "Charge Characteristics of an Oxisol and an Inceptisol from Hawaii," *Soil Sci.* 120: 49-56 (1975).
74. Gebhardt, H., and N. T. Coleman. "Anion Adsorption by Allophanic Tropical Soils: I. Chloride Adsorption," *Soil Sci. Soc. Am. Proc.* 38: 255-259 (1974).
75. Gebhardt, H., and N. T. Coleman. "Anion Adsorption by Allophanic Tropical Soils: II. Sulphate Adsorption," *Soil Sci. Soc. Am. Proc.* 38: 259-262 (1974).
76. Barrow, N. J. "Comparison of the Adsorption of Molybdate, Sulphate and Phosphate by Soils," *Soil Sci.* 109: 282-288 (1970).
77. Mekaru, T., and G. Uehara. "Anion Adsorption in Ferruginous Tropical Soils," *Soil Sci. Soc. Am. Proc.* 36: 296-300 (1972).
78. Theng, B. K. G. "Adsorption of Molybdate by Some Crystalline and Amorphous Soil Clays," *New Zealand J. Sci.* 14: 1040-1056 (1971).
79. Gonzalez, R., B. H. Appelt, E. B. Schalscha and F. T. Bingham. "Molybdate Adsorption Characteristics of Volcanic-Ash-Derived Soils in Chile," *Soil Sci. Soc. Am. Proc.* 38: 903-906 (1974).
80. Geering, H. R., E. E. Cary, L. H. P. Jones and W. H. Allaway. "Solubility and Redox Criteria for the Possible Forms of Selenium in Soils," *Soil Sci. Soc. Am. Proc.* 32: 35-40 (1968).
81. Frost, R. R., and R. A. Griffin. "Effect of pH on Adsorption of Arsenic and Selenium from Landfill Leachate by Clay Minerals," *Soil Soc. Sci. Am. Proc.* 41: 53-57 (1977).
82. Obihara, C. H., and E. W. Russell. "Specific Adsorption of Silicate and Phosphate by Soils," *J. Soil Sci.* 23: 105-117 (1972).
83. Gebhardt, H., and N. T. Coleman. "Anion Adsorption by Allophanic Tropical Soils III. Phosphate Adsorption," *Soil Sci. Soc. Am. Proc.* 38: 263-266 (1974).
84. Hingston, F. J. "Reactions between Boron and Clays," *Aust. J. Soil Res.* 2: 83-85 (1964).
85. Hatcher, J. T., C. A. Bower and M. Clark. "Adsorption of Boron by Soils as Influenced by Hydroxyl Aluminium and Surface Area," *Soil Sci.* 104: 422-426 (1967).
86. Sims, J. R., and F. T. Bingham. "Retention of Boron by Layer Silicates, Sesquioxides and Soil Materials. I. Layer Silicates," *Soil Sci. Soc. Am. Proc.* 31: 728-732 (1967).
87. Sims, J. R., and F. T. Bingham. "Retention of Boron by Layer Silicates, Sesquioxides and Soil Materials. II. Sesquioxides," *Soil Sci. Soc. Am. Proc.* 32: 364-369 (1968).
88. Sims, J. R., and F. T. Bingham. "Retention of Boron by Layer Silicates Sesquioxides and Soil Materials. III. Iron- and Aluminium-Coated Layer Silicates and Soil Materials," *Soil Sci. Soc. Am. Proc.* 32: 369-373 (1968).
89. Kafkafi, U., and B. Bar-Yosef. "The Effect of pH on the Adsorption and Desorption of Silica and Phosphate on and from Kaolinite," *Proc. Int. Clay Conf., Tokyo,* Vol. 1 691 (1969), pp. 691-696.
90. Kafkafi, U. "Hydrogen Consumption and Silica Release during Initial Stages of Phosphate Adsorption on Kaolinite at Constant pH," *Israel J. Chem.* 6: 367-373 (1968).

91. Bar-Yosef, B., U. Kafkafi and N. Lahav. "Relationships Among Adsorbed Phosphate, Silica, and Hydroxyl during Drying and Rewetting of Kaolinite Suspension," *Soil Sci. Soc. Am. Proc.* 33: 672-677 (1969).
92. Chen, Y. R., J. N. Butler and W. Stumm. "Adsorption of Phosphate on Alumina and Kaolinite from Dilute Aqueous Solutions," *J. Colloid Interface Sci.* 43: 421-436 (1973).
93. Barrow, N. J. "The Slow Reactions Between Soil and Anions: 1. Effects of Time, Temperature, and Water Content of a Soil on the Decrease in Effectiveness of Phosphate for Plant Growth," *Soil Sci.* 118: 380-386 (1974).
94. Munns, D. N., and R. L. Fox. "The Slow Reaction which Continues after Phosphate Adsorption: Kinetics and Equilibrium in Some Tropical Soils," *Soil Sci. Soc. Am. Proc.* 40: 46-51 (1976).
95. Olsen, S. R., W. D. Kemper and R. D. Jackson. "Phosphate Diffusion to Plant Roots," *Soil Sci. Soc. Am. Proc.* 26: 222-227 (1962).
96. Olsen, S. R., and F. S. Watanabe. "Diffusive Supply of Phosphorus in Relation to Soil Textural Variations," *Soil Sci.* 110: 318-327 (1970).
97. Nye, P. H. "The Rate Limiting Step in Plant Nutrient Adsorption from Soil," *Soil Sci.* 123: 292-297 (1977).
98. Tinker, P. B. H., and F. E. Sanders. "Rhizosphere Microorganisms and Plant Nutrition," *Soil Sci.* 119: 363-368 (1975).
99. Nye, P. H., and P. B. Tinker. *Solute Movement in the Soil-Root System* (Oxford: Blackwell Scientific Publications, 1977).
100. Bhat, K. K. S., P. H. Nye and J. P. Baldwin. "Diffusion of Phosphate to Plant Roots in Soil. IV. The Concentration in Distance Profile in the Rhizosphere of Roots with Root Hairs in a Low-P Soil," *Plant Soil* 44: 63-72 (1976).
101. Brewster, J. L., K. K. S. Bhat and P. H. Nye. "The Possibility of Predicting Solute Uptake and Plant Growth Response from Independently Measured Soil and Plant Characteristics. V. The Growth and Phosphorus Uptake of Rape in Soil at a Range of Concentrations and a Comparison of Results with the Predictions of a Simultaneous Model," *Plant Soil* 44: 295-328 (1976).
102. Mattingly, G. E. G. "Labile Phosphate in Soils," *Soil Sci.* 119: 369-375 (1975).
103. Barrow, N. J., and T. C. Shaw. "The Slow Reactions between Soil and Anions: 2. Effect of Time and Temperature on the Decrease in Phosphate Concentration in the Soil Solution," *Soil Sci.* 119: 167-177 (1975).
104. Barrow, N. J., and T. C. Shaw. "The Slow Reactions Between Soil and Anions: 3. The Effects of Time and Temperature on the Decrease in Isotopically Exchangeable Phosphate," *Soil Sci.* 119: 190-197 (1975).
105. Barrow, N. J., and T. C. Shaw. "The Slow Reactions Between Soil and Anions: 4. Effect of Time and Temperature on Contact between Soil and Molybdate on the Uptake of Molybdenum by Plants and on the Molybdate Concentration in the Soil Solution," *Soil Sci.* 119: 301-316 (1975).
106. Barrow, N. J., and T. C. Shaw. "The Slow Reactions between Soil and Anions. 5. Effects of Period of Prior Contact on the Desorption of Phosphate from Soils," *Soil Sci.* 119: 311-326 (1975).

107. White, R. E., and A. W. Taylor. "Reactions of Soluble Phosphate with Acid Soils: The Interpretation of Adsorption-Desorption Isotherms," *J. Soil Sci.* 28: 314-328 (1977).
108. Kuo, S., and E. G. Lotse. "Kinetics of Phosphate Adsorption and Desorption by Lake Sediments," *Soil Sci. Soc. Am. Proc.* 38: 50-54 (1974).
109. Enfield, C. G. "Rate of Phosphorus Sorption by Five Oklahoma Soils," *Soil Sci. Soc. Am. Proc.* 38: 404-407 (1974).
110. Enfield, C. G., C. C. Harlin, Jr. and B. E. Bledsoe. "Comparison of Five Kinetic Models for Orthophosphate Reactions in Mineral Soils," *Soil Sci. Soc. Am. Proc.* 40: 243-249 (1976).
111. Evans, R. L., and J. J. Jurinak. "Kinetics of Phosphate Release from a Desert Soil," *Soil Sci.* 121: 205-211 (1976).
112. Griffin, R. A., and R. G. Burau. "Kinetic and Equilibrium Studies of Boron Desorption from Soil," *Soil Sci. Soc. Am. Proc.* 38: 892-897 (1974).
113. Olsen, S. R., C. V. Cole, F. S. Watanabe and L. A. Dean. "Estimation of Available Phosphorus in Soils by Extraction with Sodium Bicarbonate," Circular 939. U. S. Department of Agriculture, Washington, D.C. (1954).
114. Bray, R. H., and L. T. Kurtz. "Determination of Total Organic and Available Forms of Phosphorus in Soils," *Soil Sci.* 59: 39-43 (1945).
115. Kilmer, V. J., and D. C. Nearpass. "The Determination of Available Sulfur in Soils," *Soil Sci. Soc. Am. Proc.* 24: 337-340 (1960).
116. Fox, R. L., R. A. Olson and H. F. Rhoades. "Evaluation of the Sulphur Status of Soils by Plant and Soil Tests," *Soil Sci. Soc. Am. Proc.* 28: 243-246 (1964).
117. Gupta, U. C., and D. C. Mackay. "Procedure for the Determination of Exchangeable Copper and Molybdenum in Podzol Soils," *Soil Sci.* 101: 93-97 (1966).
118. Barrow, N. J., and T. C. Shaw. "Sodium Bicarbonate as an Extractant for Soil Phosphate, 1. Separation of the Factors Affecting the Amount of Phosphate Displaced from Soil from those Affecting Secondary Adsorption," *Geoderma* 16: 91-107 (1976).
119. Barrow, N. J., and T. C. Shaw. "Sodium Bicarbonate as an Extractant for Soil Phosphate, II. Effect of Varying the Conditions of Extraction on the Amount of Phosphate Initially Displaced and on the Secondary Adsorption," *Geoderma* 16: 109-123 (1976).
120. Barrow, N. J. "On the Displacement of Adsorbed Anions from Soil. 2. Displacement of Phosphate by Arsenate," *Soil Sci.* 117: 28-33 (1974).
121. Barrow, N. J. "Studies on Extraction and on Availability to Plants of Adsorbed plus Soluble Sulphate," *Soil Sci.* 104: 242-249 (1967).
122. Barrow, N. J. "On the Displacement of Adsorbed Anions from Soil: 1. Displacement of Molybdate by Phosphate and by Hydroxide," *Soil Sci.* 116: 423-431 (1974).
123. Talibudeen, O. "Isotopically Exchangeable Phosphorus in Soils. III. The Fractionation of Soil Phosphorus," *J. Soil Sci.* 9: 120-129 (1958).
124. White, R. E., and A. W. Taylor. "Effect of pH on Phosphate Adsorption and Isotopic Exchange in Acid Soils at Low and High Additions of Soluble Phosphate," *J. Soil Sci.* 28: 49-61 (1977).
125. Larsen, S., and A. E. Widdowson. "Ageing of Phosphate Added to Soil," *J. Soil Sci.* 22: 5 (1971).

126. Likens, G. E., F. H. Borman, R. S. Pierce, J. S. Eaton and N. M. Johnson. *The Biogeochemistry of a Northern Hardwood Forest Ecosystem* (New York: Springer-Verlag New York, Inc., 1977).
127. Reid, G. W. "On Nutritional Pollution," *Int. J. Environ. Studies* 2: 271-275 (1972).
128. Garrell, M. H., J. C. Confer, D. Kirschner and A. W. Fast. "Effects of Hypolimnetic Aeration on Nitrogen and Phosphorus in a Eutrophic Lake," *Water Resources Res.* 13: 343-347 (1977).
129. Shukla, S. S., J. K. Syers, J. D. H. Williams, D. E. Armstrong and R. F. Harris. "Sorption of Inorganic Phosphate by Lake Sediments," *Soil Sci. Soc. Am. Proc.* 35: 244-249 (1971).
130. Lance, J. C. "Phosphate Removal from Sewage Water by Soil Columns," *J. Environ. Qual.* 6: 279-284 (1977).
131. Parizek, R. R., L. T. Kardos, W. E. Sopper, E. A. Myers, D. E. Davis, M. A. Farrell and J. B. Nesbitt. *Wastewater Renovation and Conservation*, Penn State Studies No. 23 (University Park, PA: The Pennsylvania State University Press, 1967).
132. Woolson, E. A., J. H. Axley and P. C. Kearney. "The Chemistry and Phytotoxicity of Arsenic in Soils: I. Contaminated Field Soils," *Soil Sci. Soc. Am. Proc.* 35: 938-943 (1971).
133. Woolson, E. A., J. H. Axley and P. C. Kearney. "The Chemistry and Phytotoxicity of Arsenic in Soils. II. Effects of Time and Phosphorus," *Soil Sci. Soc. Am. Proc.* 37: 254-259 (1973).
134. McClenahen, J. R. "Distribution of Soil Fluorides near an Airborne Fluoride Source," *J. Environ. Qual.* 5: 472-475 (1976).
135. Morel, F., R. E. McDuff and J. J. Morgan. "Interactions and Chemostasis in Aquatic Chemical Systems: Role of pH, pE, Solubility and Complexation," in *Trace Metals and Metal-Organic Interactions in Natural Waters*, P. C. Singer, Ed. (Ann Arbor, MI: Ann Arbor Science Publishers, Inc., 1974).
136. Morel, F., and J. J. Morgan. "A Numerical Method for Computing Equilibria in Aqueous Chemical Systems," *Environ. Sci. Technol.* 6: 58-67 (1972).

CHAPTER 3

CATION ADSORPTION BY HYDROUS METAL OXIDES AND CLAY

David G. Kinniburgh
  Hydrogeology Unit
  Institute of Geological Sciences
  Wallingford, Oxon
  United Kingdom

Marion L. Jackson
  Department of Soil Science
  University of Wisconsin
  Madison, Wisconsin

INTRODUCTION

Hydrous metal oxides are those solids that have as their sole constituents one or more metallic cations combined with the elements of water, hydrogen and oxygen. This includes the metal hydroxides, oxyhydroxides and also the metal oxides because in aqueous solution their surface chemistry invariably reveals that they are at least partially hydrated. This chapter deals principally with simple hydrous metal oxides containing only one type of metallic cation, but also reviews briefly cation adsorption on the phyllosilicate clays. Coatings of simple oxides often occur on the phyllosilicate planar surfaces.

CATION ADSORPTION BY HYDROUS METAL OXIDES

Studies of the chemistry of the hydrous metal oxides have appeared in a large number of scientific journals from a wide range of disciplines (Table I). This interest usually arises from purely practical considerations, such as the need to scavenge trace amounts of an element from solution or for the improvement of the flotation of a mineral. It is for this reason that many of the studies have involved complex, and sometimes poorly defined, chemical

Table I. Some of the Areas of Research Concerned with the Chemistry of the Hydrous Metal Oxides

Chemistry
    Analytical and radiochemistry—use in ion exchange separation, scavenging by sorption and coprecipitation
    Electrochemistry—alkaline cells, semiconductor surface chemistry
    Theoretical—electrical double layer theory

Engineering
    Mineral separation—flotation, sedimentation, activation
    Water treatment—flocculation, clarification, sedimentation, effluent decontamination, microflotation
    Catalysis—preparation and reactivity of catalysts
    Metallurgy—corrosion
    Nuclear reactors—nuclear fuels
    Materials science—ferromagnetic materials

Medicine
    Pharmaceuticals—antacids, antiperspirants

Environmental Sciences
    Soils—retention of anions, heavy metal pollution, radioactive contamination, soil acidity, soil stability, soil genesis, tropical soils, spodic horizons
    Rivers—transport of sorbed nutrients and wastes, geochemical exploration
    Lakes—nutrient retention and release, lake amelioration
    Oceans—sedimentation rate, heterocoagulation, ferromanganese nodule composition and structure, scavenging of trace metals

systems. Interest has frequently been focused on defining the behavior of the particular cation-oxide system, rather than understanding the mechanisms underlying that behavior. However, in approximately the last 15 years there has been considerable interest in the acid-base properties of the hydrous metal oxides, which has stimulated interest in models of the electrical double layer at the oxide-aqueous electrolyte interface. At the same time, there has been a growing awareness of the role that oxides can play in the environmental chemistry of many metal ions. A number of detailed studies of cation adsorption have followed which, in some cases, have combined measurements of cation adsorption with measurements of other interfacial properties such as surface area, surface charge and electrokinetic potential. There is now a wealth of information on the adsorption of cations by particular hydrous metal oxides under a wide range of conditions. The purpose of this chapter is to collect this information so as to highlight the full range of behavior found. Parts of this material have been reviewed previously [1-5]. The charge properties of the oxides or the models that have been used to explain cation adsorption will not be discussed in detail, as reviews of these topics have appeared elsewhere [6-18].

Kinetics

Adsorption of cations by hydrous metal oxides is frequently found to be extremely rapid, most of the exchange occurring within a matter of minutes [19-25]. This rapid adsorption reflects the fact that the adsorption is a surface phenomenon and that the surfaces are readily accessible to the ions in solution.

Some evidence suggests that the kinetics of adsorption depend on the state of aggregation of the primary crystallites. For example, Morgan and Stumm [26] found that Mn^{2+} sorption (and H^+ release) by dispersed colloidal MnO_2 was very rapid, but in the presence of 0.01 M $NaClO_4$ the reaction rate was considerably slower. It was suggested that in 0.01 M $NaClO_4$, the colloidal MnO_2 was flocculated and adsorption involved diffusion of Mn^{2+} to less readily available sites. Kurbatov and Wood [27] also found that the rate of uptake of Co^{2+} by coprecipitated Fe gel decreased with increased inert electrolyte concentration, but these results were not consistent with previous findings [28] and may have reflected gradual pH changes. Simon et al. [29] made a detailed study of the kinetics of Cd^{2+} sorption with freshly coprecipitated Al gel and found that equilibrium was reached in about three hours. They found that after an initial instantaneous sorption step the slower sorption could be divided into two simultaneous first-order reactions with half-lives of about 4 minutes and 40 minutes. Other workers have also described the approach to equilibrium in terms of one or more first-order reactions [24,27].

In microporous oxide systems, especially those obtained by heating, equilibrium is achieved somewhat more slowly [30]. The rate of exchange is generally controlled by the rate of ion diffusion within the particle [31,32] and this is related to the size, shape and spatial distribution of the pores. The size distribution of micropores in hydrous oxides is frequently found to be very sensitive to heat treatment. The porosity and specific surface area is generally found to reach a maximum at some particular temperature (about 400-500°C for aluminum oxides and about 200°C for iron oxides), and then the specific surface area decreases with increasing temperature as a result of sintering. In view of this change in pore structure on heating, it is not surprising that the kinetics of cation exchange [33], as well as the adsorption capacity [34], may be influenced by heat treatment. Cation adsorption by activated alumina can take many hours to reach equilibrium [30,35] and this can be attributed to rate-limiting diffusion of ions within the micropores [36], which are approximately 3 nm in diameter [37]. With these slower exchange reactions, the rate of adsorption is likely to depend on precise experimental conditions such as pH, electrolyte concentration, solid/solution ratio, temperature and amount of agitation [38].

Sometimes adsorption is found to continue slowly for a relatively long time, especially with manganese dioxide minerals. For example, McKenzie [39-41] studied the uptake of transition metal ions by a variety of synthetic manganese dioxide minerals and found that after an initial rapid uptake of metal ions there was a slow reaction which, in some cases, continued for the duration of the experiment (40 days). Such a slow reaction is indicative of more than a surface exchange reaction and in the case of cobalt may involve both oxidation of Co^{2+} to Co^{3+} and substitution of Co^{3+} for structural manganese ion [42]. Co^{2+} adsorption on manganese dioxide has also been reported to achieve equilibrium within one hour [43], which suggests that the slower reaction rate observed by McKenzie may have been the result of the high initial concentration of transition metal ion used (0.1 M).

Precise monitoring of the initial rapid rate of adsorption is frequently not possible because a finite time is required to ensure proper mixing and to separate the solid and solution phases for analysis. The separation problem may be overcome by in situ measurement of adsorption by ion-selective electrodes [24] or polarography [29]. Ion-selective electrodes have the advantage that they determine the free ion activity and can therefore also give information on the extent of cation hydrolysis [44], but their response can be slow and erratic.

In view of the importance of pH in determining the extent of adsorption, it is important in kinetic studies of adsorption to maintain a constant suspension pH, or at least to monitor the variation in suspension pH. In fact, following the drop in suspension pH after addition of a multivalent cation solution to an oxide suspension is a convenient way of studying the rate of adsorption during the first hour or so of adsorption [45]. However, in some cases, the rate of response of glass electrodes may be insufficient to give reliable pH measurements in the initial stage of adsorption [29]. In low ionic strength suspensions, the change in electrical conductivity may be used since the equivalent conductivity of H_3O^+ is considerably greater than that of other cations [46].

Little systematic work has been done on the long-term effect of aging hydrous oxide gels on their cation uptake, but uptake may continue to increase slowly over periods of months. In aging studies it is important to distinguish between those studies in which the solid and cation have remained in contact over the aging period and those involving uptake on an oxide aged in the absence of the cation. In the former case, it is possible for the cation to be incorporated into the evolving oxide structure, while in the latter case, changes in uptake only reflect changes in the number and type of adsorption sites.

Since aging involves considerable surface reorganization, it is not surprising that cation sorption itself often retards crystallization of gels. Al gels are particularly susceptible to aging effects. It was found that when Al gel was aged (crystallizing to bayerite) in the presence of an equimolar mixture of Mg, Ca, Sr and Ba nitrates, the sorption-pH curves (pH 7-11) for Ca, Sr and Ba, shifted to a higher pH whereas the Mg curve shifted to a lower pH, suggesting structural incorporation of Mg but not of the other cations [47]. Similar work with Ni, Cu, Zn, Cd and Pb on Al gel (pH 4-7) has shown that with the exception of Pb, there was a substantial shift in the sorption-pH curves to a lower pH (shifts of 0.3-1.5 pH unit), the size of the shift being Ni > Cd > Zn > Cu. These results are in qualitative agreement with the ability of Ni, Cu, Zn and Cd to form double hydroxides with aluminum.

Isotopic exchange studies confirm these findings as indicated by the curves in Figure 1. At high pH, crystallization of Al gel to bayerite (Figure 1a) resulted in increased suspension pH, but no corresponding increase in Ca sorption was found. Virtually all of the sorbed Ca remained isotopically exchangeable, indicating that it was adsorbed at the surface rather than incorporated into the bulk structure of the oxide. The lack of an increase in overall sorption is probably a result of a decrease in the surface area of the gel compensating for the increase in pH. The adsorption of Zn during aging was carried out at a lower pH of 6.2 to 6.1, at which aging was much slower. Although Zn adsorption showed only a small overall decrease with time, the isotopic exchangeability of the sorbed Zn steadily decreased until after 400 hours only 10% of the sorbed Zn was isotopically exchangeable (Figure 1b). The Zn thus was steadily incorporated into the bulk structure of the aging gel. McBride [48], using electron spin resonance (ESR) spectroscopy, has shown that the Cu^{2+} adsorbed or coprecipitated with Al gel is far less mobile than Cu^{2+} in aqueous solution (or Cu^{2+} adsorbed on clay minerals). This is consistent with some kind of structural incorporation of the adsorbed Cu^{2+}; it would be interesting to see whether the ESR spectra showed any change with time of aging. The ESR spectra also confirmed that the coprecipitated Cu^{2+} was evenly distributed in the Al gel rather than precipitated as a separate $Cu(OH)_2$ phase.

pH Dependence

Monovalent Cations

Numerous studies have shown the importance of pH as a factor controlling the extent of cation adsorption by hydrous metal oxides. For most monovalent cations, such as the alkali metal cations, adsorption is usually nonspecific

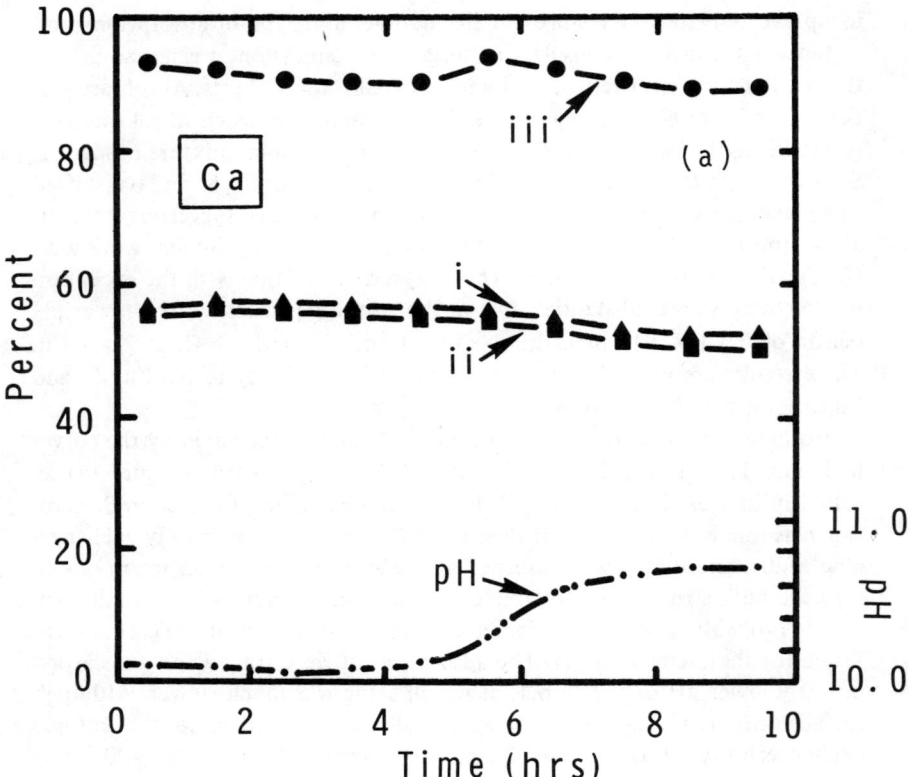

Figure 1a. Effect of time of aging on the isotopic exchangeability of coprecipitated ions. Coprecipitation of 1.0×10^{-2} M Ca with 9.3×10^{-2} M Al in $1 M$ $NaNO_3$ (pH 10.0 to 10.7). Curve i is the percent Ca or Zn coprecipitated as determined by analysis of the supernatant solution. Curve ii is the percent ^{45}Ca tracer adsorbed in 15 min after addition at various times after coprecipitation. Curve iii, calculated from curves i and ii, is the percent of the adsorbed Ca or Zn that was isotopically exchangeable in 15 min.

and therefore depends directly on the surface charge of the oxide as determined by the amount of H^+/OH^- adsorption. At the point of zero charge (pzc), there is usually very little cation sorption, but above the pzc, cations are adsorbed to counterbalance the net negative surface charge. Part of the negative surface charge is also counterbalanced by the exclusion of anions, but this usually compensates for only a small fraction of the total charge, especially at high electrolyte concentrations [49]. Some recent literature values for the pzc of various metal oxides are given in Table II. Previous tabulations were given by Parks [8,9].

The potentiometric titration method for constructing charge-pH curves [50] gives an indirect measure of the ion capacity of the electrical double layer at a given pH and electrolyte concentration. The oxide suspension is

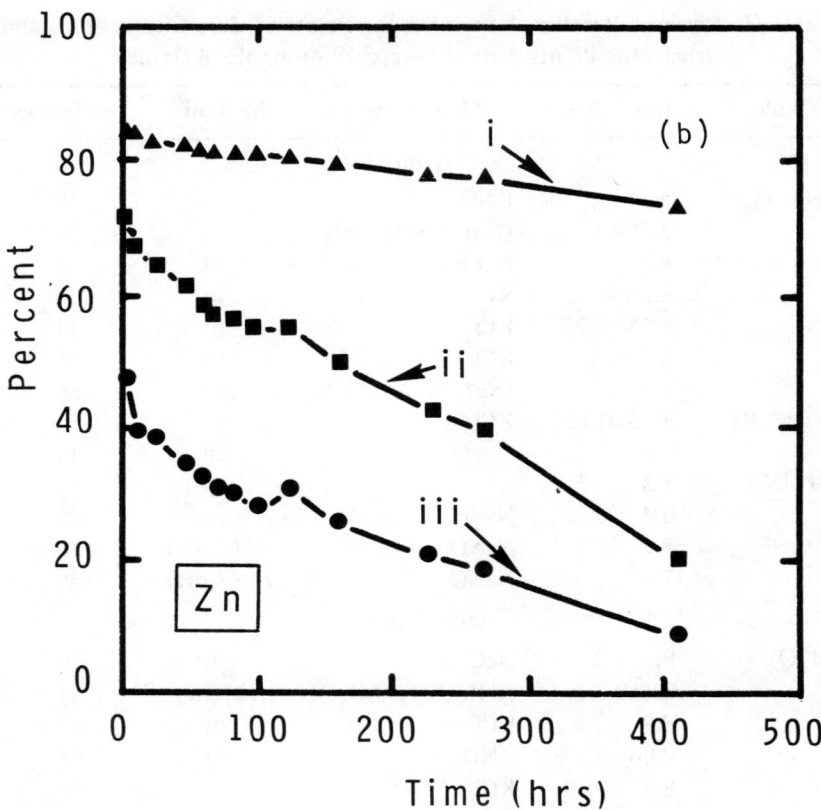

Figure 1b. Effect of time of aging on the isotopic exchangeability of coprecipitated ions. Coprecipitation of 1.0×10^{-2} M Zn with 9.3×10^{-2} M Al in $1\,M$ $NaNO_3$ (pH 6.2 to 6.1). Curve i is the percent Zn coprecipitated as determined by analysis of the supernatant solution. Curve ii is the percent ^{65}Zn tracer adsorbed in 15 min after addition at various times after coprecipitation. Curve iii, calculated from curves i and ii, is the percent of the adsorbed Zn that was isotopically exchangeable in 15 min.

titrated with either acid or base at different concentrations of indifferent electrolyte, and the pzc is located at the common intersection of the various titration curves. This titration procedure is more reliable than direct adsorption measurements at high electrolyte concentrations where the "excess salt" problem makes direct measurement of adsorption difficult. Results of the two methods are generally in reasonable agreement [51,81]. Some typical curves obtained for several oxides are shown in Figure 2. Cation adsorption increases with increasing pH and electrolyte concentration. If the electrolyte is not indifferent, i.e., either the cation or anion is specifically adsorbed, then there is no unique crossover point.

An interesting feature of the solid-solution interface common to many oxides is the high apparent surface charge density compared with other sur-

Table II. Recent Literature Values for the Points of Zero Charge (pzc) and Isoelectric Points (iep) of Several Hydrous Metal Oxides

Oxide	pzc (iep)	Electrolyte	Method[a]	Reference
		Iron		
α-Fe$_2$O$_3$:	8.5	KNO$_3$	tit	50
	8.7±0.10	(Na)Cl,NO$_3$,ClO$_4$	tit	51
	8.3	NaClO$_4$	tit	52
	8.68±0.12	KCl	tit	53
	8.45–9.27	KCl	tit	54
	8.5±0.2	KCl	tit	55
	6.0–6.6	(Na?)NO$_3$,Cl,ClO$_4$	ep	56
α-FeOOH:	7.55±0.15	KCl	tit	54
	9.0		ep	44
β-FeOOH	7.3	–	ep	56
	6.4	NaCl	ep	57
Fe gel	8.1	NaNO$_3$	tit (ΔpH)	58
	7.9	NaNO$_3$	tit (ΔpH)	59
		Aluminum		
Al$_2$O$_3$:	9.2±0.2	NaCl	stp	60
	9.1±0.1	(K)Cl,NO$_3$,ClO$_4$	tit,ep,subs	61
	8.9	KCl	tit	62
	9.06	KNO$_3$	tit	63
	8.5	KCl	tit	64
	8.3	KCl,KNO$_3$	ep	65
	8.3	NaClO$_4$	tit	66
AlOOH	9.2	NaNO$_3$	ep	67
Al gel	9.0–9.2	NaCl	ep	68
	9.3	deionized	ep	69
	~9.4	NaNO$_3$	tit (ΔpH)	58
		Silicon		
SiO$_2$	~3.5	KCl	tit	70
	1.8–2.0	KCl,KClO$_4$	ep,subs	71
	~3	KCl	tit	72
	~2	KNO$_3$	ep	73
		Titanium		
TiO$_2$	7.15	KCl	tit	74
	5.9–6.0	NaClO$_4$	tit	75
	5.6±0.2	KNO$_3$	ep	73
	6.39	NaClO$_4$	tit	76

Table II, continued

Oxide	pzc (iep)	Electrolyte	Method[a]	Reference
TiO$_2$	5.5	KNO$_3$	tit,ep	77
	5.8	–	ep	44
		Manganese		
α-MnO$_2$	4.6±0.2	–	ep	78
	4.5±0.5	–	subs	78
β-MnO$_2$	7.3±0.2	–	ep	78
	7.3±0.2	–	subs	78
γ-MnO$_2$	5.6±0.2	–	ep	78
	5.5±0.2	–	subs	78
δ-MnO$_2$	1.5	–	ep	78
	2.8±0.3	NaClO$_4$	tit	26
	2.25	NaCl	ep	79
	∼2	NaNO$_3$	ep	80
Mn(II)–	∼2	–	ep	78
manganite	1.8+0.5	–	subs	78

[a] tit = potentiometric titration (pzc); stp = streaming potential (iep); ep = electrophoretic mobility (iep); subs = rate of subsidence (iep).

faces. These high surface charge densities are not usually accompanied by particularly high electrokinetic potentials and the reasons for this have been the subject of much discussion [11,14,15,17,84,85]. Two explanations have been put forward: (1) the presence of a porous surface gel coating capable of adsorbing ions to a depth of a few nm (i.e., increased surface area for adsorption); and (2) the formation of ion pairs between discrete surface sites and adsorbed counterions so that some of the counterions are adsorbed closer to the surface than the shear plane. There is little experimental work on the surface structure of oxides, although it is well known that ground quartz (SiO$_2$) has a disturbed surface layer that is more soluble than the bulk. Etching experiments on vitreous silica rods have shown that the penetration depth of Na$^+$ is less than 0.3 nm [86]; however, similar experiments with another type of silica (microporous silica) have confirmed that Na$^+$, but not Br$^-$ or N$_2$, can diffuse slowly into the bulk structure [87]. There is no direct experimental evidence of any gel coatings on colloidal-sized crystalline oxides, and for these the surface ion pair hypothesis seems more plausible. In certain circumstances, where there is specific adsorption of both cation and anion, the combined effects of their adsorption can effectively mask the underlying surface charge almost completely over a wide pH range. This has been observed for Al$_2$O$_3$ in MgSO$_4$ electrolyte [65].

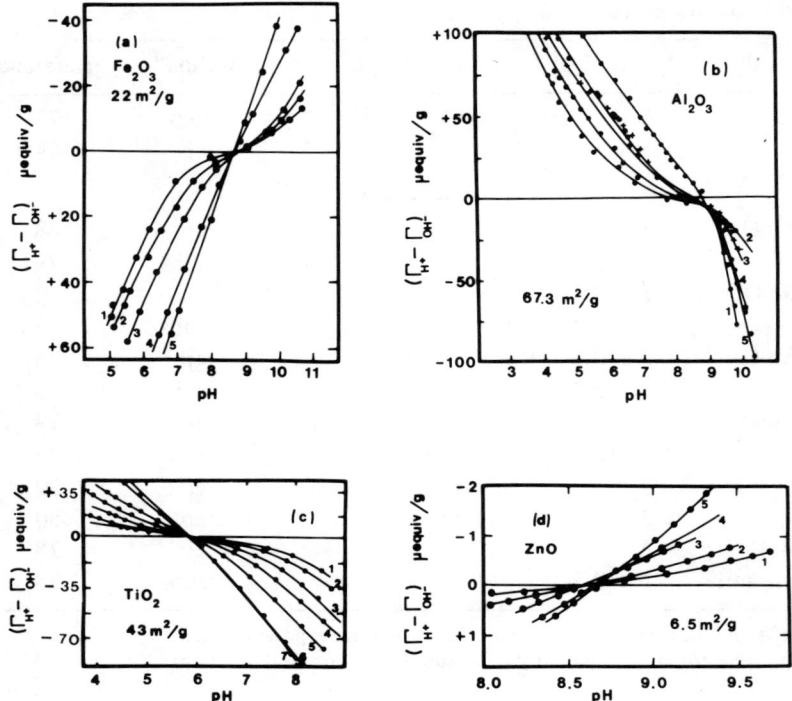

Figure 2. Charge-pH curves for various oxides (of specific surface area indicated) over a range of background electrolyte concentrations: (a) Fe_2O_3 (hematite) in KNO_3: 1 = 10^{-4} M; 2 = 10^{-3} M; 3 = 10^{-2} M; 4 = 10^{-1} M; 5 = 1 M [50]; (b) Al_2O_3 in KCl: 1 = 10^{-4} M; 2 = 10^{-3} M; 3 = 10^{-2} M; 4 = 10^{-1} M; 5 = 1 M [62]; (c) TiO_2 (rutile) in $NaNO_3$: 1 = 10^{-3} M; 2 = 3 x 10^{-3} M; 3 = 10^{-2} M; 4 = 3 x 10^{-2} M; 5 = 10^{-1} M; 6 = 1 M; 7 = 2 M [82]; (d) ZnO in $NaNO_3$: 1 = 3 x 10^{-4} M; 2 = 2 x 10^{-3} M; 3 = 1.3 x 10^{-2} M; 4 = 3.5 x 10^{-2} M; 5 = 9.7 x 10^{-2} M [83].

Where specific adsorption of monovalent cations occurs, the pH dependence of adsorption depends on the particular mechanism involved. But as with nonspecific adsorption, an increase in pH generally leads to an increase in adsorption, although the increases are not as great as found with multivalent cations [88].

Divalent Cations

It is well known that storage of solutions containing trace concentrations of multivalent cations can lead to a considerable loss in concentration as a result of the cations being adsorbed on the walls of the vessel [89]. This loss is usually undesirable and can be prevented by acidification, for example by the addition of 5 ml/l of concentrated (16 M) nitric acid [90]. The tendency for multivalent cations to be adsorbed on solid surfaces has also been put to use for scavenging impurities from solution and for the selective preconcentration of a number of trace elements. The hydrous metal oxides, particularly those of iron, aluminum and manganese, have been used most frequently for these purposes, and the efficiency of removal of the ions from solution has invariably been found to be strongly pH dependent. A rapid increase in uptake of the metal ion usually occurs over a narrow pH range (Figure 3). This is particularly true for the strongly hydrolyzable cations [1,91-96], and has led many workers to view this uptake as an H^+-M^{2+} exchange reaction with the protons derived from the weakly acidic surface-OH groups.

Kurbatov and co-workers [45] were the first to apply the law of mass action to this ion exchange equilibrium and this has proved to be a useful empirical approach. It is useful for comparing adsorption of different ions under otherwise identical conditions where the contribution of nonspecific adsorption is negligible. These conditions are frequently encountered since multivalent cation adsorption is often carried out in the presence of a high concentration of background electrolyte at a pH below the pzc, where the surface has a net positive charge. In such cases, adsorption is usually better correlated with the hydrolysis characteristics of the adsorbing ion rather than with the charge properties of the solid. The ion exchange approach also emphasizes the importance of the distribution ratio and the usefulness of the log D against pH plot, where D is equal to [% sorbed/(100 − % sorbed)] and is a solid:solution distribution ratio. This plot, which we shall call a Kurbatov plot, is useful for two reasons. Firstly, it is frequently linear for trace divalent cation adsorption, so enables the sigmoid-shaped percent adsorption against pH curve to be transformed to a simple two-parameter equation (Figure 4). Secondly, one of these parameters, the slope of the plot, gives an indication of the stoichiometry of the H^+-M^{n+} exchange reaction providing adsorption is at very low surface coverages and low solution concentrations ($< 10^{-6} M$). No other method is suitable for determination of this stoichiometry under these conditions. Nonetheless, caution must be exercised in the interpretation of the slope of the log D against pH plot in terms of the stoichiometry of the H^+-M^{n+} exchange reaction because this interpretation has not been firmly established theoretically and is known to be suspect under certain conditions [97-99]. At best, this interpretation is only valid in the linear region of the adsorption isotherm (i.e., at very low concentrations), although it has frequently been applied outside this region. At higher concentrations, even though the plots may still be linear, they tend to underestimate the true H^+/M^{n+} stoichiometry of the exchange reaction (Figure 4).

The plot has given reasonably good straight lines for divalent cation uptake by MnO_2 [20,22,100], goethite [101], Al_2O_3 [30] and Fe gel [45,102]. Often, for the adsorption of alkaline earth cations the slope is between 0.8

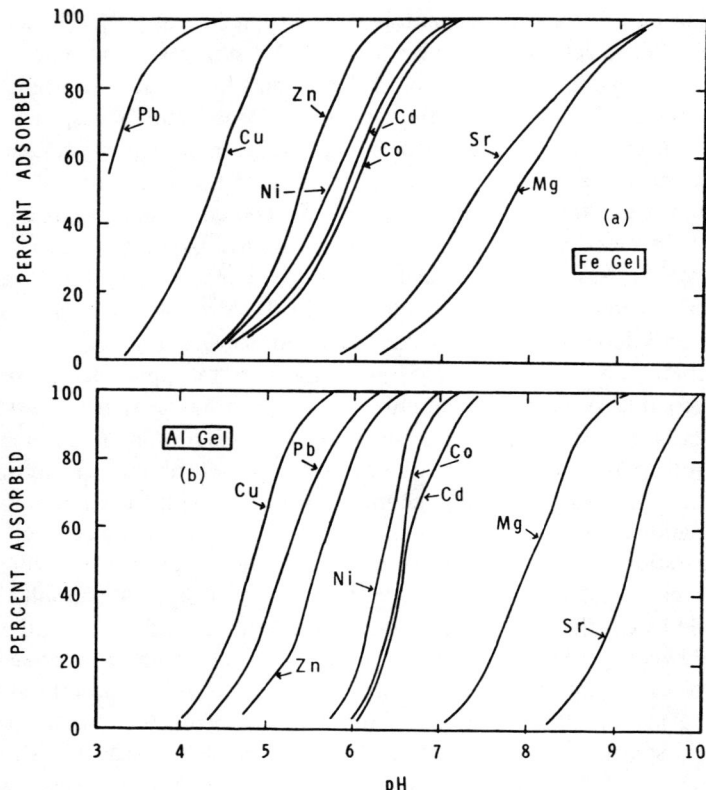

Figure 3. Adsorption-pH curves for eight divalent cations, each present at 1.25×10^{-4} M, from a 1 M $NaNO_3$ background electrolyte: (a) fresh Fe gel (0.093 M Fe); (b) fresh Al gel (0.093 M Al) [47].

and 1.2, which indicates that exchange is closer to being equimolar than equivalent [26,35,58,64,102-105]. For transition metal ions, the slope is usually greater than this [106,107], suggesting a difference in H^+/M^{n+} exchange stoichiometry. If at very low cation concentrations the slope of the plot is 1.0, then an increase in pH of one pH unit will increase adsorption by $10^{1.0}$ or 10 times, whereas a slope of 1.5 indicates an increase in adsorption of $10^{1.5}$ or 32 times.

The second parameter of the Kurbatov plot, the intercept at log D = 0, is the pH at which 50% of the original cation concentration is adsorbed, pH_{50}. For any particular solid:solution ratio, the pH_{50} reaches a limiting value as the total divalent cation concentration in the system is decreased. This value is related to the free energy of adsorption and is a convenient parameter for comparing the relative selectivity of adsorption of a number of cations [47,

CATION ADSORPTION

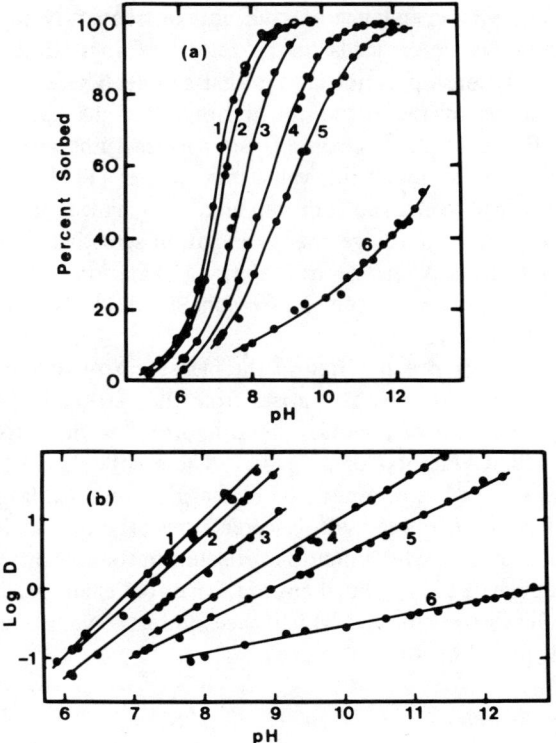

Figure 4. Strontium sorption by coprecipitation with 0.1 M Fe gel in 1 M NaNO$_3$ as a function of suspension pH and initial Sr concentration: (a) percent sorbed against pH curves at a range of initial Sr concentrations: 1 (open circles) = 10^{-7} M; 1 (full circles) = 8 × 10^{-6} M; 2 = 8 × 10^{-5} M; 3 = 10^{-3} M; 4 = 5 × 10^{-3} M; 5 = 10^{-2} M; 6 = 5 × 10^{-2} M; (b) the above curves expressed in terms of a distribution coefficient [102].

101]. The limiting pH_{50} decreases by about one pH unit for a tenfold increase in solid:solution ratio and is generally in the region of one to four pH units less than the first hydrolysis constant, pK_1, of the adsorbing cation.

Tri- and Quadrivalent Cations

While most quantitative studies of cation adsorption on hydrous metal oxides have dealt with the adsorption of divalent cations, tri- and quadrivalent cations can also be very strongly adsorbed. The ion exchange model has been extended to the coprecipitation of a variety of tri- and quadrivalent cations, such as Y^{3+}, Ce^{3+}, $RuNO_3^{4+}$ and Zr^{4+}, with hydrous metal oxides

[108-112]. The pH dependence of such uptake frequently suggests that uptake of an n-valent cation leads to the release of approximately $(n-1)$ H^+, although the relationship is not always that simple. These results can be interpreted in terms of the predominant uptake of the hydrolyzed species, $M(OH)_{n-1}^+$. For each M^{n+} adsorbed, the surface gains approximately one positive charge, and although this value may be less [113], it is rarely more. Many of these highly charged ions can undergo changes in oxidation state and this usually leads to a large change in cation selectivity. This can be useful in quantitatively separating ions in the two oxidation states. For example, Np(VI) has been separated from Np(V) on Si gel [114] and Cr(III) from Cr(VI) on Fe gel [115].

Considerable caution is required in the measurement of the adsorption of tri- and quadrivalent cations. This arises from their strong hydrolysis and the tendency to precipitate as colloidal metal hydroxides, or in special cases, to form the so-called "radiocolloids" [116]. These colloids are frequently not visible to the unaided eye, especially in the presence of a larger amount of another solid. Many of these hydrolysis reactions are slow and lead to a variety of polynuclear ions, which may be reflected in the slow attainment of adsorption equilibrium [117]. Furthermore, small pH changes can lead to relatively large changes in uptake, which necessitates very careful pH measurements for reliable adsorption-pH curves.

H^+/M^{n+} Stoichiometry

Since specific adsorption of multivalent cations almost always involves proton exchange, an important characteristic of this adsorption process is the number of protons released, or hydroxide ions adsorbed, for each cation adsorbed. This is referred to as the H^+/M^{n+} stoichiometry. At low cation concentrations and high solid:solution ratios, the steepness of the rise in the adsorption-pH curves is an indication of this stoichiometry, as discussed previously. However, a more direct method of determining this ratio is by the back-titration of the H^+ released, or, equivalently, by determining the acid-base curves in the presence and absence of the adsorbed ion. The adsorption against pH curve must also be determined either simultaneously or separately under similar conditions. Both of these methods are suitable for determining the desorption stoichiometry also, but this has not been reported. Some results for the H^+/M^{n+} adsorption stoichiometry obtained by this method are shown in Table III. There is obviously no single value that is characteristic of a particular cation or of a particular solid. However, the general tendency for the alkaline earth cations to have lower values than the transition metal ions is reasonably clear. This also tends to be the order of adsorption selectivity of these cations, and it is possible that the affinity of an ion is related in a general

Table III. H^+ Released for Each Metal Ion Adsorbed as Derived from Direct Back-Titration

Solid	Metal ion	n	Reference
δ-MnO_2	Pb^{2+}	1.4	118
	Cd^{2+}	1.3	–
	Zn^{2+}	1.1	–
	Tl^+	0.38	–
δ-MnO_2	Mn^{2+}	1.0–1.7[a]	26
	Co^{2+}	2.10±0.26	99
	Zn^{2+}	2.10±0.05	119
SiO_2	Ca^{2+}	1.0	105
	Co^{2+}	1.1–1.7[a]	120
	Fe^{2+}	1 –2[a]	121
MnO_2	Mg^{2+}	0.05–0.25	43
	Ca^{2+}	0.12–0.35	–
	Sr^{2+}	0.45	–
	Ba^{2+}	0.50	–
	Co^{2+}	1.03	–
	Mn^{2+}	0.90	–
	Ni^{2+}	1.18	–
	Zn^{2+}	0.14	–
Al_2O_3	Ca^{2+}	1.0–1.3[b]	64
	Pb^{2+}	1.5[b]	66
α-FeOOH	Cu^{2+}	1.5–2.0[a]	122
α-FeOOH	Cu^{2+}	2.4±0.5	123
	Pb^{2+}	2.0±0.5	–
	Cd^{2+}	2.2±0.6	–
	Co^{2+}	2.3±0.7	–
	Zn^{2+}	2.2±0.3	–
Fe gel	Zn^{2+}	1.65	130
	Ca^{2+}	0.95	–
	Pb^{2+}	1.2–1.6[a]	124

[a] Increase with pH.
[b] No trend with pH.

way to its ability to displace protons from the surface, or, alternatively, that the more strongly adsorbed ions (those adsorbed at lower pH) tend to hydrolyze more readily. In some cases, an increasing trend in the H^+/M^{n+} stoichiometry with increasing pH has been observed. This might reflect a change in the dominant species adsorbed as the concentration of the more hydrolyzed species in solution increases. The fact that the H^+/M^{2+} exchange stoichiometry is usually less than two for divalent cation adsorption means that the surface charge becomes increasingly positive, which is reflected in a change in the electrokinetic properties of the interface (Figure 5). In some cases, specific adsorption can reverse the sign of the effective charge of the surface. On weak acid surfaces having relatively few displaceable H^+, such as sulfides, the pH dependence of adsorption is less than on oxides, and this is also reflected in a lower H^+/M^{n+} stoichiometry. For example, the H^+/Ca^{2+} stoichiometry on many oxides is close to one (Table III), whereas on ZnS the value is about 0.3 [125].

Adsorption on manganese dioxide, especially cobalt adsorption, may involve a greater or lesser part of the charge balance being satisfied by the desorption of Mn^{2+}, either derived from adsorbed Mn^{2+} or from the displacement of structural manganese [43,99,126]. Loganathan et al. [80] found that for each mole of Co(II) adsorbed by δ-MnO_2 at pH 4, about 1 mole of H^+ and 0.5 mole of Mn^{2+} were displaced. By contrast, Zn^{2+} adsorption displaced less than 10% of the manganese that an equivalent amount of cobalt adsorption

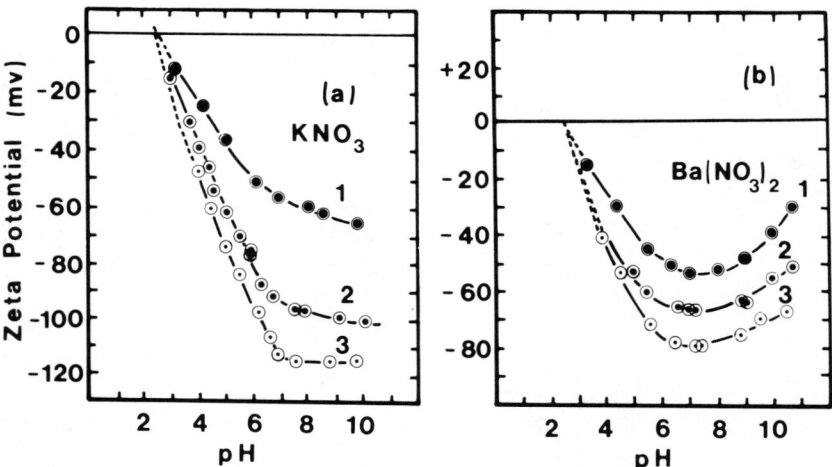

Figure 5. Change in zeta potential of vitreous silica as a function of pH and electrolyte concentration in the presence of (a) KNO_3, an indifferent electrolyte (1 = 10^{-2} M; 2 = 10^{-3} M; 3 = 10^{-4} M); (b) $Ba(NO_3)_2$ (1 = 3.3 x 10^{-4} M; 2 = 10^{-4} M; 3 = 3.3 x 10^{-5} M). Ba^{2+} (or $BaOH^+$) unlike K^+, is specifically adsorbed at high pH and this decreases the magnitude of the zeta potential in this region [138].

displaced. There is no evidence for this type of exchange being important with other oxides.

A direct consequence of the strong pH dependence of cation adsorption is the finding that there is usually a decrease in the suspension pH when an increment of an adsorbing cation is added to an oxide suspension. This drop in pH, which is very rapid, reflects the net release of H^+ and is an indication of the amount of specific adsorption. If it is reasonable to assume a constant H^+/M^{n+} stoichiometry for different parts of the adsorption isotherm, then by successively back-titrating the H^+ released after the addition of increments of M^{n+} it is possible to construct an adsorption isotherm for the specifically adsorbed ions. This approach has the advantages of being rapid and only requiring two burets and a pH meter. It is particularly suitable for studying the adsorption close to the adsorption maximum, where the usual approach of monitoring the drop in cation concentration in solution is insensitive and susceptible to analytical errors. The results using this approach (Figure 6) are in reasonable agreement with the batch approach for Ca^{2+} adsorption on Fe gel, although the H^+/Ca^{2+} stoichiometry for best fit seems to vary between about 0.9 and 1.2, compared with a value of 0.95 obtained by direct measurement. An ion-selective electrode might also be useful because adsorption could be measured simultaneously and the H^+/M^{n+} stoichiometry monitored continuously. The batch approach includes nonspecific adsorption, which might contribute significantly to the overall adsorption at high M^{n+} concentrations.

Relationship between Adsorption, Hydrolysis and Precipitation

While the general characteristics of uptake of multivalent cations by metal oxide dispersions are reasonably well known, the precise mechanisms responsible for this uptake are not easy to identify unambiguously [44,127]. The rapid increase in percentage uptake over a narrow pH range indicates that uptake is accompanied by the net release of H^+, or uptake of OH^-, but it is not clear whether the protons released come from surface-OH groups as a result of ion exchange reactions or from the preferential adsorption of hydroxy complexes from solution, or both [59,66]. As the suspension pH is increased toward the first hydrolysis constant of the cation, the concentration of hydroxy complexes increases rapidly, but the actual hydroxy species formed depend very much on the particular cation and its concentration, and sometimes on the precise conditions of formation. Adsorption is often significant when only an extremely small fraction of the cation is hydrolyzed in solution. While this does not rule out adsorption of $MOH^{(n-1)+}$ species, it implies that they are strongly preferred to the free ions. At high cation concentrations, polynuclear cations may be formed [128,129]. Even at low concentrations of divalent cations, the MOH^+ species may be of only minor, transitional

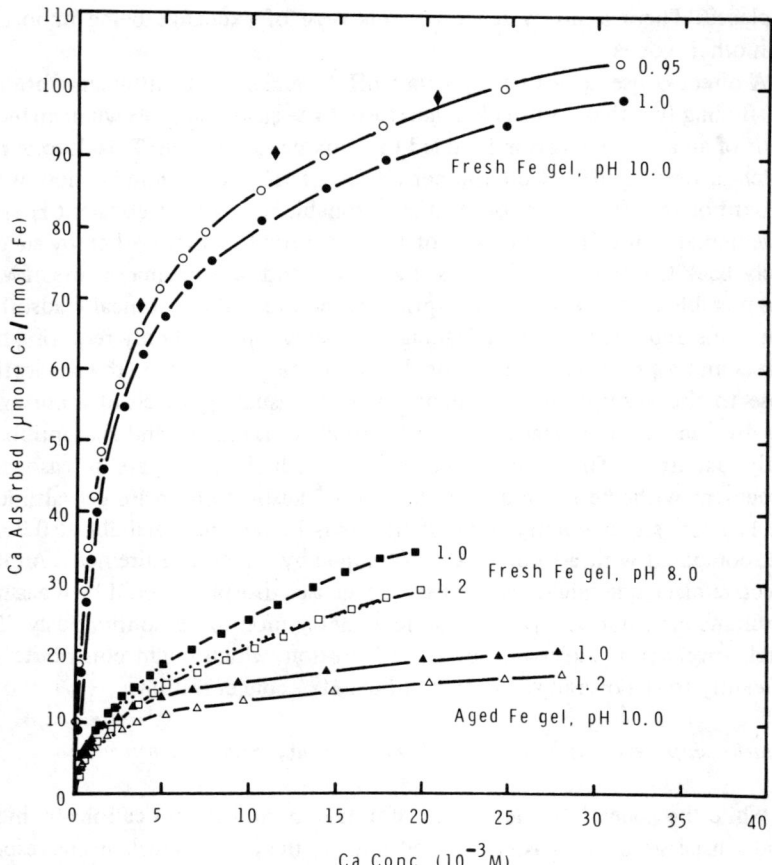

Figure 6. Ca adsorption on fresh and aged Fe gel calculated from the amount of OH^- required to maintain a constant pH during the titration of Fe gel with $Ca(NO_3)_2$. The values of the H^+/Ca^{2+} stoichiometry assumed are indicated. The experimentally determined curve (fresh gel, pH 8.0) is indicated by the dotted line, while three experimental points (diamonds) for fresh Fe gel (pH 10.0) are also included. The aged Fe gel had been aged for 35 months in $1\ M$ $NaNO_3$ during which time the pH decreased from pH 7.0 to pH 3.2.

importance compared with $M(OH)_2^0(aq)$. And when metal hydroxide precipitation does occur, the solids initially precipitated frequently have a wide range of particle sizes and solubilities. Many metal hydroxides can precipitate at substoichiometric OH^-/M^{2+} ratios by precipitation as a basic salt or as an indefinite range of solids with loosely associated anions to balance their net positive surface charge. Frequently these solids are only metastable and slowly either dissolve or age to more stable crystalline forms. This makes experimental results of cation uptake less precise at high concentrations and results in an intermediate region between simple adsorption at low concentrations and hydroxide precipitation at high concentrations. At high concentrations it is difficult to determine whether the cation is adsorbed or precipitated. The percent uptake-pH curves frequently show two features that suggest that adsorption is the dominant mechanism. Firstly, the curves shift to a lower pH as the initial cation concentration is decreased, which is the opposite of what would be expected for hydroxide precipitation. Secondly, the tailing of the curves on the low pH side of the inflection point (Figures 3 and 4) is characteristic of adsorption but not of precipitation as predicted from a strict solubility product approach. The close relationship between adsorption and precipitation is also reflected in the fact that the "adsorption" isotherms usually do not show an inflection at the onset of precipitation, at least not until bulk precipitation is appreciable.

Because of the similarities between adsorption, coprecipitation and precipitation, it is not surprising that cation selectivity is also reasonably well correlated with the solubility of the corresponding metal hydroxide [47,101]. However, it seems to be a general observation that adsorption invariably occurs before the initiation of bulk precipitation from solution [47,130-132]. It has been suggested that the solubility product of metal hydroxides is lowered at solid-solution interfaces as a result of the low dielectric constant of the water at the interface [73], although presumably this is also the case even when pure solids are precipitated. The hydrated metal oxide surface might act as a template for the formation of a surface coating of the hydroxide of the sorbed cation. The ease with which this could be done would depend on the similarities between the structures of the corresponding hydroxides and would essentially involve the extension of the oxide surface. Repulsion of the multivalent cations making up this surface coating would be minimized by the interpenetration of OH^-, as in the solid hydroxides. Parfitt [133] has discussed the surface nucleation and precipitation of $Al(OH)_3$ and $Si(OH)_4$ on TiO_2 particles with particular reference to the effect of these coatings on the electrophoretic mobility and heat of immersion of the particles. Strongly adsorbed surface coatings can drastically change the surface chemistry of minerals, and widespread use of this fact is made in the mineral processing industry. Even a fairly low surface coverage of adsorbed aluminum can susstantially reduce the solubility of silica [134]. Adsorbed ions may also hinder the crystallization of gels [135].

Under some circumstances, one metal oxide may facilitate the apparent "precipitation" (coagulation) of another oxide particle by virtue of their opposite surface charges, a process called heterocoagulation. Even at trace concentrations of divalent cations, where separate solid phase formation is not possible, the cation may still be present as a colloid, perhaps adsorbed on dust particles. Benes and Kopicka [136] found that in their 5 x $10^{-7} M$ 'Cd$(NO_3)_2$' solutions the Cd was present as Cd^{2+} below pH 5, but at a pH greater than 5, the Cd was mostly adsorbed on colloidal impurities in the solution.

Spectral information on the structure of the adsorbed cation-metal oxide complex is largely lacking although recent IR and ESR spectral studies promise to lead to interesting results [48,137]. The extent of hydration and location of the adsorbed ions remains uncertain and these are important parameters in theoretical models of the adsorption mechanism. The specifically adsorbed ions are believed by some to retain their entire primary hydration sheath [96,138] and this is supported by experimental studies based on visible absorption spectra. Although these and other data [88] seem to support the proposition it certainly has not been proven conclusively. There is some evidence that adsorbed ions can coordinate directly with the surface oxygens [139], at least in some cases, but more convincing evidence is required on this point. The formation of double hydroxides involves a form of dehydration of the cations, and naturally occurring oxide minerals frequently contain a variety of isomorphous (impurity) cations within their structures [140,141].

Not much work has been done on cation adsorption in nonaqueous media, but it is interesting that Sr removal by Al_2O_3 was considerably greater in a 90% MeOH-10% H_2O mixture than in a pure aqueous system [141]. A similar trend was said to have been found for Sr adsorption by Fe gel and a manganese oxide in a variety of nonaqueous media.

Cation Selectivity

The selectivity of an exchanger is a measure of its relative affinity for different ions. It is determined by both the properties of the ion and the exchanger, with the solvent playing an important intermediary role. As a result, the selectivity sequence is an important clue to the microenvironment of the adsorbed ion. It is also of more direct importance because it controls the extent of uptake of a particular cation and the effect of one cation on the uptake of another.

Monovalent Cations

There is usually little specific interaction between oxide surfaces and monovalent cations, so selectivity differences are usually quite small. However, definite differences do exist. For example, two extreme selectivity se-

CATION ADSORPTION

quences have been found with the alkali metal cations: $Li^+ < Na^+ < K^+ < Rb^+ < Cs^+$ and its reverse, $Cs^+ < Rb^+ < K^+ < Na^+ < Li^+$. The first sequence is the "normal" sequence and is found with most strong acid ion exchange resins [142], clay minerals and a number of hydrous metal oxides (Table IV), especially SiO_2. This sequence has been explained qualitatively by assuming that the binding force is derived from the coulombic interaction between the ion and the surface and that the adsorbed ions retain their primary hydration sheath. The ion with the smallest hydrated radius (Li^+) will be able to approach the surface most closely and will therefore be held most strongly. The larger cations, such as Cs^+, are more polarizable and more readily allow distortions to their hydration spheres than smaller, strongly hydrated cations, such as Li^+. On microporous solids, such as silica gel, steric factors may strongly influence the selectivity in favor of the hydrated ions.

However, selectivity sequences of the reverse order have been found on a number of exchangers (Table IV), and this sequence seems particularly prevalent on weak acid exchangers, such as hydrous metal oxides. The reason for this reversed selectivity sequence is not certain, and it is probable that several factors combine to effect the reversal. One interpretation involves the effect of the solid on the water adjacent to the oxide surfaces. It is well known that this water is strongly bound to the surface [152-155], perhaps through formation of hydrogen bonds between a single adsorbed water molecule and two underlying surface hydroxyl groups [156,157]. It has been suggested that the ordered water close to the surface would favor adsorption of ions that can maintain this order [82]. This reasoning has been used to explain the greater specific adsorption on rutile of Li^+, a 'structure maker,' compared with Cs^+, a

Table IV. Selectivity Sequences for the Alkali Metal Cations on Various Hydrous Metal Oxides

Sequence	Oxide	Reference
$Cs > Rb > K > Na > Li$	Si gel	143
$K > Na > Li$	Si gel	144
$Cs > K > Li$	SiO_2	145
$Cs > K > Na > Li$	SiO_2	146
$Li > Na > K$	SiO_2	147
$K > Na > Li$	Al_2O_3	148
$Li > K \sim Cs$	Fe_2O_3	55
$Li > Na > K \sim Cs$	Fe_2O_3	149
$Cs \sim K > Na > Li$	Fe_3O_4	150
$Li > Na > Cs$	TiO_2	75
$Li > Na > K$	$Zr(OH)_4$	151

'structure breaker.' Several authors have identified the normal sequence with structure-breaking surfaces and the reverse sequence with structure-promoting surfaces and have developed models of cation selectivity somewhat similar to Eisenmann's model for cation selectivity [104,149,158]. The ability of a surface to order the interfacial water molecules has been related to the polarity and field strength of the solid. The higher the polarity of the solid, that is, the higher the pzc, the greater the structural order of the interfacial water. This implies that a solid having a high pzc should prefer Li^+ compared with a solid having a low pzc. While there is some evidence to support this [149], the relationship is far from perfect. Certainly SiO_2 has a low pzc and often obeys the normal sequence (Table IV). This line of reasoning may help to explain why solids having the same chemical composition, but different structures, sometimes exhibit different selectivity sequences. For example, MnO_2 can exist in many structural forms, each having a different polarity and pzc [78]. For $\delta\text{-}MnO_2$, the selectivity sequence is $Cs^+ > Na^+$ [104], whereas for $\beta\text{-}MnO_2$, the sequence is $Na^+ > Cs^+$. This relates to a pzc of 1.5 for $\delta\text{-}MnO_2$ and 7.3 for $\beta\text{-}MnO_2$ and is qualitatively in agreement with the preceding arguments.

While in general there is little evidence of specific adsorption of monovalent cations on hydrous metal oxides [91,92], there are exceptions. Indeed, the ion-pairing model [159] implies that specific adsorption of simple electrolyte ions, such as Na^+, Cl^-, NO_3^-, etc., is the rule rather than the exception. Certainly Li^+ at high concentrations can be specifically adsorbed by hematite [49]. Hydrous antimony pentoxide (HAP) selectively adsorbs a number of monovalent cations, especially Na^+, and it has been used in neutron activation analysis to remove unwanted $^{24}Na^+$ activity [160,161]. Ag^+ has also been shown to be selectively adsorbed by Si gel [113], Fe gel [162,163] and MnO_2 [22, 164]. This is probably related to the exceptionally high polarizability of Ag^+ and its tendency to hydrolyze. Under certain conditions, reduction of Ag^+ to Ag^0 by Fe^{2+} may be a significant scavenging mechanism for ferric hydroxide [165]. Tl^+ is selectively adsorbed on MnO_2 [118].

Divalent Cations

There is at present no unifying principle governing the selectivity of oxides for divalent cations. Consider, for example, the selectivity sequence of the alkaline earth cations, Mg^{2+}, Ca^{2+}, Sr^{2+} and Ba^{2+} on various hydrous metal oxides (Table V).

There is apparently little consistency in the selectivity sequences, even for similar materials. Most noticeable are the differences between the Fe gel and Fe_2O_3 and between the two Al_2O_3 samples. In some cases the selectivity differences only become apparent at high concentrations [55,166]. The high selectivity shown by hematite ($\alpha\text{-}Fe_2O_3$) for Mg^{2+} and Li^+ at high concentrations has been attributed to the penetration of these ions into the solid surface [55] because of the similarity in size between Fe^{3+} (0.064 nm), Mg^{2+} (0.065 nm) and Li^+ (0.068 nm). This mechanism implies that these strongly adsorbed ions are at least partially dehydrated. The selectivity sequence may also vary with pH. For example, on Fe gel, Sr > Mg changes to Mg > Sr at high pH [47]. Also, Cd > Zn on MnO_2 at pH 4.0 but Zn > Cd at pH 6.0 and 8.0 [169]. Differences in the H^+/M^{n+} stoichiometry for different cations could lead to reversals in selectivity with changing pH since the ion with the higher H^+/M^{n+} stoichiometry would tend to be increasingly favored at higher pH.

Al gel shows a relatively high affinity for Mg^{2+} (Table V), which is consistent with the fact that of the alkaline earth cations, Mg^{2+} forms the most stable double hydroxide, and the selectivity is therefore probably related to the ability of the ions to fit into structural holes at the surface or to form an incipient surface coating of $Mg(OH)_2$.

Divalent transition and heavy metal cations are invariably considerably more strongly adsorbed than the alkaline earth cations, but their selectivity sequences vary (Table VI). While there is a broad relationship between relative affinity and the tendency to hydrolyze [47,113,123,128], there are also many exceptions, as revealed by the different sequences found on different solids.

Manganese oxide minerals behave significantly differently from many of the other oxide minerals in several respects (see the section on H^+/M^{n+} stoichiometry). Co(II) is particularly strongly taken up by manganese oxides at low pH, and it has been suggested that Co^{2+}, perhaps after surface oxidation to Co^{3+}, is able to replace Mn^{3+} in the crystal structure of various manganese oxide minerals [39,99]. This substitution has been considered unlikely be-

Table V. Selectivity Sequences for the Adsorption or Coprecipitation of Alkaline Earth Cations on Various Hydrous Metal Oxides

Sequence	Oxide	Reference
Ba > Ca	Fe gel	166
Ba > Ca > Sr > Mg	Fe gel	47
Mg > Ca > Sr > Ba	α-Fe_2O_3	55
Mg > Ca > Sr > Ba	Al gel	47
Mg > Ca > Sr > Ba	Al_2O_3	64
Ba > Sr > Ca	α-Al_2O_3	35
Ba > Sr > Ca > Mg	MnO_2	100
Ba > Sr > Ca > Mg	δ-MnO_2	43
Ba > Ca > Sr > Mg	SiO_2	167
Ba > Sr > Ca	SiO_2	168

Table VI. Selectivity Sequences for the Adsorption or Coprecipitation of a Variety of Divalent Transition and Heavy Metal Cations on Several Hydrous Metal Oxides

Sequence	Oxide	Reference
Pb > Zn > Cd	Fe gel	118
Zn > Cd > Hg	Fe gel	170
Pb > Cu > Zu > Ni > Cd > Co	Fe gel	47
Cu > Zn > Co > Mn	α-FeOOH	101
Cu > Pb > Zn > Co > Cd	α-FeOOH	123
Cu > Zn > Ni > Mn	Fe_3O_4	150
Cu > Pb > Zn > Ni > Co > Cd	Al gel	47
Cu > Co > Zn > Ni	MnO_2	171
Co > Cu > Ni	MnO_2	172
Pb > Zn > Cd	MnO_2	118
Co $\simeq$ Mn > Zn > Ni	MnO_2	43
Cu > Zn > Co > Ni	δ-MnOOH	41
Co > Cu > Zn > Ni	α-MnO_3	41
Co > Zn	δ-MnO_2	99
Zn > Cu > Co > Mn > Ni	Si gel	173
Zn > Cu > Ni $\simeq$ Co > Mn	Si gel	174
Cu > Zn > Co > Fe > Ni > Mn	SnO_2	175

cause of the large discrepancy in ionic radius between octahedrally coordinated low-spin Co^{3+} (0.052 nm) and the high-spin Mn^{3+} (0.065 nm) found in manganese(IV) oxide structures [42]. Instead, Burns [42] proposed that after surface oxidation of Co^{2+} to Co^{3+}, the Co^{3+} could substitute for Mn^{4+} (0.054 nm), an ion of similar size. This strong affinity of Co for manganese minerals is also found in manganese nodules from soils and sediments [176, 177]. The iron phase of these nodules may also be significant, but by contrast with its marked affinity for manganese minerals, Co^{2+} does not show exceptional affinity for either α-FeOOH (goethite) or Fe gel (Table VI and Figure 3), probably because Co^{2+} is not oxidized under these conditions [45].

Coprecipitation and Formation of Double Hydroxides

A recurring characteristic of many of the hydrous metal oxides is their high degree of structural disorder, especially when freshly precipitated by addition of a strong base to a metal salt solution. This frequently results in gelatinous precipitates of variable composition, which show diffuse characteristics when examined by such methods as X-ray diffraction, infrared spectroscopy, and differential thermal and thermogravimetric analysis. This lack of long-range order is manifest in the large specific surface area of these solids and their high chemical reactivity. The ability of these materials to adsorb cations from solution has been discussed in some detail, but the frequent similarity between cation adsorption and coprecipitation [58,162, 166,178,181] is somewhat surprising. This similarity suggests that coprecipitation frequently occurs by an adsorption mechanism, although large differences are sometimes found [182]. Differences seem to arise when the concentration of the minor component is relatively high and when the two cations are sufficiently similar for isomorphous substitution to occur. The frequent coprecipitation of impurities was of particular importance in early work concerned with quantitative gravimetric analysis, where very pure precipitates were required.

While most adsorption studies have involved the adsorption or coprecipitation of a minor component onto, or with, a major component, a number of studies have also been made where the concentration of cations were of the same order of magnitude. Feitknecht and co-workers [183-185] performed a series of studies of the coprecipitation of di- and trivalent metal hydroxides, the products of which having been called "double hydroxides." The term "double" was derived from the belief that the structure of these compounds was composed of two layers: a trioctahedral layer containing the divalent cations interleaved with disordered dioctahedral layers containing the trivalent cation. These compounds were prepared by precipitation from solutions containing a mixture of a divalent cation (Ca^{2+}, Mg^{2+}, Ni^{2+} or Mn^{2+}) and a trivalent cation (Al^{3+}, Fe^{3+} or Cr^{3+}). Frequently, characteristic X-ray patterns were obtained, but chemical analysis also revealed the presence of variable amounts of Cl^-, the dominant anion, suggesting a variable, positive surface charge. However, these variations in chemical composition did not lead to detectable differences in cell dimensions [183,184].

More recently, the synthesis [186] and structure [187] of coprecipitated Mg-Al hydroxycarbonates has been worked out. Two particularly stable structures formed that have Mg/Al ratios of 5:1 and 2:1, and these consist of brucite-like layers in which Al^{3+} replaces up to a maximum of about one in three Mg^{2+}. The resulting positive charge is balanced by interlayers containing sheets of anions, such as OH^- and CO_3^{2-}, with water molecules in some or all of the sites remaining unoccupied by anions [187]. The extent of Al^{3+} for Mg^{2+} substitution was reflected by changes of the basal spacings, larger amounts of substitution giving smaller basal spacings [186]. This structure for the double hydroxides differs significantly from that proposed by Feitknecht, and more recently by Ross and Kodama [188], in that both the di- and trivalent cations are situated in a single layer, rather than in two distinct layers [189].

Soil goethites (α-FeOOH) may contain up to 15 mol % substitution of Fe by Al [190]; more recently, a wide range of $(Fe,Al)_2O_3$ mixed oxides have been found in soils and synthesized [191]. These mixed oxides have been studied by a variety of physical techniques, such as X-ray diffraction, infrared analysis, differential thermal analysis and Mössbauer spectroscopy, which allow the extent of substitution to be estimated [186,192-195].

Weathering of New Idria serpentinite in arid land of California has produced analogous Mg^{2+} and Fe^{3+} brucite-like layers arranged as one, two and three layers (pyroaurite, coalingite, and coalingite-K, respectively [196]) interstratified with CO_3^{2-} and H_2O interlayers [197].

It is apparent from all of these studies that the coprecipitation of two cations, one in trace amounts and the other in large excess, is one end of a continuum which extends to the formation of discrete, crystalline compounds.

Concentration Dependence

The relationship between the amount of a substance adsorbed at constant temperature and its concentration in the equilibrium solution is called the adsorption isotherm. The adsorption isotherm is important from both a theoretical and a practical point of view. Its shape is related to the energy of adsorption and the number of adsorption sites, but the relationship is not necessarily simple. The adsorption isotherm contains the information necessary to determine the effect of the oxide on buffering additions and removals of cations from solution. If adsorption isotherms are known at several pH values, the effect of a change of pH can also be estimated.

The principle governing adsorption at low solution concentrations is Henry's law, which gives a linear adsorption isotherm passing through the origin. This linearity is normally observed only at very low concentrations, although linear isotherms have been found at higher concentrations [198]. In the region of the adsorption isotherm in which Henry's law is obeyed, the fraction of the total amount that is adsorbed is independent of the initial concentration (Figure 4).

At higher concentrations, adsorption is below what is expected from Henry's law, a result arising partly by the filling up of a significant number of the total possible adsorption sites. This results in a smaller chance of a solution ion that impinges on the surface of finding a vacant site, and therefore being adsorbed. The Langmuir isotherm takes this into account and has been widely used in both gas adsorption and solution adsorption studies. For example, the Langmuir adsorption isotherm has been used to describe the concentration dependence of cation uptake by MnO_2 [22,26,99,100,164], by TiO_2 [103], by Al gel [199,200], by Al_2O_3 [201,202], by Cr gel [182] and by Fe_3O_4 [150,203].

The adsorption maximum increases with pH, as illustrated with some results for the specific adsorption of Ca^{2+} by γ-Al_2O_3 (Figure 7). The Langmuir constant, which is a measure of the free energy of adsorption, was not affected by pH in this case (Figure 7). While the Langmuir isotherm frequently fits the data quite well over a fairly small concentration range, it usually breaks down over a larger range. This is also true with gas adsorption on solids and can be explained in terms of lateral interactions of adsorbed species, multilayer adsorption (i.e., precipitation in the case of the solid-solution interface) and surface heterogeneity [204]. The change in surface charge resulting from adsorption makes subsequent adsorption more difficult, so the free energy of adsorption varies with surface coverage. Several models of oxide surfaces that take this into account have been presented [17,64,205,206]. Intrinsic cation-surface affinity constants can be found by extrapolating the conditional affinity constants to zero surface charge (potential) [104].

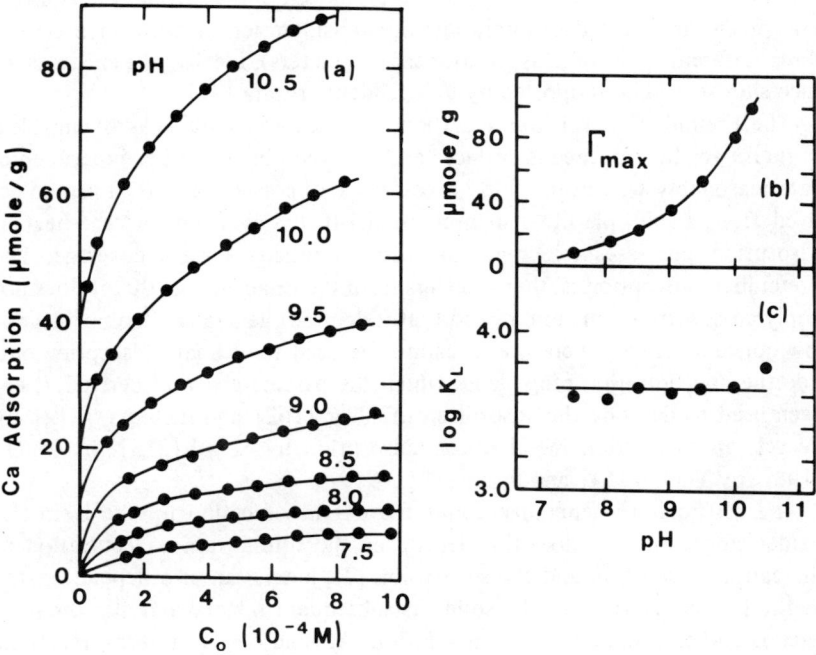

Figure 7. (a) Adsorption isotherms for Ca adsorption on Al_2O_3 as a function of suspension pH; (b) the variation of the two Langmuir adsorption parameters, the maximum adsorption capacity, Γ_{max}; and (c) the Langmuir adsorption constant, K_L [64].

Inflections in titration or adsorption curves [207] have been interpreted by some workers in terms of the polyfunctional weak acid character of some hydrous metal oxides; for example, TiO_2 [103,208], MnO_2 [20,209] and SiO_2 [210]. Such characteristics are usually sensitive to the experimental conditions used, especially the ionic strength and rate of titration. There is little agreement over the number or kind of different surface functional groups involved and in many cases, when ion pairing of the electrolyte ions at the surface ("surface complexation") is taken into account, then various titration curves may be adequately described without invoking surface heterogeneity [17]. However, at the molecular level, surface heterogeneity is probably the rule rather than the exception, even for pure minerals [211]. This arises from the structural diversity of crystal faces, and the presence of exposed edges, corners and defects. With aged gels there is probably a broad range of surface structures with varying degrees of structural order and cation affinity. Over a limited concentration range, the adsorption isotherm may be dominated by the filling of a single type of adsorption site, but if the free energies of adsorption for the different sites are quite close, then filling of several sites will occur simultaneously and some type of multiple-site model would be required. The precision of the data rarely justifies consideration of more than two or three different types of sites (four or six parameters). Ideally, the existence of such sites should be supported by independent criteria [99].

The Freundlich adsorption equation is particularly suitable as an empirical isotherm for heterogeneous surfaces and is often able to fit the experimental data reasonably well over a fairly broad range of concentrations. It can be derived from a multiple-site Langmuir model if it is assumed that the heat of adsorption decreases exponentially with increasing surface coverage. The Freundlich adsorption isotherm suffers from the disadvantage that it does not imply an adsorption maximum and that it does not generally become linear at low concentrations. Therefore, it cannot be used to estimate adsorption outside the concentration range from which the parameters were derived. It has been used to describe the adsorption of silver [162] and mercury [212] by Fe gel, and transition metal cation adsorption by Si gel [213], by MnO_2 [198], by ThO_2 [214] and by Fe gel [179].

It is difficult to generalize about the extent of cation uptake by metal oxides since not only does this depend on the equilibrium concentration of the cation in solution and the suspension pH, but it can also depend on the method of preparation of the solid, its subsequent history and also the presence of other interacting ions in solution. At high concentrations, reactions such as surface-induced hydroxide precipitation may be important, and these can obscure the extent of the underlying adsorption reaction. A number of studies have looked at the adsorption and coprecipitation (scavenging) of low concentrations of cations by Fe gels (Table VII), and these results give a use-

Table VII. Divalent Cation Adsorption/Coprecipitation by Iron Hydrous Oxide Gel at $10^{-7}\ M\ M^{2+}$

Cation	pH	Method[a]	Background Electrolyte	Uptake (μmol M^{2+}/ mmol Fe)	Reference
Ca^{2+}	7.0	S	1 M NaNO$_3$	0.00077	58
	8.0	S	1 M NaNO$_3$	0.0068	58
Sr^{2+}	7.0	S	1 M NaNO$_3$	0.00085	58
	7.0	S	?	0.00075	98
	7.0	C	1 M NaNO$_3$	0.0010	102
	8.0	C	8×10^{-3} M NH$_4$Cl	0.0066	106
	8.0	S	1 M NaNO$_3$	0.0076	58
	8.0	C	1 M NaNO$_3$	0.010	102
Ba^{2+}	7.0	S	1 M NaNO$_3$	0.0042	58
	7.5	C	8×10^{-3} M NH$_4$Cl	0.038	106
	8.0	S	1 M NaNO$_3$	0.051	58
Co^{2+}	6.2	C	2×10^{-2} M NH$_4$Cl	0.028	45
	7.1	C	2×10^{-2} M NH$_4$Cl	0.20	45
	7.0	S	$\sim 1 \times 10^{-2}$ M NH$_4$Cl	0.16	95
	7.0	C	dil. NH$_4$Cl	0.24	215
	7.0	C	8×10^{-3} M NH$_4$Cl	0.19	106
	8.0	C	2×10^{-2} M NH$_4$Cl	2.4	45
Zn^{2+}	5.0	S	1 M NaNO$_3$	0.0011	130
	6.0	S	1 M NaNO$_3$	0.048	130
	7.0	C	0.01 M KNO$_3$	0.58	170
Cu^{2+}	5.0	S	0.1 M NaNO$_3$	0.025	59
	6.0	S	0.1 M NaNO$_3$	1.1	59

[a] S = adsorption; C = coprecipitation.

ful indication of the amount of adsorption likely to be found at low concentrations. Below about $10^{-6}\ M\ M^{2+}$ Henry's law is usually reasonably well obeyed, and we have made use of this in extrapolating all of the results to a common equilibrium concentration of $1.0 \times 10^{-7}\ M\ M^{2+}$. The specific surface areas of the gels are not available, but 320 m^2/g (35 m^2/mmol Fe) is probably a reasonable estimate for most Fe gels, although it might be as high as 600 m^2/g [59]. Not surprisingly, the adsorption densities are invariably low; for example, 0.001 μmol M^{2+}/mmol Fe is equivalent to an average area of 5.8 $\times$ 10^4 nm^2/ion (320 m^2/g) or an average separation of 240 nm on the surface. To put the cation adsorption on Fe gel into perspective, we have found that about 200 μmol of surface-OH/mmol Fe are titrated between pH 3 and 10 in 1 M NaNO$_3$, and Davis and Leckie [59] have estimated a total site density for Fe gel of about 900 μmol/mmol Fe. If the formula Fe$_2$O$_3$.2FeOOH.2.6 H$_2$O for Fe gel (ferrihydrite) recently proposed by Russell [215] is correct, then there should be a total of 500 μmol OH/mmol Fe. Because the majority of these OH groups are on the surface, their number decreases with aging [216]. The agreement between different sources (Table VII) is reasonable in view of the different conditions under which adsorption took place. The results highlight the importance of pH since an increase in pH from 5.0 to 6.0 increases Cu^{2+} and Zn^{2+} adsorption about 40 times.

We have found that when Fe gel ages, the shape of the adsorption isotherm can change considerably. As expected, at high concentrations adsorption by aged Fe gel was less than that by fresh Fe gel (on a weight for weight basis), and this is probably due to the overall reduction in specific surface area of the gel. Shuman [217] found that the Zn adsorption maxima of Fe and Al gels decreased roughly in proportion to the decrease in surface area on aging. However, at low concentrations ($\sim 10^{-7} M$), we found a substantial *increase* in Zn adsorption by aged Fe gels, even when expressed on a weight for weight basis and when the aged gel showed a substantial overall reduction in H$^+$/OH$^-$ titration capacity [107] (also see Figure 6, aged gel). This indicates the formation of a small number of active sites in the relatively crystalline phase of the aged gels (goethite) and it appears that the disordered fresh gel is characterized by a large number of low-affinity sites, whereas the aged gel is characterized by a smaller number of these sites and, in addition, a number of high-affinity sites. Fully crystalline hematite and goethite were less reactive than the aged gels on a weight for weight basis, perhaps reflecting a surface area effect. This complicated effect of aging deserves more attention since it is important when considering the environmental significance of oxides in cation retention.

The extent of adsorption on crystalline oxides is more variable, and substantial disagreement can occur even when adsorption is compared on a surface area basis. For example, contrasting results were obtained for Co(II) adsorption on quartz [120,132].

At high equilibrium concentrations, adsorption densities can get much higher than those shown in Table VII. Particularly with gels, it is frequently difficult to determine a true adsorption maximum because adsorption is obscured by hydroxide precipitation [130]. For the transition and heavy metal cations, uptakes of 100 μmol M^{2+}/mmol Fe are possible at an equilibrium concentration of $10^{-3} M$ [218]. Even for Ca^{2+}, we have found that reasonably high uptakes are possible; for example, 20 and 80 μmol Ca^{2+}/mmol Fe at an equilibrium concentration of $10^{-2} M$ Ca^{2+} and pH 8 and 10, respectively. Such high uptakes can change the overall surface chemistry of the oxides considerably. In the case of the transition and heavy metal cations, where the H^+/M^{2+} adsorption stoichiometry is closer to two (equivalent to $M(OH)_2$ adsorption), the new surface will still show weak acid properties but largely reflecting the new surface coating rather than the underlying oxide substrate [73,127,133]. With cations such as Ca^{2+}, where the H^+/M^{2+} stoichiometry is closer to one, there could be a significant reduction in the net number of surface-OH groups.

Effect of Complex Formation

An alternative explanation for the fact that the H^+/M^{n+} adsorption stoichiometry is frequently less than n is that the adsorbed species may not be M^{n+}, but a lower charged ion pair or complex ion. The possibility that hydroxo complexes are preferentially adsorbed has been discussed already, but a number of adsorption experiments have been carried out in systems where appreciable complex formation involving other ligands occur and this can give rise to peculiar adsorption isotherms [219-221]. The effect of complex formation on the overall adsorption of a metal ion will depend on the stability of the complex formed and the relative affinity of the surface for the free and complexed forms. If the stability of the cation-solid surface complex is large compared with the stability of the cation-solution ligand complex(es), then complex formation in solution will have little effect on the overall extent of adsorption. This can be true even if a large proportion of the cation is in a complexed state in solution in the absence of the solid. Calculation of the species present must be made in much the same way as for complex equilibria in solution, with the exception that the adsorption isotherm for each adsorbed species must also be included [73,220]. Steric and charge factors will mean that not all of the possible surface-complex cation interactions will be important and, in practice, it will probably be unnecessary to consider the whole matrix of possible interactions.

One system in which the effect of complex formation was studied involved uptake of transition metal cations from NH_3/NH_4Cl solutions [178, 179]. This was of interest because of the possible analytical separation of these ions from the alkaline earth cations by selective uptake from a NH_3/NH_4Cl system. The transition metal cations tend to form ammine complexes which, although they are cations, are not strongly taken up by Fe gel, especially at high NH_4Cl concentrations and high pH [178,179,222-225]. It is probable that the adsorbed species in these systems is the aquo complex, rather than the ammine complex. With this assumption, Simon et al. [225] calculated the stability constants and average coordination number of cadmium ammine complexes from the effect of increasing NH_4Cl concentrations on decreasing the amount of Cd coprecipitated with Al gel. Conversely, if the stability of the complex in solution is known, the free energy of adsorption can be calculated. This is only possible if there is a single dominant complex species formed in solution. The preference of the solids for the aquo ion rather than the ammine complex is probably related to the large size of the ammine complex and the relatively weak adsorption of the NH_3 ligand. Similarly, when other ligands, both inorganic and organic, form stable complexes in solution but themselves are not strongly adsorbed, then the presence of these ligands will tend to decrease adsorption. The exceptional stability of Hg-Cl complexes leads to a decrease in Hg adsorption in the presence of Cl^- [127,212,226,227]. However, with Cl^- concentrations greater than about 0.1 M, Zn adsorption on Fe gel was slightly increased, indicating possible adsorption of $ZnCl^+$ [228]. Trace concentrations of cyanide ($\sim 10^{-6} M$) prevented Ag^+ adsorption by Fe gel [221].

Various organic ligands, such as EDTA, citrate, cysteine and glutamate, can be coadsorbed during cation adsorption and, in some cases, the adsorbed ligand can assist in cation adsorption, particularly when the adsorbed ligand has a strongly complexing functional group directed outward to the solution [221]. EDTA decreased Cu adsorption on silica, but the decrease was not as great as expected if only free Cu^{2+} had been adsorbed. This suggested that adsorption of one or more Cu(en) complexes was involved [229]. On the other hand, EDTA led to an overall increase in Ag^+ adsorption on silica [221].

With cations that form very stable hydroxy complexes, increasing pH can actually lead to a decrease in adsorption because of the formation of soluble hydroxy metal complexes. Maxima in adsorption-pH curves have been observed for mercury, ruthenium, cerium, cobalt and zinc [10,91,92,212,223, 225,227]. These maxima are usually found only at a pH greater than 10 and at low concentrations because the solubility product of the metal hydroxide must not be exceeded. Under certain circumstances, the maxima in the adsorption-pH curves may arise from causes other than hydroxy metal complex formation, such as dissolution or recrystallization of the solid at high pH, enhancement of ligand formation as in NH_4^+/NH_3 systems (Figure 8), or the formation of a finely dispersed hydroxide [117].

Temperature Dependence

A study of the temperature dependence of adsorption reactions gives valuable information about the enthalpy and entropy changes during adsorption. Greater adsorption is often found at lower temperatures [45,97,162], but the differences are usually small. Sometimes quite large differences are found, in-

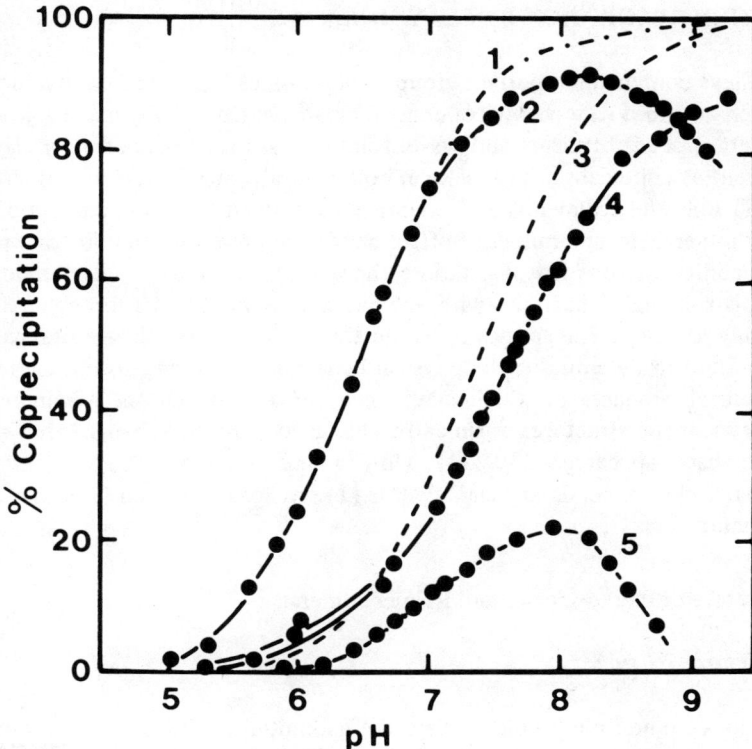

Figure 8. Coprecipitation of 10^{-3} M Cd by Al gel in the presence and absence of ammonia as a function of suspension pH: 1 = 10^{-1} M Al + KOH; 2 = 10^{-1} M Al + NH$_4$OH; 3 = 10^{-2} M + KOH; 4 = 10^{-2} M Al + NH$_4$OH + 0.06 M NH$_4$Cl; 5 = 10^{-2} M Al + NH$_4$OH + 0.5 M NH$_4$Cl [225].

cluding increases in adsorption with increasing temperature [91,178,179,198]. There seems to be no particular trend in these differences, and caution is required in interpreting many of them since temperature changes can affect several factors at the same time. For example, an increase in temperature may increase the rates of the adsorption, hydrolysis and recrystallization reactions. Also, in some of the earlier work, it is difficult to know how well the pH was controlled. Temperature changes can also change the dissociation constant of water and alter the potential of the reference electrode.

CATION ADSORPTION BY CLAY MINERALS

Clays contain an important group of phyllosilicate (layer silicate) minerals, which adsorb a wide variety of constituents from naturally occurring groundwater, applied fertilizers and waste solutions. A major part of most clays is present as colloidal-sized particles in both natural mineral deposits and virtually all soils and sediments. Soil scientists have investigated ion adsorption by clay minerals to determine the effect of adsorption on the mobility and plant availability of ions in soils. Before the adsorption concept was developed, scholars thought that salts with soluble cations would pass through soils as readily as water. The species of cation that occupy the exchange sites can be important in determining the physical behavior of soils and also of the many industrial products in which clays occur or are introduced. A number of reviews of the structure, origin and colloidal and cation exchange behavior of clays have appeared [230-234]. Only a brief summary is given here. In general, clay minerals are classified as (1) phyllosilicates and (2) other constituents of clays.

Structure of Phyllosilicates and Related Minerals

Structural Scheme

The common phyllosilicate clays are kaolinite, smectites (the montmorillonite and saponite subgroups), vermiculites, micas and chlorites. All of these phyllosilicates are built up from a combination of two types of sheet structures—tetrahedral and octahedral. The tetrahedral sheet is a two-dimensional sheet of composition $Si_4O_6(OH)_4$ in which each Si is at the center of a tetrahedron of four oxygen or hydroxide ions. The octahedral sheet consists of closely packed oxygen or hydroxide ions in which Al, Fe or Mg ions are in octahedral coordination. When Al only is present, it is only necessary that two-thirds of the possible octahedral positions be filled in a gibbsite-like layer of composition $Al_4(OH)_{12}$. When Mg is present in place of Al, all octahedral positions are filled, giving a brucite-like layer of composition $Mg_6(OH)_{12}$. These two types of octahedral filling are referred to as dioctahedral and trioctahedral, respectively. By interposition of oxygen ions of the tetrahedral sheet for hydroxide ions of the octahedral sheets, the two sheets are condensed to form unit layers. Such layers are built up in a series of ways to produce the smallest phyllosilicate entities that may exist independently. The Si tetrahedra are arranged so that their tips are all pointing towards the octahedral sheet.

In micas, vermiculites and smectites, the tetrahedral:octahedral sheet ratio is 2:1, with the octahedral sheet sandwiched between the two tetrahedral sheets. Thus, these mineral groups have similar top and bottom planar Si-O-Si surfaces. In chlorite, a brucite- or gibbsite-like sheet is interposed between two mica-like layers, giving a tetrahedral:octahedral sheet ratio of 2:1:1. The 2:1 mineral structure extends to offsetting double or triple rows of tetrahedral chains linked through oxygen atoms. Interchain adsorption spaces are a result in palygorskites (attapulgites) and sepiolites.

In kaolinite and halloysite (dioctahedral minerals), the tetrahedral and octahedral sheets are combined in a ratio of 1:1. Thus, these minerals have two different kinds of planar surfaces—one Si-O-Si and one Al-OH. Hydrogen bonding between the —OH of one layer and —O of the adjacent dehydrated layers gives the kaolinite structure of which the unit layers are not readily separated by H_2O. Curvature of similar structural layers gives an interlayer water molecule sheet in halloysite. Both minerals are penetrated by high concentrations of polar salts such as CsCl and, to some extent, KOAc, or by polar organic compounds such as hydrazine or dimethyl sulfoxide [235].

The unit layers of different mineral species have the ability to stack on top of one another within a single particle. This stacking is usually irregular [236], as seen by high-resolution electron microscopy, but regular repeating sequences are common in certain mineral specimens. Stacking of layers of different minerals, both irregular and regular, is called interstratification. Further refinements involve the particular stacking bonding and isomorphous substitutional (charge-imparting) arrangements that occur within and between similar unit layers.

Isomorphous Substitution and Cation Exchange Capacity

Most phyllosilicate minerals have permanent negative surface charges arising from nonstoichiometric isomorphous substitution of cations within their structures. Commonly, Al^{3+} substitutes for Si^{4+} in the tetrahedral sheet and Mg^{2+} for Al^{3+} in the dioctahedral sheet. Other cations may proxy with similar results, e.g., Fe^{3+} for Si^{4+}. Either type of isomorphous substitution gives a net negative charge of unity. This charge may be compensated by back-substitution in the layers (+1) or by interlayer cations. When the layer charge density is high (about 2.5 meq/g), each ditrigonal cavity in the interlayer space bounded by the Si-O-Si tetrahedral sheets has a charge of (−1) and each is filled with K^+ in K-mica. Penetration of water and cations between the layers is slow and difficult. Charged interlayer hydroxyl sheets of chlorite make hydration and cation penetration difficult or impossible. If the substitution is near zero, the layers are also difficult to separate by water and cations, as in kaolinite (dioctahedral), or impossible, as in serpentine (trioctahedral). Similar nonexpansive properties occur with the unsubstituted 2:1 minerals, pyrophyllite (dioctahedral) and talc (trioctahedral). Interlayer attraction may occur between layers and the lack of interlayer cations with their tendency to hydrate hinders expansibility.

The expandable phyllosilicates, such as montmorillonite and vermiculite, have a charge of 0.8-1.7 meq/g. The negative layer charges are exactly balanced by the adsorption (or desorption) of counterions, and therefore the phrase cation exchange is often used in place of the more general phrase cation adsorption. At low electrolyte concentrations, the charge of phyllosilicate layers is satisfied by the accumulation of cations in a diffuse layer at the surface and exclusion of anions from this surface region. Anion exclusion decreases with increasing electrolyte concentration and is small for concentrations greater than $10^{-2} N$. The quantity of exchangeable cations held is usually expressed as the cation exchange capacity (CEC), e.g., 125-170 meq/100 g for vermiculite and 80-125 meq/100 g for montmorillonite. Expansion is facilitated by exchangeable cations having a tendency to be highly hydrated in solution. Li^+ is notable in this respect [235,237].

Structure of Other Clays

Amorphous Materials and Imogolite

Materials of short-range order have a network of Si-O, Al-O, Fe(III)-O, etc., linkages with mainly irregular radial distribution in space. Most frequently, they result from partial leaching of volcanic glass, but can result from partial leaching (weathering) of crystalline materials. Allophane has a hydrous spheroidal microstructure [238]. Imogolite has the structure of a gibbsite-like tube linked internally by silica tetrahedra [239]. Isomorphous cation substitution, edge effects and hydrous oxide coatings [240,241] provide cation exchange and adsorption properties. The cation exchange properties of these poorly ordered aluminosilicates have been studied much less than their well-crystallized counterparts. They probably behave more like the variable charge colloids (simple oxides) than the constant charge phyllosilicate minerals.

Oxides

Hydrous Al, Fe, Mn, Ti and other cationic oxides-hydroxides occur in clays as coatings on phyllosilicates [240,241] and as free gels and crystals. They account for some of the cation adsorption properties of soils.

Cation Exchange by Ferrialuminosilicates

Most cation exchange studies with ferrialuminosilicate clays have involved cation exchange on smectites, particularly montmorillonite, and, to a lesser extent, on kaolinite, vermiculite, biotite, muscovite, chlorite, allophane and imogolite.

Montmorillonite clays have been widely studied for their electrical double layer properties. Because of the isomorphous substitutional origin of their layer charge, they behave predominantly as constant-charge surfaces, in contrast to the variable charge surfaces of the simple oxides. For a given montmorillonite, the layer charge may vary ±15% within the layer sequence [242]. However, the exposed octahedral edges of montmorillonite and other phyllosilicate minerals act to some extent as variable-charge surfaces. These edges become positively charged below pH 9 or so, depending on their octahedral composition. Interaction of the positively charged edges and negatively charged planar surfaces plays an important part in the physical behavior of many clays. The pH-dependent charge properties of the clay minerals are evident from potentiometric titrations of clay suspensions, and, as might be expected, the shape of the titration curve depends strongly on the concentration of background electrolyte. Complications arise in the interpretation of these curves because of the presence of Al^{3+} and nonexchangeable $Al_n(OH)_{3n-x}^{x+}$ polymers on the clay surfaces [240,241,243-245]. With montmorillonite, the edges probably account for about 10% of the total external surface area, but with the other phyllosilicates the proportion is much less. With allophane and imogolite the pH-dependent charge of edges and surfaces is much greater.

Cation retention on ferrialuminosilicate clays lacking isomorphous substitutional charge decreases with electrolyte concentration and varies with the species of cation and pH. The concept of a CEC is therefore useful only when the precise conditions of its measurements are also stated. Because cations hydrolyze to a varying extent, complete washing out of the saturating cation is advisable, and 0.01 N final concentration of divalent cations [246] or 0.02 N KCl in 80% methanol (with weighing of the excess salt) is recommended [247] for phyllosilicates. Use of a 0.05 N final salt concentration and estimation of excess salt by weighing has been recommended for allophane and imogolite [238].

The clay minerals show a relatively high affinity for polyvalent over monovalent cations when salts of the same molarity are present in a single solution. The high affinity of micas for Cs^+, Rb^+ and K^+ results from the ability of the large monovalent cations to lose their water of hydration and form polar bonds with the structural oxygen of clays. These cations enter the interlayer spaces at the wedge-shaped boundaries between mica and vermiculite at the

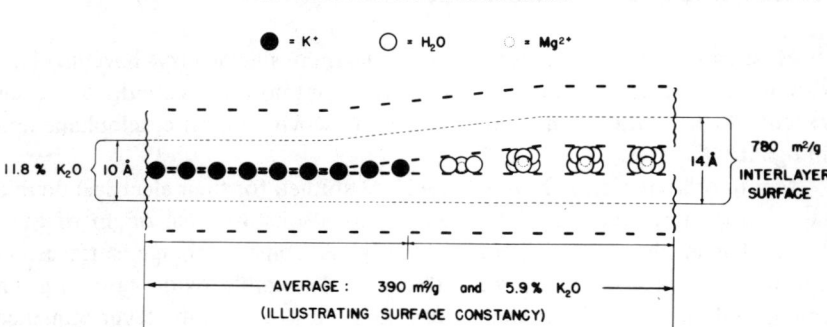

Figure 9. Cleavage margin of mica core surrounded by vermiculite showing a constancy of coverage by K^+ or water and hydrated exchangeable cations [248,307].

edges [248] (Figure 9) of weathered mica. Among the monovalent cations, adsorption is greater for the larger, less-hydrated cations such as Cs^+ than for smaller ions such as K^+. This affinity for Cs^+ is most pronounced at the periphery of micaecous mineral [249-251] particles. Hydrated multivalent cations can occupy only the outer periphery [252] (Figure 10) of the frayed edges of crystals [237].

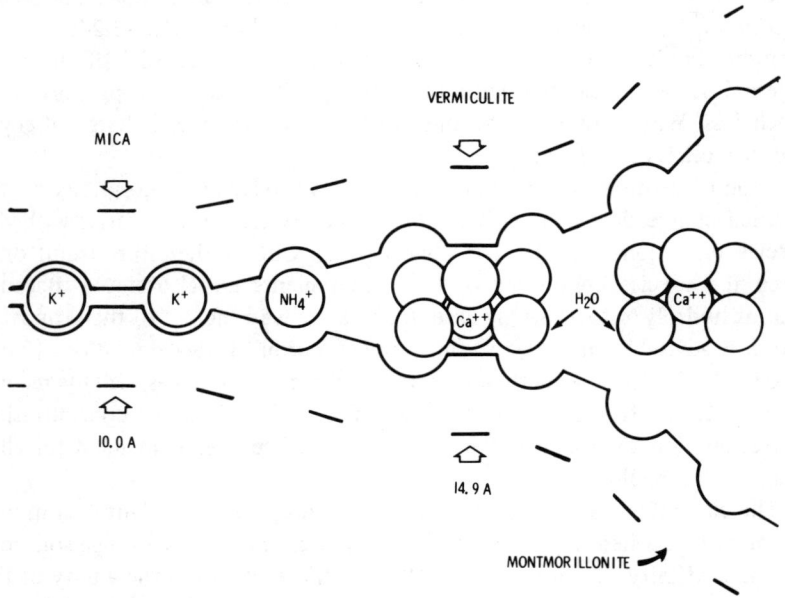

Figure 10. Mica weathering to a vermiculite and montmorillonite around the rim of a mica particle showing steric hindrance to entrance of hydrated cations near the K^+ interlayer [252].

CATION ADSORPTION

The phyllosilicate minerals usually do not show strong specific adsorption for multivalent cations, which can probably be attributed to the relative inactivity of the siloxane (Si-O-Si) structure found on the planar surfaces of most of these minerals. However, specific interactions may occur at low concentrations.

Homovalent Exchange

The mutual exchange of similarly charged (homovalent) cation species can be interpreted in a more straightforward manner than heterovalent exchange, so gives a better insight into the structural causes of cation exchange selectivity in clays. The results are conveniently reported in terms of exchange isotherms, with reduced axes showing the equivalent fraction of one of the ions in the adsorbed state against its equivalent fraction in solution. The ion exchange reaction may be written as follows:

$$A^{n+} + B^{n+} - \text{clay} = B^{n+} + A^{n+} - \text{clay} \qquad (1)$$

and an empirical selectivity coefficient defined by

$$K_c = \frac{(a_{B^{n+}})N_{A^{n+}}}{(a_{A^{n+}})N_{B^{n+}}} \qquad (2)$$

where $N_{A^{n+}}$ and $N_{B^{n+}}$ are the equivalent fractions of A^{n+} and B^{n+} on the clay and $(a_{A^{n+}})$ and $(a_{B^{n+}})$ are the activities of A^{n+} and B^{n+} in solution. When the activity coefficients of A^{n+} and B^{n+} in solution are similar, it is possible to substitute the activity ratio by the concentration ratio. K_c usually varies with the extent of exchange, but when K_c is known over the entire range of compositions from pure A to pure B, it is possible to calculate an average thermodynamic equilibrium constant [253]. Changes in selectivity can then be interpreted in terms of changes in the activity coefficients of the adsorbed ions, but this is merely a formal exercise and is of no predictive value. For practical purposes, K_c is more useful.

The alkali metal cations are the most widely studied series of cations and, in nearly all cases, the order of selectivity is $Cs^+ > Rb^+ > K^+ > Na^+ > Li^+$, which shows that the clay minerals favor the cation with the smaller *hydrated* radius since the order of increasing hydrated ionic radii is the reverse of this series [254]. Cs^+, Rb^+ and K^+ have a low energy of hydration in the adsorbed state and this allows close approach to the silicate layers, an energetically favorable situation for polar bonding. The hydration of cations can be determined directly from the basal X-ray diffraction spacings of the homoionic clays and also from the effect of water activity on alkali metal ion selectivity [255]. NH_4^+ also behaves as a weakly hydrated cation and like Cs^+, Rb^+ and K^+, can become irreversibly "fixed" (exchangeable with difficulty) in vermiculite and micaceous minerals.

The initial cation saturation can have a significant effect on cation uptake, especially when there is a large difference in the hydration characteristics of the two ions involved. This is particularly true for the micaceous minerals and often leads to hysteresis in the exchange isotherm. For example, Sawhney [250] found that appreciably more Cs^+ was adsorbed and fixed by biotite, muscovite, illite and vermiculite when the sample was Ca^{2+} saturated than when K^+ saturated. This probably was a result of the greater accessibility of the hydrated interlayer sites in the Ca^{2+} saturated minerals, rather than a reflection of intrinsic differences in K^+-Ca^{2+}-Cs^+ exchange selectivity. With montmorillonite and kaolinite, Sawhney found the reverse sequence, namely more Cs^+ was adsorbed by the K^+ saturated minerals than by the Ca^{2+} saturated minerals. Hydrated ions may become trapped in vermiculite when there is layer contraction brought about by an interlayer collapsing ion.

Differences in selectivity for clay minerals from different sources are most noticeable when Cs^+ is involved in the exchange. For example, values of K_c for Na^+ for Cs^+ exchange by montmorillonite varied from 6 to 40 as the Cs^+ coverage decreased [254]. In another montmorillonite, K_c remained more or less constant at 25-30 over a wide range of Cs^+ coverage [256]. The K^+ and Cs^+ selectivity on montmorillonite increases sharply at low surface coverages ($< 3\%$ saturation), indicating that there are a small number of sites highly specific for K^+ and Cs^+ [257]. These sites may not be characteristic of montmorillonite, but of small amounts of micaceous vermiculite impurities that are frequently present in naturally occurring montmorillonites. This is supported by the finding that the K^+ selectivity of the < 0.08-μm particle size fraction of several reference and soil montmorillonites was significantly lower than that of the 2- to 0.2-μm fraction at low K^+ saturations [258]. Fluorhectorite also shows high Cs^+ selectivity, although in this case micaceous impurities are presumably not involved [259]. High-charge interlayer areas could be involved.

Approximate values of K_c for other exchange pairs on montmorillonite are Na^+ for Rb^+ (4-18), Na^+ for K^+ (2-4), Na^+ for Li^+ (0.7-1.4) and NH_4^+ for Rb^+ (3.5) [254,256,260]. The same selectivity sequence for alkali metal cation adsorption is found on vermiculite as for montmorillonite, but the Li^+ for Na^+ K_c on vermiculite ranged from 6 to 22, compared with a value close to 1 or 2 for montmorillonite, indicating relatively greater Na^+ selectivity on vermiculite compared with montmorillonite [261]. There are few accurate exchange data available for kaolinite, but Gast [254] found that there was as much difference in selectivity coefficients between two different montmorillonites as between a kaolinite and the montmorillonites. The selectivity sequence found on kaolinite is $Li^+ < Na^+ < K^+ < Cs^+$, but the origin of the small surface charge is uncertain. Cation adsorption is pH and concentration dependent, like that on the hydrous oxides. It may be complicated by hydrolysis of cations such as Li^+ and Na^+ and by competition with aluminum dissolved from clay [262,263].

Somewhat surprisingly, there is little evidence for dependence of cation exchange selectivity or of fixation on the location of charge (tetrahedral versus octahedral substitution). The fixation of K^+ by montmorillonites (after drying) increased with total layer charge density, but no relationship between K^+ fixation and the structural source of charge could be found [264]. However, there is evidence that the structural origin of charge can affect the orientation of hydrated cations adsorbed in the interlayer spaces of phyllosilicates [265].

Cation exchange with the alkaline earth cations does not show as much variation as found with the alkali metal cations because individual differences in hydration energy are not as great as for the alkali metal ions. The alkaline earth cations remain hydrated when adsorbed, and the usual selectivity sequence for montmorillonites is $Ba^{2+} > Sr^{2+} > Ca^{2+} > Mg^{2+}$ [266,267]. This sequence is also found on vermiculite [261,268]. There is little information on divalent cation exchange selectivity on kaolinite; however, $Ca^{2+} > Mg^{2+}$ was found on a kaolinitic soil clay [269].

Under most conditions, the exchange selectivity of divalent transition and heavy metal cations over the alkaline earth cations is not great. This contrasts with the situation found on the hydrous oxides. For example Mn^{2+} selectivity was found to be between Mg^{2+} and Ca^{2+} on montmorillonite [270]. Cu^{2+} showed no preference over Mg^{2+} on montmorillonite, although it showed a large preference over Na^+ [271]. Selectivity can vary with exchange site occupancy. At low surface loadings of Cu^{2+} on montmorillonite, Ca^{2+} was preferred to Cu^{2+}; however, at high Cu^{2+} loadings, there was a selectivity reversal [272]. The preference at low Cu^{2+} coverages has been reversed by raising the pH [273]. Increased uptake of Cu^{2+} by montmorillonite has been found in the presence of Cu(II) tetrammine complex at high pH [274], but this was not a simple ion exchange reaction since at high tetrammine concentrations uptake amounted to twice the cation exchange capacity of the clay. Less than 10% of the adsorbed Cu(II) at this high saturation was isotopically exchangeable. Also, Cu(II) tetrammine complexes were found to be irreversibly adsorbed by montmorillonite [259]; Cd^{2+} showed only a small preference over Ca^{2+} on montmorillonite, mica or kaolinite [275]; and Pb^{2+} was usually favored by only a factor of two or three over Ca^{2+}. The equilibrium Ca^{2+} concentration must therefore be considered when interpreting Pb^{2+} adsorption (exchange) on clays [276].

Heterovalent Exchange

Heterovalent cation exchange by clay minerals has attracted much attention, and many attempts have been made to find a model that explains the behavior over a wide range of experimental conditions [277]. While a number of models are satisfactory over a limited range of experimental conditions, no universally acceptable model is available. This is perhaps not surprising in view of the complexities of the exchange processes.

For mono-divalent exchange, the reaction may be written as follows:

$$A^{2+} + 2B^+\text{-clay} = 2B^+ + A^{2+}\text{-clay} \tag{3}$$

and an empirical selectivity coefficient defined by the following:

$$K_c = \frac{(a_{B^+})^2 N_{A^{2+}}}{(a_{A^{2+}}) N_{B^+}} \tag{4}$$

where (a_{A^+}) and $(a_{B^{2+}})$ are the activities of A^+ and B^{2+} in the equilibrium solution and $N_{A^{2+}}$ and N_{B^+} are the equivalent fractions of A^+ and B^{2+} in the adsorbed state, respectively. A similar equation can be written for mono-trivalent exchange. K_c is not dimensionless as it is in homovalent exchange, and selectivity is strongly dependent on the total cation concentration. When the activity coefficients of the cations in the mixed electrolyte are not known or cannot be reasonably calculated, an uncorrected selectivity coefficient, K'_c, can be defined in a similar manner to K_c above but using molarities rather than activities. Then K_c may be determined without a knowledge of the activity coefficients by extrapolating K'_c to zero ionic strength [278]. This extrapolation also has the advantage that it tends to minimize hysteretic effects which are often found in heterovalent exchange. Extrapolation is particularly convenient when the averaged thermodynamic exchange constant, K, is to be calculated using the Gaines-Thomas approach because it gives K_c directly for the desired standard state, namely a homoionic clay in equilibrium with an infinitely dilute solution of the corresponding salt. There are a number of examples of this thermodynamic approach to mono-divalent exchange [257,259, 265,270,279].

While selectivity is governed by the same factors as in homovalent exchange, an additional factor may be important in heterovalent exchange, namely, the spatial distribution of charged sites. Multivalent cations must simultaneously satisfy the charge of two or more adjacent exchange sites since isomorphous substitution usually produces singly charged sites. While the spatial distribution of charge at the surface of the clay minerals is not known for certain, it is likely that the effect of the discrete charge-producing site in the octahedral and tetrahedral sheets will still be evident at the surface, and it would be expected that divalent cations would favor sites that are closer together. McBride [271] has discussed this point and has suggested that this might explain the high Cu(II) preference of Na montmorillonite compared with Mg montmorillonite. Similarly, when exchange occurs in the interlayer spaces of vermiculite or montmorillonite domains, the spatial relationship of the surface charge of adjacent layers might be important. It is perhaps significant that in Na^+-Ca^{2+} montmorillonite suspensions at low Na^+ loadings the interlayer regions were predominantly Ca^{2+} saturated, while the external surfaces were predominantly Na^+ saturated [280].

An indication of the mono-divalent cation selectivity is given by the fraction of sites occupied by the monovalent cation in equilibrium with a solution containing equinormal concentrations of the mono- and divalent cations. For example, with a total cation concentration of 0.01 N, the following monovalent fractional coverages, N^+, have been found for montmorillonite at 25-30°C for the following ion pairs: NH_4^+-Mg^{2+} (0.32), NH_4^+-Ca^{2+}, (0.27), NH_4^+-Sr^{2+} (0.25), NH_4^+-Ba^{2+} (0.20), Na^+-Ba^{2+} (< 0.01), NH_4^+-Mn^{2+} (0.26), K^+-Ca^{2+} (0.18) [266,270,279]. The K^+-Ca^{2+} selectivity of a number of clay minerals at 0.01 N total electrolyte concentration gave the following values for N^+: montmorillonite (0.10), vermiculite (0.19), biotite, 2-0.2 μm (0.49), biotite, 5-2 μm (0.48), muscovite, 0.2-0.08 μm (0.56), and muscovite, 2-0.2 μm (0.63) [246]. The relatively high K^+ selectivity of the mica minerals is evident. In general, selectivity for the divalent cation increases with a decrease in total equivalent concentration. It also increases with temperature. Also, for a given type of mineral, divalent cation selectivity should increase with surface charge density, although this is not always true [259].

Most of the studies of heterovalent exchange by clay minerals have involved mono-divalent cation exchange. Exchange involving trivalent cations is difficult to study because the acid conditions required to prevent hydrolysis are often sufficient to lead to the partial decomposition of the clay mineral. In acid conditions, Al^{3+} is usually strongly preferred to mono- and divalent cations [281,282], although at high concentrations (1 N) K^+ is preferred to Al^{3+} on montmorillonite [283]. When exchange of aluminum is considered at pH > 3, the significance of hydrolyzed polynuclear Al species must be considered, since these are strongly adsorbed on clay surfaces [245,284]. The size of the adsorbed polynuclear species is variable, although the most probable range in acid solution is $Al_6(OH)_{12}^{6+}$ to $Al_{13}(OH)_{32}^{7+}$, with an OH:Al ratio of about 2.5, or larger species. It is nonexchangeable with Ca^{2+} or Ba^{2+} and can contribute to the pH-dependent charge of the clay [244,285]. The preferential adsorption of OH-Al species with an OH:Al ratio of 2.5 takes place from solutions with OH:Al ratios considerably less than 2.5, which means that the clays appear to promote hydrolysis of aluminum [286]. Uptake of OH-Al by clays is probably described best in terms of Langmuir-type adsorption models [287,288].

In solutions in which ion-pair or complex formation occurs, the exchange of all cationic species must be considered if the exchange constants established are to be independent of the anion present. Ion-pairing is most likely to occur at high salt concentrations and in the presence of such anions as CO_3^{2-}, HCO_3^- and SO_4^{2-} [289].

Adsorption Involving Structural Cations

Proton Adsorption by Silicates

Protons from acids or hydrolytic reactions of adsorbed cations penetrate into the surface few atomic sheets of most silicates, with a release of structural cations [286,290-292]. This is usually designated as weathering of minerals rather than ion exchange, since it is generally irreversible at soil temperatures and pressures. The reactions are, however, reversible at higher temperatures and pressures. Such reactions have been designated as "spontaneous interchange reactions" [293]. Silicic acid (pK_1 = 9.5) inclusions in the silicates, including clays, act as proton sinks [286,294]. Proton adsorption creates a high acidity at the solid-liquid interface, and Si-O-Si and Si-O-M groups (where M is Mg, Al or Fe) dissociate forming small polymers and monomers that are either adsorbed on the surface or diffuse into the bulk solution [292-295]. In trioctahedral minerals, the protons exchange octahedral Fe(II) or Mg-ions or tetrahedral Al ions. In dioctahedral phyllosilicates, the central oxygen must rotate so that a proton may be attached to the free pair of electrons [292]. The protonation to form silicic acid in silicates has been shown to be a first-order reaction, with the rate being directly proportional to the surface charge density and to the MgO content of micaceous minerals [294,296].

A major study of proton-silicate reactions has elucidated the basic chemistry of soil acidity and liming [286,291,297,298] in a kind of cyclic growth of knowledge of the role of protons [297] and aluminum [298,299] in soil acidity [300]. Soil acidity and its neutralization by liming is of considerable agricultural importance [301] and has been categorized into five groups according to the acid strength of the proton donor sites [245], which include $Al(OH_2)_6^{3+}$, polynuclear hydroxy aluminum species [245,248] and humus carboxyl groups. Further pedological significance of proton interchange for Mg^{2+} of silicates lies in the high percentage of Mg^{2+} saturation in solonetz soils [302] and soils developed on serpentinite rocks [303].

Release of Native Potassium from Micas

Micas are "nonexpanding" minerals because the high layer charge (240-250 meq/100 g) results in locking the interlayer cations, usually K^+, in the highly charged interlayer space. Ordinarily, rapid cation exchange therefore takes place only on the external surfaces, at defects and in the expanded wedge zones at the edges of the flakes. The release of native K from micas is an important weathering reaction that occurs in soils and can eventually result in the transformation of mica to vermiculite and smectite by charge lowering through proton adsorption (Si-O → SiOH) and release of tetrahedral aluminum (and iron) [304]. Further epitaxial crystallization to chlorite and kaolinite occurs [305]. Muscovite often transforms to dioctahedral chlorite, then kaolinite. The weathering of micas to vermiculite and other clays has been reviewed in detail [230,248,305-309] and is summarized in the following extended form [248,307]:

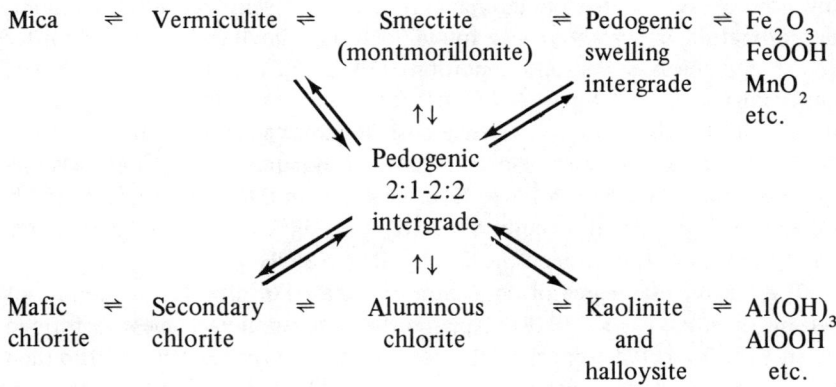

In the laboratory, the key to exchange of K^+ from micas is that the solution concentration of K^+ should be kept low [310] either by the use of sodium tetraphenylboron, which removes K^+ from solution by precipitation, or by frequent renewal of the replacing solution. Ba^{2+} is a particularly effective inorganic cation for replacing interlayer K^+, but concentrated Na^+ solutions [243] are also effective, especially if refluxed with the mineral. The release of K is inhibited when there is above a certain critical threshold concentration of K^+ in solution. This threshold K^+ concentration depends on the particular mica and the nature and relative concentration of the replacing cation, but is of the order of $10^{-4} M$ (4 μg/ml) for biotites and $10^{-5} M$ (0.4 μg/ml) for muscovites in the presence of $1 M$ NaCl [311]. H^+ adsorption, unlike that of other cations, can release K^+ along with other structural cations. Even moderately acidic concentrations ($\sim 10^{-3} M$) lead to partial structural disintegration and dissolution of the mica [312]. Strong acids leave only silicate relics [313, 314].

The rate of K^+ release from biotites (trioctahedral) is many times greater than from muscovites (dioctahedral), and within the trioctahedral micas the rate of K^+ release tends to decrease markedly with increasing F^- for OH^- substitution and with increasing oxidation of octahedral Fe^{2+}. The unifying theme here seems to be the orientation of the O-H bond of some of the octahedral $-OH$ groups or, more specifically, the separation (repulsion) between the H^+ of the $-OH$ group and the adjacent interlayer, K^+. In biotites (symmetrical octahedral occupancy), the $-OH$ bond is perpendicular to the basal plane, whereas in muscovites (asymmetrical ocathedral occupancy), the $-OH$ bond is inclined giving a greater $H^+...K^+$ separation, less repulsion of interlayer K^+ and slower K^+ release. Similarly, when F^- substitutes for OH^-, there is no H^+ to repel the K^+ and consequently the interlayer K^+ is bound more strongly [311].

Oxidation of Fe(II) to Fe(III) in biotite was believed to be the key to lowering of layer charge and interlayer hydration of vermiculite [306]. Later, deprotonation of hydroxyl was found to occur simultaneously in chlorites [315] and micas under acid conditions [241,316,317]. Even under neutral conditions, the deprotonation of OH occurred with oxidation of Fe(II) in micas [309]. Also, the iron valence change is reversible without concurrent change in cation exchange capacity [241]. Nevertheless, most naturally occurring vermiculites have a lower layer charge than their parent micas [304]. Concurrent ejection of structural Fe(III) and Mg^{2+} lowers the layer charge [309] and leads to proton adsorption by Si-O → SiOH.

The greater the degree of octahedral Fe oxidation, the slower the rate of release of interlayer K^+ [318]. Apparently, octahedral vacancies are formed when Fe^{3+} is ejected from the octahedral sheet, and the resulting biotite then has inclined O-H bonds and behaves in spots like a dioctahedral mica. This may help explain the long-known fact that the rate of release of the last of the K^+ that can be released from fine-grained mica (< 1 μm) is slower than from coarser-grained fractions of the same micas. A greater proportion of the octahedral iron is likely to have been oxidized and ejected and OH orientation altered from $90°$ in the fine-grained mica fractions.

The layer charge decrease of dioctahedral micas (little or no Fe^{2+}) is well known and may occur as a result of proton adsorption [319]. When K^+ is removed from muscovite in the laboratory, however, there is little immediate loss of layer charge, and the resulting vermiculite has an exceptionally high cation capacity (> 200 meq/100 g) [320]. The rate of cation exchange in these dioctahedral vermiculites is much slower than in usual lower-charged, vermiculites because of the more restricted interlayer expansion of the former [321].

The release of K^+ from micas is not quickly reversible at 25°C and 1 atm. A number of K^+-depleted phlogopite micas were collapsed after treatment with K^+ solutions, but the apparent basal spacings were about 0.003 nm greater than in the original unaltered micas [322]. Furthermore, K was released from the reconstituted micas more readily than from the original micas.

Adsorption of Trace Amounts of Cations

Of considerable practical significance is the fact that clays can adsorb trace amounts of divalent cations from solutions containing a large excess of monovalent cations. The adsorption and fixation of small quantities of Cs^+ by micaceous minerals has been mentioned previously, but the adsorption of trace amounts of hydrated cations is not so well documented. Such adsorption might be anticipated since the exposed octahedral edges of the clay minerals could act in much the same way as the simple oxides.

Early work indicated that the sorption of certain transition metals could exceed the cation exchange capacity of the clay. Since this was observed at a pH where cation hydrolysis was suspected, sorption of hydrolyzed ions, such as $CuOH^+$, was implicated. While there may have been some ambiguity between true adsorption and hydroxide precipitation at the high cation concentrations used in these early studies, more recent work has confirmed pH-dependent cation adsorption at low cation concentrations where precipitation is unlikely [323-325]. Traces of Co^{2+} were adsorbed by montmorillonite in the presence of 0.1 N $CaCl_2$. The amounts increased steadily above pH 5; however, not all of the adsorbed Co^{2+} was desorbed by 2.5% acetic acid. This nonexchangeable fraction increased steadily over 30 days, suggesting that Co^{2+} was somehow becoming incorporated into the crystal structure. Adsorption of Co^{2+} traces by montmorillonite in approximately 0.01 N NaCl [323] resulted in approximately 100 times more Co^{2+} adsorption than in 0.1 N $CaCl_2$. These results suggest that the background electrolyte is important even for trace adsorption. Mg^{2+} also interfered with trace adsorption of Cu^{2+} and Co^{2+} on micaceous clay, but adsorption was still significant in the presence of high background electrolyte concentrations [325]. At an equilibrium Co^{2+} (and Zn^{2+}) concentration of $10^{-7} M$ in seawater ($\sim$ 0.7 M NaCl at pH 8.2), about 2 µg/g of Co and 23 µg/g of Zn were adsorbed on a micaceous clay [326]. These reactions involve proton release rather than proton adsorption.

Most of the trace adsorption studies on clay minerals have concerned Cu^{2+}, Co^{2+} or Zn^{2+}, but the adsorption of trace amounts of Cd^{2+} by Na^+ montmorillonite has been studied [327]. In an essentially salt-free system at pH 6.5-7.0, montmorillonite adsorbed 25 µg/g of Cd^{2+} at an equilibrium Cd^{2+} concentration of only 1 µg/l ($10^{-8} M$ Cd^{2+}). In line with other evidence quoted above, this adsorption decreased markedly in the presence of a background electrolyte, especially NaCl, in which Cd^{2+}-Cl^- complexes are formed. There was evidence, however, that some of the high-affinity Cd^{2+} adsorbing sites were not affected by Na^+ competition.

A large number of organic and inorganic ligands prevented uptake of Cu(II) and Zn(II) by kaolinite, montmorillonite and illite when the metal was present solely as an anionic complex [328-330]. However, uptake of protonated Cu(II) and Zn(II) N-containing complexes was significant, particularly with montmorillonite. Clays appear to promote the precipitation of metal hydroxides at low pH and inhibit the dissolution of amphoteric hydroxides, such as $Zn(OH)_2$, at high pH.

The location of the highly selective adsorption sites is often thought to be at the mineral edges, but the evidence for this is only circumstantial [331]. It has been suggested [332] that transitional metal cations can be adsorbed at the edges of phyllosilicates and, in so doing, extend the octahedral sheet [332]. These ions may, in turn, adsorb silicic acid, which would then complete the lateral extension of the layer structure. This suggestion was supported by the stoichiometry of transition metal ion adsorption and by the observation that silicic acid increased Zn adsorption on montmorillonite [333]. However, ESR spectra of Cu adsorbed on montmorillonite have indicated that the edge sites of montmorillonite may not be the principal sites of Cu adsorption [334]. Detailed studies of the pH-dependence of adsorption of trace quantities of cations on the clay minerals may help resolve the issue. Caution is required in interpreting such studies, however, since clay minerals are often coated with various quantities of the simple hydrous oxides, particularly those of aluminum and iron [314]. These have not always been removed beforehand, so adsorption may not be directly related to the phyllosilicate per se but to the hydrous oxide coatings. Even initially pure Na phyllosilicates can accumulate small amounts of these coatings after prolonged storage in salt-free solutions. Hydrolysis of Na^+ clays causes H^+ release from H_2O and initiates proton adsorption and structural decomposition. The rate of such decomposition of some phyllosilicates increased with octahedral substitution [335].

Under natural conditions, adsorbed anions such as phosphate, silicate and carbonate, as well as organic and metal oxide coatings, can interfere with trace cation adsorption on clays. While this may complicate interpretation of the results, it seems desirable that adsorption on at least some of these "dirty" clays should be studied under as near natural soil conditions as possible. An untreated Georgia kaolinite adsorbed about 0.1 meq Cu^{2+}/100 g (30 μg/g) at an equilibrium concentration of $5 \times 10^{-5} M$ Cu^{2+} (pH 5-6), but when the clay had been pretreated with acidified (pH 3.4) NaCl to remove adsorbed hydroxy Al ions, adsorption increased eightfold [334]. Aluminum vacancies may have been created in the kaolin structure by such an acid treatment and these may have been partly responsible for the increased adsorption. Small amounts of native organic matter associated with bentonites may be significant in adsorption of traces of cations such as Cu^{2+}, for which organic matter selectivity is high [336].

Cation Interchange in Burial Diagenesis

The cation exchange capacity of phyllosilicates is subject to drastic changes as the weathering reactions are reversed during diagenesis through burial in sedimentary formations [337]. With increasing depth of burial, montmorilmorillonite is altered through mixed-layer assemblages to mica and chlorite with accompanying massive cation interchanges between structures. Most of the expansible layers are lost by 12,000 ft of burial. Kaolinite is destroyed, tetrahedral aluminum and K^+ are increased, and extensive migration of ions occurs. Excess of water over that required for hydroxyl formation creates "superpressure." There is a great migration of ions, gases such as methane, and sometimes petroleum [292]. The CEC of the minerals becomes low because the interlayer charges of mica and chlorite are not accessible for cation exchange. At greater depths of burial (increased temperatures and pressures), there is a granitization of the clays. The weathering reactions which lower the mica charge and make the structural layers cleave and exchange interlayer cations, are reversible under burial diagenesis of sediments [337].

CONCLUSIONS

In aqueous solution, the surfaces of metal oxides, hydroxides and hydrous oxides are made up of OH groups capable of showing weak acid behavior, and the extent to which they are protonated determines the surface charge. At a pH above the point of zero charge (pzc), the solid has a net negative surface charge, which increases with pH and electrolyte concentration. This negative surface charge is largely balanced by the adsorption of cations and, for simple 1:1 electrolytes at moderate concentrations (< 0.1 M), the surface charge, and, hence, cation adsorption, is more or less independent of the cation. At higher electrolyte concentrations, especially at high negative surface charges, small differences in effective surface charge are often found, and the surface charge for the alkali metal cations often follows the lyotropic series $Li^+ < Na^+ < K^+ < Rb^+ < Cs^+$, or its reverse.

With divalent cations, cation adsorption involves specific interactions with the surface and adsorption shows a much greater dependency on solution pH than can be explained by the straightforward development of the underlying surface charge. Indeed, divalent cation adsorption can take place in the presence of a large background concentration of a 1:1 electrolyte. For iron and aluminum oxides in particular, adsorption usually takes place even when the surfaces have a net positive surface charge. With silica, adsorption usually takes place on a negatively charged surface; however, adsorption on silica does not appear to be as strong as on the other common oxide surfaces, even when compared on a surface area basis. In general, the greater the tendency of the divalent cation to hydrolyze, the greater the adsorption. The selectivity sequences vary with different oxides and can also vary with cation concentration and pH. Significant divalent cation adsorption frequently occurs when only a very small fraction of the cation in solution is hydrolyzed. In the case of the alkaline earth cations, approximately one H^+ is released (or one OH^- coadsorbed) for each M^{2+} adsorbed; however, for the transition and heavy metal cations, this stoichiometry coefficient is usually between 1.5 and 2.0. The source of the H^+ released is uncertain. It could come either from a bulk or surface water molecule as a result of the preferential adsorption of hydrolyzed species such as MOH^+, or, alternatively, it could be released by direct exchange from surface-OH groups. Detailed spectroscopic (IR, ESR) and isotopic exchange (^2H, ^3H) studies should resolve this question, at least for large surface occupancies.

The pH at which divalent cation adsorption becomes significant varies with the particular cation and solid involved, the solid:solution ratio, the specific surface area of the solid, the total cation concentration, and the concentration of other interacting species. Therefore, it is obviously difficult to generalize. However, for iron and aluminum oxides, the critical pH is about 6.5-9 for the alkaline earth cations, 3-5 for Cu^{2+}, Pb^{2+} and Hg^{2+}, 5-6.5 for Zn^{2+}, Co^{2+}, Ni^{2+} and Cd^{2+}, and 6.5-7.5 for Mn^{2+}. The critical pHs tend to be a bit higher for silica and lower for manganese oxides. Adsorption densities vary greatly, but in the case of iron hydrous oxide gels at an equilibrium concentration of 10^{-7} M M^{2+} and pH 7, for example, typical adsorption densities are 0.001 μmol/mmol Fe (0.4 μg/g or 3×10^{-11} mol/m^2) for Ca^{2+} and about 0.2 μmol/mmol Fe (130 μg/g or 6×10^{-9} mol/m^2) for Co^{2+}. Adsorption, unlike hydroxide or carbonate precipitation, tends to become more, rather than less, important at low cation concentrations. The separation of the percent adsorption against pH curves gives an indication of the possible competition between different ions. For example, the wide separation between the percent adsorption against pH curves for the alkaline earth cations compared with the transition and heavy metal ions indicates that Ca^{2+} would not be expected to interfere greatly with transition and heavy metal ion adsorption, even when the calcium concentration is several orders of magnitude greater than that of the transition or heavy metal cation. Hydrolysis, including the formation of polynuclear species, and hydroxide precipitation dominate the behavior of trivalent cations such as Fe^{3+} and Al^{3+}.

Since much of the present interest in cation adsorption by the hydrous oxides is concerned with their possible role in controlling the transport of trace metals in the environment, there is a need for more information at environmentally realistic concentrations, i.e., $< 10^{-7}$ M. Also, the reversibility of adsorption, perhaps after a long period of aging, warrants more detailed study because the ability of adsorbed ions to be desorbed by changing conditions is pertinent to the environmental question.

Cation adsorption by phyllosilicate clays differs from that on the simple oxides because for the clays, the surface charge (usually negative) is largely controlled by the amount of isomorphous substitution within the clay structure, rather than by the adsorption of H^+ or OH^- from solution. Cation exchange is therefore usually a more appropriate term than cation adsorption. Furthermore, although often extensive, the planar surfaces of most phyllosilicates (largely Si-O-Si surfaces) are relatively inactive compared with Al-OH or Fe-OH surfaces of iron and aluminum oxides, and even the Al-OH surface of kaolinite does not seem to show marked reactivity. Rather, cation selectivity on clays tends to be dominated by the energetics of fitting exchangeable cations into interlayer spaces, which is best exemplified by the exchange behavior of the alkali metal cations. Cs^+, Rb^+ and K^+, which are weakly hydrated, tend to be favored over the more strongly hydrated Na^+ and Li^+ ions. With divalent cations, differences in hydration energies are smaller and consequently differences in selectivity are smaller. With many clays, in contrast with the oxides, there is little difference in selectivity between the alkaline earth cations and the transition and heavy metal cations, at least at high concentrations. Most cation exchange studies on clays have been performed with relatively high cation concentrations and there is a need to investigate trace cation uptake by "clean" and "dirty" clays in more detail. A "method of additions" approach might be a direct and useful approach for understanding the role of the various components in determining cation uptake in more complex mixed clay-oxide-organic matter systems.

ACKNOWLEDGMENTS

This research was supported in part by the Department of Soil Science, School of Natural Resources, College of Agricultural and Life Sciences, University of Wisconsin, Madison, projects 1123 and 1336; in part by the National Science Foundation EAR76-19783-JACKSON; and in part by the Ecological Sciences Branch, Division of Biomedical and Environmental Research, U.S. Department of Energy Contract EY-76-S-02-1515 (paper COO-1515-83); through an International Consortium for Interinstitutional Cooperation in the Advancement of Learning (ICICAL). The authors wish to thank Ms. Chrystie Jackson for typing the manuscript.

REFERENCES

1. Zhabrova, G. M., and E. V. Egorov. "Sorption and Ion Exchange on Amphoteric Oxides and Hydroxides," *Russ. Chem. Rev.* 30: 338-346 (1961).
2. Churms, S. C. "Inorganic Ion-Exchangers, Part IV," *S. Afr. Ind. Chem.* 19: 87-92, 96 (1965).
3. Jenne, E. A. "Controls on Mn, Fe, Co, Ni, Cu, and Zn Concentrations in Soils and Water: the Significant Role of Hydrous Mn and Fe Oxides," in *Trace Inorganics in Water,* Adv. Chem. Ser. 73 (Washington, DC: American Chemical Society, 1967), pp. 337-387.
4. Jenne, E. A. "Trace Element Adsorption by Sediments and Soils—Sites and Processes," in *Molybdenum in the Environment,* Vol. 2, W. R. Chappell and K. K. Petersen, Eds. (New York: Marcel Dekker, Inc., 1977), pp. 425-553.
5. Veselý, V., and V. Pekárek. "Synthetic Inorganic Ion-Exchanger - I. Hydrous Oxides and Acidic Salts of Multivalent Metals," *Talanta* 19: 219-262 (1972).
6. Bar-Yosef, B., A. M. Posner and J. P. Quirk. "Zinc Adsorption and Diffusion in Goethite Pastes," *J. Soil Sci.* 26: 1-21 (1975).
7. Pyman, M. A. F., J. W. Bowden and A. M. Posner. "The Movement of Titration Curves in the Presence of Specific Adsorption," *Aust. J. Soil Res.* 17: 191-195 (1979).
8. Parks, G. A. "The Isoelectric Points of Solid Oxides, Solid Hydroxides, and Aqueous Hydroxo Complex System," *Chem. Rev.* 65: 177-198 (1965).
9. Parks, G. A. "Aqueous Surface Chemistry of Oxides and Complex Oxide Minerals," in *Equilibrium Concepts in Natural Water Systems,* Adv. Chem. Ser. 67 (Washington, DC: American Chemical Society, 1967), pp. 121-160.
10. Fuerstenau, D. W. "Interfacial Properties in Mineral/Water Systems," *Pure Appl. Chem.* 24: 135-165 (1970).
11. Lyklema, J. "The Electrical Double Layer on Oxides," *Croat. Chem. Acta* 43: 249-260 (1971).
12. Bowden, J. W., A. M. Posner and J. P. Quirk. "Ionic Adsorption on Variable Charge Mineral Surfaces. Theoretical-Charge Development and Titration Curves," *Aust. J. Soil Res.* 15: 121-136 (1977).
13. Breeuwsma, A. *Adsorption of Ions on Hematite (α-Fe_2O_3).* A Colloid-Chemical Study. Ph.D. Thesis, Agricultural University, Wageningen, Netherlands (1973).
14. Perram, J. W., R. J. Hunter and H. J. L. Wright. "The Oxide-Solution Interface," *Aust. J. Chem.* 27: 461-475 (1974).
15. Lyklema, J. "The Structure of the Electrical Double Layer on Porous Surfaces," *J. Electroanal. Chem.* 18: 341-348 (1968).
16. Leckie, J. O., and R. O. James. "Control Mechanisms for Trace Metals in Natural Waters," in *Aqueous-Environmental Chemistry of Metals,* A. J. Rubin, Ed. (Ann Arbor, MI: Ann Arbor Science, 1974), pp. 1-76.
17. Davis, J. A., R. O. James and J. O. Leckie. "Surface Ionization and Complexation at the Oxide/Water Interface. I. Computation of Electrical Double Layer Properties in Simple Electrolytes," *J. Colloid Interface Sci.* 63: 479-499 (1978).

18. Davis, J. A. "Adsorption of Trace Metals and Complexing Ligands at the Oxide-Water Interface," Ph.D. Thesis, Stanford University, Stanford, CA (1977).
19. Ahrland, S., I. Grenthe and B. Norén. "The Ion Exchange Properties of Silica Gel. I. The Sorption of Na^+, Ca^{2+}, Ba^{2+}, UO_2^{2+}, Gd^{3+}, Zr(IV) + Nb, U(IV) and Pu(IV)," *Acta Chem. Scand.* 14: 1059-1076 (1960).
20. Kolařík, Z. "Sorption radioaktiver Isotopen an Niederschlägen. VII. Sorption von Strontium am Mangan (IV)-Hydroxyd," *Colln Czech. Chem. Commun.* 27: 951-958 (1962).
21. Ahmed, S. M., and D. Maksimov. "Studies of the Oxide Surfaces at the Liquid-Solid Interface. Part II. Fe Oxides," *Can. J. Chem.* 46: 3841-3846 (1968).
22. Posselt, H. A., F. J. Anderson and W. J. Weber. "Cation Sorption on Colloidal Hydrous Manganese Dioxide," *Environ. Sci. Technol.* 2: 1087-1093 (1968).
23. Vydra, F. "Use of Rotating Disc Electrodes for the Study of the Kinetics of Sorption Processes," *J. Electroanal. Chem.* 25: 13-15 (1970).
24. Rophael, M. W., and M. A. Malati. "Characterization of Manganese Dioxides. I. The Rate and Activation Energy of Adsorption of Calcium Ions by β-Manganese Dioxide," *Chem. Ind.* (1972), pp. 768-769.
25. Zasoski, R. J., and R. G. Burau. "A Technique for Studying the Kinetics of Adsorption in Suspensions," *Soil Sci. Soc. Am. J.* 42: 372-374 (1978).
26. Morgan, J. J., and W. Stumm. "Colloid Chemical Properties of Manganese Dioxide," *J. Colloid Sci.* 19: 347-359 (1964).
27. Kurbatov, M. H., and G. B. Wood. "Rate of Adsorption of Cobalt Ions on Hydrous Ferrics Oxide," *J. Phys. Chem.* 56: 698-701 (1952).
28. Kurbatov, M. H. "Rate of Adsorption of Barium Ions in Extreme Dilution, by Hydrous Ferric Oxide," *J. Am. Chem. Soc.* 71: 858-863 (1949).
29. Simon, J., W. Schulze and M. Völtz. "Eine kinetische Untersuchung der Cd^{2+}-Sorption an gelartigen Aluminiumhydroxid," *Z. Anorg. Allg. Chem.* 394: 233-242 (1972).
30. Bonner, W. P., H. A. Bevis and J. J. Morgan. "Removal of Strontium from Water by Activated Alumina," *Health Phys.* 12: 1691-1703 (1966).
31. Nancollas, G. H., and R. Paterson. "The Kinetics of Ion Exchange on Zirconium Phosphate and Hydrous Zirconia," *J. Inorg. Nucl. Chem.* 22: 259-268 (1961).
32. Heitner-Wirguin, C., and A. Albu-Yaron. "Hydrous Oxides and Their Cation-Exchange Properties," *J. Appl. Chem.* 15: 445-448 (1965).
33. Heitner-Wirguin, C., and A. Albu-Yaron. "Hydrous Oxides and Their Cation-Exchange Properties. II. Structure and Equilibrium Experiments," *J. Inorg. Nucl. Chem.* 28: 2379-2384 (1966).
34. Kraus, K. A., H. O. Phillips, T. A. Carlson and J. S. Johnson. "Ion Exchange Properties of Hydrous Oxides," *Proc. 2nd Int. Conf. Peaceful Uses Atomic Energy*, Geneva, 28: 3-19 (1958).
35. Belot, Y., C. Gailledreau and R. Rzekiecki. "Retention of Strontium-90, Calcium-45, and Barium-140 by Aluminum Oxide of Large Area," *Health Phys.* 12: 811-823 (1966).
36. Lippens, B. C., and J. H. de Boer. "Studies on Pore Systems in Catalysts. III. Pore-Size Distribution Curves in Aluminum Oxide," *J. Catalysis* 3: 44-49 (1964).

37. Bowen, J. H., R. Bowrey and A. S. Malin. "A Study of the Surface Area and Structure of Activated Alumina by Direct Observation," *J. Catalysis* 7: 209-216 (1967).
38. Churms, S. C. "The Effect of pH on the Ion-Exchange Properties of Hydrated Alumina. Part II. Kinetics of Exchange," *J. S. Afr. Ind. Chem. Inst.* 19: 108-114 (1966).
39. McKenzie, R. M. "The Sorption of Cobalt by Manganese Minerals in Soils," *Aust. J. Soil Res.* 5: 235-246 (1967).
40. McKenzie, R. M. "The Reaction of Cobalt with Manganese Dioxide Minerals," *Aust. J. Soil Res.* 8: 97-106 (1970).
41. McKenzie, R. M. "The Sorption of some Heavy Metals by the Lower Oxides of Manganese," *Geoderma* 8: 29-35 (1972).
42. Burns, R. G. "The Uptake of Cobalt into Ferromanganese Nodules, Soils, and Synthetic Manganese(IV) Oxides," *Geochim. Cosmochim. Acta* 40: 95-102 (1976).
43. Murray, J. W. "The Interaction of Metal Ions at the Manganese Dioxide Solution Interface," *Geochim. Cosmochim. Acta* 39: 505-519 (1975).
44. James, R. O., P. J. Stiglich and T. W. Healy. "Analysis of Models of Adsorption of Metal Ions at Oxide/Water Interfaces," *Disc. Faraday Soc.* 59: 142-156 (1975).
45. Kurbatov, M. H., G. B. Wood and J. D. Kurbatov. "Isothermal Adsorption of Cobalt from Dilute Solutions," *J. Phys. Chem.* 55: 1170-1182 (1951).
46. Bunzl, K., W. Schmidt and B. Sansoni. "Kinetics of Ion Exchange in Soil Organic Matter. IV. Adsorption and Desorption of Pb^{2+}, Cu^{2+}, Cd^{2+}, Zn^{2+} and Ca^{2+} by Peat," *J. Soil Sci.* 27: 32-41 (1976).
47. Kinniburgh, D. G., M. L. Jackson and J. K. Syers. "Adsorption of Alkaline Earth, Transition, and Heavy Metal Cations by Hydrous Oxide Gels of Iron and Aluminum," *Soil Sci. Soc. Am. J.* 40: 796-799 (1976).
48. McBride, M. B. "Retention of Cu^{2+}, Ca^{2+}, Mg^{2+} and Mn^{2+} by Amorphous Alumina," *Soil Sci. Soc. Am. J.* 42: 27-31 (1978).
49. Breeuwsma, A., and J. Lyklema. "Physical and Chemical Adsorption of Ions in the Electrical Double Layer on Hematite (α-Fe_2O_3)," *J. Colloid Interface Sci.* 43: 437-448 (1973).
50. Parks, G. A., and P. L. de Bruyn. "The Zero Point of Charge of Oxides," *J. Phys. Chem.* 66: 967-973 (1962).
51. Albrethsen, A. E. "An Electrochemical Study of the Ferric Oxide-Solution Interface," Ph.D. Thesis, Massachusetts Institute of Technology, Cambridge, MA (1963).
52. Onoda, G. Y., and P. L. de Bruyn. "Proton Adsorption at the Ferric Oxide/Aqueous Solution Interface. I. A Kinetic Study of Adsorption," *Surf. Sci.* 4: 48-63 (1966).
53. Smith, G. W., and T. Salman. "Zero-Point of Charge of Hematite and Zirconia," *Can. Metal. Quart.* 5: 93-107 (1966).
54. Atkinson, R. J., A. M. Posner and J. P. Quirk. "Adsorption of Potential-Determining Ions at the Ferric Oxide Aqueous Electrolyte Interface," *J. Phys. Chem.* 71: 550-558 (1967).
55. Breeuwsma, A., and J. Lyklema. "Interfacial Electrochemistry of Hematite (α-Fe_2O_3)," *Disc. Faraday Soc.* 52: 324-333 (1971).
56. Matijević, E., and P. Scheiner. "Ferric Hydrous Oxide Sols. III. Preparation of Uniform Particles by Hydrolysis of Fe(III)-Chloride, -Nitrate,

and -Perchlorate Solutions," *J. Colloid Interface Sci.* 63: 509-524 (1978).
57. Rubio, J., and E. Matijević. "Interactions of Metal Hydrous Oxides with Chelating Agents," *J. Colloid Interface Sci.* 68: 408-421 (1979).
58. Kinniburgh, D. G., J. K. Syers and M. L. Jackson. "Specific Adsorption of Trace Amounts of Calcium and Strontium by Hydrous Oxides of Iron and Aluminum," *Soil Sci. Soc. Am. Proc.* 39: 464-470 (1975).
59. Davis, J. A., and J. O. Leckie. "Surface Ionization and Complexation at the Oxide/Water Interface. II. Surface Properties of Amorphous Iron Oxyhydroxide and Adsorption of Metal Ions," *J. Colloid Interface Sci.* 67: 90-107 (1978).
60. Robinson, M., J. A. Pask and D. W. Fuerstenau. "Surface Charge of Alumina and Magnesia in Aqueous Media," *J. Am. Ceram. Soc.* 47: 516-520 (1964).
61. Yopps, J. A., and D. W. Fuerstenau. "The Zero Point of Charge of Alpha Alumina," *J. Colloid Sci.* 19: 61-71 (1964).
62. Sadek, H., A. K. Helmy, V. M. Sabet and T. F. Tadros. "Adsorption of Potential-Determining Ions at the Aluminum Oxide Aqueous Interface and the Point of Zero Charge," *J. Electroanal. Chem.* 27: 257-266 (1970).
63. Tewari, P. H., and A. W. McLean. "Temperature Dependence of Point of Zero Charge of Alumina and Magnetite," *J. Colloid Interface Sci.* 40: 267-272 (1972).
64. Huang, C. P., and W. Stumm. "Specific Adsorption of Cations on Hydrous γ-Al_2O_3," *J. Colloid Interface Sci.* 43: 409-420 (1973).
65. Nechaev, É. A., and T. B. Golovanova. "Specific Ion Adsorption and the Structure of the Electrical Double Layer on Aluminum Oxide," *Russian J. Colloid Sci.* 36: 311-314 (1974).
66. Hohl, H., and W. Stumm. "Interaction of Pb^{2+} with Hydrous γ-Al_2O_3," *J. Colloid Interface Sci.* 55: 281-288 (1976).
67. Alwitt, R. S. "The Point of Zero Charge of Pseudoboehmite," *J. Colloid Interface Sci.* 40: 195-198 (1972).
68. Deželić, N., H. Bilinski and R. H. H. Wolf. "Precipitation and Hydrolysis of Metallic Ions-IV. Studies on the Solubility of Aluminum Hydroxide in Aqueous Solution," *J. Inorg. Nucl. Chem.* 33: 791-798 (1971).
69. Brace, R., and E. Matijević. "Aluminum Hydrous Oxide Sols - I. Spherical Particles of Narrow Size Distribution," *J. Inorg. Nucl. Chem.* 35: 3691-3705 (1973).
70. Bolt, G. H. "Determination of the Charge Density of Silica Sols," *J. Phys. Chem.* 61: 1166-1169 (1957).
71. Healy, T. W., R. O. James and R. Cooper. "The Adsorption of Aqueous Co(II) at the Silica-Water Interface," in *Adsorption from Aqueous Solution*, Adv. Chem. Ser. No. 79 (Washington, DC: American Chemical Society, 1968), pp. 62-73.
72. Tadros, T. F., and J. Lyklema. "Adsorption of Potential-Determining Ions at the Silica-Aqueous Electrolyte Interface and the Role of Some Cations," *J. Electroanal. Chem.* 17: 267-275 (1968).
73. James, R. O., and T. W. Healy. "Adsorption of Hydrolyzable Metal Ions at the Oxide-Water Interface. II. Charge Reversal of SiO_2 and TiO_2 Colloids by Adsorbed Co(II), La(III), and Th(IV) as Model Systems," *J. Colloid Interface Sci.* 40: 53-64 (1972).

74. Smith, G. W., and T. Salman. "The Adsorption of Dehydroabietylamine Acetate on Synthetic Rutile," *Can. Metal Quart.* 6: 167-179 (1967).
75. Bérubé, Y. G., and P. L. de Bruyn. "Adsorption at the Rutile Solution Interface. I. Thermodynamic and Experimental Study," *J. Colloid Interface Sci.* 27: 305-318 (1968).
76. Schindler, P. W., and H. Gamsjäger. "Acid-Base Reactions of the TiO_2 (Anatase)-Water Interface and the Point of Zero Charge of TiO_2 Suspensions," *Kolloid-Z. Z. Polymere* 250: 759-763 (1972).
77. Cornell, R. M., A. M. Posner and J. P. Quirk. "A Titrimetric and Electrophoretic Investigation of the pzc and the iep of Pigment Rutile," *J. Colloid Interface Sci.* 53: 6-13 (1975).
78. Healy, T. W., A. P. Herring and D. W. Fuerstenau. "The Effect of Crystal Structure on the Surface Properties of a Series of Manganese Dioxide," *J. Colloid Interface Sci.* 21: 435-444 (1966).
79. Murray, J. W. "The Surface Chemistry of Hydrous Manganese Dioxide," *J. Colloid Interface Sci.* 46: 357-371 (1974).
80. Loganathan, P., R. G. Burau and D. W. Fuerstenau. "Influence of pH on the Sorption of Co^{2+}, Zn^{2+} and Ca^{2+} by a Hydrous Manganese Oxide," *Soil Sci. Soc. Am. J.* 41: 57-62 (1977).
81. Hingston, F. J., A. M. Posner and J. P. Quirk. "The Role of the Proton in Determining Adsorption Envelopes," *J. Soil Sci.* 23: 177-192 (1972).
82. Bérubé, Y. G., and P. L. de Bruyn. "Adsorption at the Rutile-Solution Interface. II. Model of the Electrochemical Double Layer." *J. Colloid Interface Sci.* 28: 92-105 (1968).
83. Blok, L., and P. L. de Bruyn. "The Ionic Double Layer at the ZnO/Solution Interface. I. The Experimental Zero Point of Charge," *J. Colloid Interface Sci.* 32: 518-526 (1970).
84. Wright, H. J. L., and R. J. Hunter. "Adsorption at Solid-Liquid Interfaces. I. Thermodynamics and the Adsorption Potential," *Aust. J. Chem.* 26: 1183-1189 (1973).
85. Wright, H. J. L., and R. J. Hunter. "Adsorption at Solid-Liquid Interfaces. II. Models of the Electrical Double Layer at the Oxide Solution Interface," *Aust. J. Chem.* 26: 1191-1206 (1973).
86. Smit, W., C. L. M. Holten, H. N. Stein, J. J. M. de Goeij and H. M. J. Theelen. "A Radiotracer Determination of the Adsorption of Sodium Ion in the Compact Part of the Double Layer of Vitreous Silica," *J. Colloid Interface Sci.* 63: 120-128 (1978).
87. Smit, W., C. L. M. Holten, H. N. Stein, J. J. M. de Goeij and H. M. J. Theelen. "A Radiotracer Determination of the Sorption of Sodium Ions by Microporous Silica Films," *J. Colloid Interface Sci.* 67: 397-407 (1978).
88. Anderson, J. H. "The Local Environment of Co(II), Cu(II), and Cr(III) Supported on Silica Gel," *J. Catalysis* 28: 76-82 (1973).
89. Smith, A. E. "A Study of the Variation with pH of the Solubility and Stability of Some Metal Ions at Low Concentrations in Aqueous Solution. Part I," *Analyst* 98: 65-68 (1973).
90. Shendrikar, A. D., V. Dharmarajan, H. Walker-Merrick and P. W. West. "Adsorption Characteristics of Traces of Barium, Beryllium, Cadmium, Manganese, Lead and Zinc on Selected Surfaces," *Anal. Chim. Acta* 84: 409-417 (1976).

91. Pushkarev, V. V. "Adsorption of Radioactive Isotopes on Ferric Hydroxide," *Russian J. Inorg. Chem.* 1: 176-185 (1956).
92. Voznesensskii, S. A., V. A. Pushkarev and V. F. Bagretsov. "Sorption of Radioactive Isotopes by Aluminum Hydroxide," *Russian J. Inorg. Chem.* 3: 363-368 (1958).
93. Fuerstenau, M. C., D. A. Elgillani and J. D. Miller. "Adsorption Mechanisms in Nonmetallic Activation Systems," *Trans. AIME* 247: 11-14 (1970).
94. Strohal, P., K. Molnar and I. Bacic. "Preconcentration of Trace Elements by Aluminum Hydroxide," *Mikrochim. Acta* 586-590 (1972).
95. Šipalo-Žuljević, J., and R. H. H. Wolf. "Sorption of Lanthanum(III), Cobalt(II) and Iodide Ions at Trace Concentrations on Ferric Hydroxide," *Mikrochim. Acta* 315-320 (1973).
96. James, R. O., and T. W. Healy. "Adsorption of Hydrolyzable Metal Ions at the Oxide-Water Interface. III. A Thermodynamic Model of Adsorption," *J. Colloid Interface Sci.* 40: 65-81 (1972).
97. Schulze, W., and M. Scheffler. "Sorptionseffekte an Eisen(III)-hydroxid-Fällungen. I. Sorption von Sr^{2+} Ionen bei Fällung mit Verschiedenen Basen," *Z. Anal. Chem.* 226: 395-401 (1967).
98. Schulze, W., and M. Scheffler. "Sorptionseffekte an Eisen(III)-hydroxid-Fällungen. II. Über den Sorptionsmechanismus von Verschiedenartigan Kationen," *Z. Anal. Chem.* 229: 161-169 (1967).
99. Loganathan, P., and R. G. Burau. "Sorption of Heavy Metal Ions by a Hydrous Manganese Oxide," *Geochim. Cosmochim. Acta* 37: 1277-1293 (1973).
100. Gabano, J. P., P. Étienne and J. F. Laurent. "Étude des Proprietes de Surface du Bioxyde de Manganese," *Electrochim. Acta* 10: 947-963 (1965).
101. Grimme, H. "Die Adsorption von Mn, Co, Cu and Zn durch Goethit aus verdünnten Lösungen," *Z. Pflanzenernähr. Düng. Bodenkunde* 121: 58-65 (1968).
102. Kolařík, Z. "Sorption Radioaktiver Isotopen an Niederschlägen. VI. System Eisen(III)-Hydroxyd-Strontiumnitratlösung und die allgemeinen Gesetzmässigkeiten der Sorption am Eisen(III)-Hydroxyd," *Colln Czech. Commun.* 27: 938-950 (1962).
103. Herrmann, M., and H. P. Boehm. "Saure Hydroxylgruppen auf der Oberfläche," *Z. Anorg. Allg. Chem.* 368: 73-86 (1969).
104. Stumm, W., C. P. Huang and S. R. Jenkins. "Specific Chemical Interaction Affecting the Stability of Dispersed Systems," *Croat. Chem. Acta* 42: 223-245 (1970).
105. Iler, R. K. "Coagulation of Colloidal Silica by Calcium Ions, Mechanisms, and Effect of Particle Size," *J. Colloid Interface Sci.* 53: 476-488 (1975).
106. Duval, J. E., and M. H. Kurbatov. "The Adsorption of Cobalt and Barium Ions by Hydrous Ferric Oxide at Equilibrium," *J. Phys. Chem.* 56: 982-984 (1952).
107. Kinniburgh, D. G., K. Sridhar and M. L. Jackson. "Specific Adsorption of Zinc and Cadmium by Iron and Aluminum Hydrous Oxides," in *The Fifteenth Hanford Life Sciences Symposium on Biological Implications of Metals in the Environment*, USERDA, Natl. Technical Information Service, Springfield, VA (1977), pp. 231-239.

108. Kolařík, Z., and V. Kouřim. "Sorption Radioaktiver Isotopen an Niederschlägen. IV. Sorption des Yttriums an Eisen(III)-Hydroxyd," *Colln Czech. Chem. Comm.* 26: 1082-1091 (1961).
109. Kolařík, Z. "Sorption Radioaktiver Isotopen an Niederschlägen. VIII. Sorption von Spurenmengen Yttrium und Cer am Mangan(IV)-hydroxyd," *Colln Czech. Chem. Comm.* 27: 1333-1336 (1962).
110. Kolařík, Z., and J. Szlaur. "Sorption Radioaktiver Isotopen an Niederschlägen. IX. Sorption kleiner Zirkonium-, Ruthenium- und Uranmengen mittels Mangan(IV)-Hydroxyd. Einfluss der Komplexbildung auf das Sorptionsgleichgewicht," *Colln Czech. Chem. Comm.* 27: 1993-2004 (1962).
111. Kolařík, Z., and J. Szlaur. "Sorption radioakitver Isotopen an Niederschlägen. X. Sorption von Cerspuren mittels Eisen(III)-hydroxyds," *Colln Czech. Chem. Comm.* 28: 2818-2821 (1963).
112. Sýkora, S., and Z. Kolařík. "Sorption Radioaktiver Isotopen an Niederschlägen. XI. Sorptionseigenschaften von Zinn(IV)-hydroxyd," *Colln Czech. Chem. Comm.* 29: 1350-1359 (1964).
113. Dugger, D. L., J. H. Stanton, B. N. Irby, B. L. McConnell, W. W. Cummings and R. W. Maatman. "The Exchange of Twenty Metal Ions with the Weakly Acidic Silanol Groups of Silica Gel," *J. Phys. Chem.* 68: 757-760 (1964).
114. Inoue, Y., and O. Toshiyama. "Determination of the Oxidation States of Neptunium at Tracer Concentrations by Adsorption on Silica Gel and Barium Sulfate," *J. Inorg. Nucl. Chem.* 39: 1443-1447 (1977).
115. Chuecas, L., and J. P. Riley. "The Spectrophotometric Determination of Chromium in Sea Water," *Anal. Chim. Acta* 35: 240-276 (1966).
116. Kepák, F. "Adsorption and Colloidal Properties of Radioactive Elements in Trace Concentrations," *Chem. Rev.* 71: 357-370 (1971).
117. Stryker, L. J., and E. Matijević. "Adsorption of Hydrolyzed Hafnium Ions on Glass," in *Adsorption from Aqueous Solution,* Advances in Chemistry Ser. No. 79 (Washington, DC: American Chemical Society, 1968), pp. 44-61.
118. Gadde, R. R., and H. A. Laitinen. "Studies of Heavy Metal Adsorption by Hydrous Iron and Manganese Oxides," *Anal. Chem.* 46: 2022-2026 (1974).
119. Kozawa, A. "On an Ion Exchange Property," *J. Electrochem. Soc.* 106: 552-556 (1959).
120. Clauss, C. R. A., and K. Weiss. "Cobalt Ion Exchange of Crystalline Quartz," *J. Colloid Interface Sci.* 61: 577-581 (1977).
121. Schindler, P. W., B. Furst, R. Dick and P. U. Wolf. "Ligand Properties of Surface Silanol Groups. I. Surface Complex Formation with Fe^{3+}, Cu^{2+}, Cd^{2+}, and Pb^{2+}," *J. Colloid Interface Sci.* 55: 469-475 (1976).
122. Quirk, J. P., and A. M. Posner. "Trace Element Adsorption by Soil Minerals," in *Trace Elements in Soil-Plant-Animal Systems,* D. J. D. Nicholas and A. R. Egan, Eds. (New York: Academic Press, 1975), pp. 97-107.
123. Forbes, E. A., A. M. Posner and J. P. Quirk. "The Specific Adsorption of Divalent Cd, Co, Cu, Pb, and Zn on Goethite," *J. Soil Sci.* 27: 154-166 (1976).
124. Gadde, R. R., and H. A. Laitinen. "Study of the Sorption of Lead by Hydrous Ferric Oxide," *Environ. Lett.* 5: 223-235 (1973).

125. Moignard, M. S., R. O. James and T. W. Healy. "Adsorption of Calcium at the Zinc Sulphide-Water Interface," *Aust. J. Chem.* 30: 733-740 (1977).
126. Murray, J. W. "The Interaction of Cobalt with Hydrous Manganese Dioxide," *Geochim. Cosmochim. Acta* 39: 635-647 (1975).
127. Forbes, E. A., A. M. Posner and J. P. Quirk. "The Specific Adsorption of Inorganic Hg(II) Species and Co(III) Complex Ions on Goethite," *J. Colloid Interface Sci.* 49: 403-409 (1974).
128. Baran, V. "Hydroxyl Ion as a Ligand," *Coord. Chem. Rev.* 6: 65-93 (1971).
129. Baes, C. F., and R. E. Mesmer. *The Hydrolysis of Cations* (New York: Wiley, 1976), pp. 287-294.
130. Kinniburgh, D. G., and M. L. Jackson. "Calcium and Zinc Adsorption by Iron Hydrous Oxide Gel," in preparation.
131. Tewari, P. H., A. B. Campbell and W. Lee. "Adsorption of Co^{2+} by Oxides from Aqueous Solution," *Can. J. Chem.* 50: 1642-1648 (1972).
132. James, R. O., and T. W. Healy. "Adsorption of Hydrolyzable Metal Ions at the Oxide-Water Interface. I. Co(II) Adsorption on SiO_2 and TiO_2 as Model Systems," *J. Colloid Interface Sci.* 40: 42-52 (1972).
133. Parfitt, G. D. "Precipitation of Hydrolysis Products onto Oxide Surfaces," *Croat. Chem. Acta* 45: 189-194 (1973).
134. Iler, R. K. "Effect of Adsorbed Alumina on the Solubility of Amorphous Silica in Water," *J. Colloid Interface Sci.* 43: 399-408 (1973).
135. Nalović, L., and M. Pinta. "Comportement du fer en présence des éléments de transition. Etude Experiments Précipitation, Déshydration, Dissolution," *Comp. Rend. Acad. Sci. Paris* 274: 628-631 (1972).
136. Benes, P., and K. Kopicka. "The State and Adsorption Behaviour of Traces of Cadmium in Aqueous Solutions," *J. Inorg. Nucl. Chem.* 38: 2043-2048 (1976).
137. Parfitt, R. L., and J. D. Russell. "Adsorption on Hydrous Oxides. IV. Mechanisms of Adsorption of Various Ions on Goethite," *J. Soil Sci.* 28: 297-305 (1977).
138. Wiese, G. R., R. O. James and T. W. Healy. "Discreteness of Charge and Solution Effects in Cation Adsorption at the Oxide/Water Interface," *Disc. Faraday Soc.* 52: 302-311 (1971).
139. Burwell, R., R. G. Pearson, G. L. Haller, P. B. Tjok and S. P. Chock. "The Adsorption and Reaction of Coordination Complexes on Silica Gel," *Inorg. Chem.* 4: 1123-1128 (1965).
140. McKenzie, R. M. "An Electron Microprobe Study of the Relationship Between Heavy Metals and Manganese and Iron in Soils and Ocean Floor Nodules," *Aust. J. Soil Res.* 13: 177-188 (1975).
141. Carlson, L., T. Koljonen, P. Lahermo and R. J. Rosenberg. "Case Study of a Manganese and Iron Precipitate in a Ground-Water Discharge in Somero, Southwestern Finland," *Bull. Geol. Soc. Finland* 49: 159-173 (1977).
142. Bessonov, V. A., and N. N. Krylova. "Extraction of Strontium from Methanol with Aluminum Hydroxide," *Russian J. Colloid Sci.* 36: 767-772 (1974).
143. Helferrich, F. *Ion Exchange* (New York: McGraw-Hill Book Co., 1962).
144. Tien, H. T. "Interaction of Alkali Metal Cations with Silica Gels," *J. Phys. Chem.* 69: 350-352 (1965).

145. Altug, I., and M. L. Hair. "Cation Exchange in Porous Glass," *J. Phys. Chem.* 71: 4260-4263 (1967).
146. Abendroth, R. P. "Behaviour of a Pyrogenic Silica in Simple Electrolytes," *J. Colloid Interface Sci.* 34: 591-596 (1970).
147. Bartell, F. E., and Y. Fu. "Adsorption from Aqueous Solutions by Silica," *J. Phys. Chem.* 33: 676-687 (1929).
148. Churms, S. C. "The Effect of pH on the Ion-Exchange Properties of Hydrated Alumina. Part I. Capacity and Selectivity," *J. S. Afr. Chem. Inst.* 19: 98-107 (1966).
149. Dumont, F., and A. Watillon. "Stability of Ferric Oxide Hydrosols," *Disc. Faraday Soc.* 52: 352-360, 375-376 (1971).
150. Venkataramani, B., K. S. Venkateswarlu and J. Shankar. "Sorption Properties of Oxides. III. Iron Oxides," *J. Colloid Interface Sci.* 67: 187-194 (1978).
151. Britz, D., and G. H. Nancollas. "Thermodynamics of Cation Exchange of Hydrous Zirconia," *J. Inorg. Nucl. Chem.* 31: 3861-3868 (1969).
152. Farmer, V. C. "Water on Particle Surfaces," in *The Chemistry of Soil Constituents,* D. J. Greenland and M. H. B. Hayes, Eds. (London: Wiley Inc., 1978), pp. 405-448.
153. Anderson, J. H., and K. A. Wickersheim. "Near Infrared Characterization of Water and Hydroxyl Groups on Silica Surfaces," *Surf. Sci.* 2: 252-260 (1964).
154. Boutin, H., and H. Prask. "Study of Water Vapor Adsorbed on Gamma-Alumina and Silica by Slow Neutron Inelastic Scattering," *Surf. Sci.* 2: 261-266 (1964).
155. Jurinak, J. J. "Interaction of Water with Iron and Titanium Oxide-Surfaces: Goethite, Hematite, and Anatase," *J. Colloid Sci.* 19: 477-487 (1964).
156. Morimoto, T., M. Nagao and F. Tokuda. "The Relation between the Amounts of Chemisorbed and Physisorbed Water on Metal Oxides," *J. Phys. Chem.* 73: 243-248 (1969).
157. McCafferty, A., and A. C. Zettlemoyer. "Adsorption of Water Vapour on α-Fe_2O_3," *Disc. Faraday Soc.* 52: 239-254 (1971).
158. Eisenmann, G. "Theory of Membrane Electrode Potentials: an Examination of the Parameters Determining the Selectivity of Solid and Liquid Ion Exchangers and of Neutral Ion Sequestering Molecules," in *Ion Selective Electrodes,* R. A. Durst, Ed., Natl. Bureau of Standards Special Publication 314, Washington, DC: U.S. Government Printing Office (1969).
159. Yates, D. E., S. Levine and T. W. Healy. "Site-Binding Model of the Electrical Double Layer at the Oxide/Water Interface," *Chem. Soc. Faraday Trans.* I. 70: 1807-1818 (1974).
160. Gills, T. E., W. F. Marlow and B. A. Thomson. "Determination of Trace Elements in Glass by Activation Analysis Using Hydrated Antimony Pentoxide for Sodium Removal," *Anal. Chem.* 42: 1831-1833 (1970).
161. Ralston, H. R., and E. S. Sato. "Sodium Removal as an Aid to Neutron Activation Analysis," *Anal. Chem.* 43: 129-131 (1971).
162. Dyck, W. "Adsorption and Coprecipitation of Silver on Hydrous Ferric Oxide," *Can. J. Chem.* 46: 1441-1444 (1968).
163. Davis, J. A., and J. O. Leckie. "Effect of Adsorbed Complexing Ligands on Trace Metal Uptake by Hydrous Oxides," *Environ. Sci. Technol.* 12: 1309-1315 (1973).

164. Anderson, B. J., E. A. Jenne and T. T. Chao. "The Sorption of Silver by Poorly Crystalline Manganese Oxides," *Geochim. Cosmochim. Acta* 37: 611-622 (1973).
165. Hem, J. D. "Reactions of Metal Ions at Surfaces of Hydrous Iron Oxide," *Geochim. Cosmochim. Acta* 41: 527-538 (1977).
166. Kurbatov, J. D., J. L. Kulp and E. Mack. "Adsorption of Strontium and Barium Ions and Their Exchange on Hydrous Ferric Oxide," *J. Am. Chem. Soc.* 67: 1923-1929 (1945).
167. Tadros, T. F., and J. Lyklema. "The Electrical Double Layer on Silica in the Presence of Bivalent Counter-ions," *J. Electroanal. Chem.* 22: 1 (1969).
168. Malati, M. A., and S. F. Estefan. "The Role of Hydration in the Adsorption of Alkaline Earth Ions onto Quartz," *J. Colloid Interface Sci.* 24: 306-307 (1966).
169. Zasoski, R. J., and R. G. Burau. "Sorption and Sorptive Interaction of Cd and Zn on Hydrous Manganese Oxide," *Agronomy Abstr.* 122 (1977).
170. Bruninx, E. "The Coprecipitation of Zn, Cd and Hg with Ferric Hydroxide," *Philips Res. Repts.* 30: 177-191 (1975).
171. Kozawa, A. "On an Ion Exchange Property," *J. Electrochem. Soc.* 106: 552-556 (1959).
172. Murray, D. J., T. W. Healy and D. W. Fuerstenau. "The Adsorption of Aqueous Metal on Colloidal Hydrous Manganese Oxide," in *Adsorption from Aqueous Solution,* Adv. Chem. Ser. No. 79 (Washington, DC: American Chemical Society, 1968), pp. 74-81.
173. Vydra, F., and J. Galba. "Sorption von Metallkomplexen an Silicagel. V. Sorption von Hydrolysenprodukten des Co^{2+}, Mn^{2+}, Ni^{2+}, Zn^{2+} an Silicagel," *Colln Czech. Chem. Comm.* 34: 3471-3478 (1969).
174. Taniguechi, K., M. Nakajima, S. Yoshida and K. Tarama. "The Exchange of the Surface Protons in Silica Gel with Some Kinds of Metal Ions," *Nippon Kagaku Zasshi* 91: 525-529 (1970).
175. Donaldson, J. D., and M. J. Fuller. "Ion Exchange Properties of Tin (IV) Materials—I. Hydrous Tin (IV) Oxide and its Cation Exchange Properties," *J. Inorg. Nucl. Chem.* 30: 1083-1092 (1968).
176. McKenzie, R. M. "An Electron Microprobe Study of the Relationship Between Heavy Metals and Manganese and Iron in Soils and Ocean Floor Nodules," *Aust. J. Soil Res.* 13: 177-188 (1975).
177. Van der Weijden, C. H. "Some Geochemical Controls on Ni and Co Concentrations in Marine Ferromanganese Deposits," *Chem. Geol.* 18: 65-80 (1976).
178. Kolthoff, I. M., and B. Moscovitz. "Studies on Coprecipitation and Aging. XI. Adsorption of Ammonia Copper Ion on and Coprecipitation with Hydrous Ferric Oxide. Aging of the Precipitate," *J. Phys. Chem.* 41: 629-644 (1937).
179. Kolthoff, I. M., and L. G. Overholser. "Studies on Aging and Coprecipitation. XXVIII. Adsorption of Divalent Ions on and Coprecipitation with Ortho Ferric Hydroxide in Ammoniacal Medium," *J. Phys. Chem.* 43: 767-780 (1939).
180. Kolařík, Z., and V. Kouřim. "Sorption Radioaktiver Isotopen an Niederschlägen. II. Strontium und Yttriumsorption an Eisen(III)- und Aluminiumhydroxyd," *Colln Czech. Chem. Comm.* 25: 1000-1007 (1960).

181. Schönfeld, T., and C. Friedmann. "Aufnahme von Suprenmengen Antimon (Radioantimon) durch einige Metalloxihydrat-Niederschlägen," *Monat. Chemie* 101: 1518-1531 (1970).
182. Simon, J., W. Schulze and R. Reinke. "Sorptionseffekte an Metall (III)-hydroxid-Fällungen. III. Sorption and Mitfällung von Cd^{2+}-Ionen an frischgefälltem Chrom(III)-hydroxid," *Z. Anal. Chem.* 264: 4-7 (1973).
183. Feitknecht, W., and M. Gerber. "Zur Kenntnis der Doppel-hydroxide und basischen Doppelsalze. III. Über Magnesium-Aluminum Doppelhydroxyd," *Helv. Chim. Acta* 25: 131-137 (1942).
184. Feitknecht, W., and F. Held. "Über Magnesium-Aluminiumdoppelhydroxyd und Hydroxydoppelchlorid," *Helv. Chim. Acta* 27: 1495-1501 (1944).
185. Feitknecht, W. "Ordnungsvorgänge bei kolloiddispersen Hydroxyden und Hydroxysalzen," *Kolloid-Z.* 136: 52-66 (1954).
186. Gastuche, M. C., G. Brown and M. M. Mortland. "Mixed Magnesium-Aluminium Hydroxides. I. Preparation and Characterization of Compounds Formed in Dialyzed Systems," *Clay Minerals* 7: 177-192 (1967).
187. Brown, G., and M. C. Gastuche. "Mixed Magnesium-Aluminium Hydroxides. II. Structure and Structural Chemistry of Synethetic Hydroxy-carbonates and Related Mineral Compounds," *Clay Minerals* 7: 198-201 (1967).
188. Ross, C. J., and H. Kodama. "Properties of Synthetic Magnesium-Aluminium Carbonate Hydroxide and its Relationship to Magnesium-Aluminium Double Hydroxide, Manasseite and Hydrotalcite," *Am. Mineral* 52: 1036-1047 (1967).
189. Taylor, H. F. W. "Crystal Structures of Some Double Hydroxide Minerals," *Mineral Mag.* 39: 377-389 (1973).
190. Norrish, K., and R. M. Taylor. "The Isomorphous Replacement of Iron by Aluminium in Soil Goethite," *J. Soil Sci.* 12: 294-306 (1961).
191. Schwertmann, U., R. W. Fitzpatrick and J. le Roux. "Al Substitution and Differential Disorder in Soil Hematites," *Clays Clay Minerals* 25: 373-374 (1977).
192. de Villiers, J. M., and T. H. Van Rooyen. "Solid Solution Formation of Lepidocrocite-Boehmite and its Occurrence in Soil," *Clay Minerals* 7: 229-235 (1967).
193. Jónás, K., and K. Solymár. "Preparation, X-ray, Derivatographic and Infrared Study of Aluminium-Substituted Goethites," *Acta Chim. Acad. Sci. Hung.* 66: 383-394 (1970).
194. Biais, R., A. Bonnemayre, X. Gramont, M. Michel, H. Gibert and C. Janot. "Etude des Substitutions Al-Fe dans des Oxydes et Hydroxydes de Synthèse. Préparation de Diaspore Ferrifère," *Bull. Soc. Fr. Mineral. Cristallog.* 95: 308-321 (1972).
195. Korecz, L., I. Kurecz, G. Menczel, E. Papp-Molnar, E. Pungor and K. Burger. "Mössbauer Investigation of Iron-Aluminium Mixed Oxides," *Talanta* 19: 1599-1604 (1972).
196. Mumpton, F. A., H. W. Jaffe and C. S. Thompson. "Coalingite, a New Mineral from the New Idria Serpentinite, Fresno and San Benito Counties, California," *Am. Mineral* 50: 1893-1913 (1965).

197. Pastor-Rodriguez, J., and H. F. W. Taylor. "Crystal Structure of Coalingite," *Mineral. Mag.* 38: 286-294 (1971).
198. Yousef, A. A., M. A. Arafa and M. A. Malati. "Adsorption of Sulphite, Oleate and Manganese(II) Ions by β-Manganese Dioxide and its Activation in Flotation," *J. Appl. Chem. Biotechnol.* 21: 200-207 (1971).
199. Simon, J., W. Schulze and M. Völtz. "Sorptionseffekte an Metall(III)-Hydroxid-Fällungen. I. Sorption und Mitfällung von Cd^{2+} Ionen bei frischgefälltem Aluminiumhydroxid," *Z. Anal. Chem.* 257: 108-111 (1971).
200. Simon, J., W. Schulze and L-W. Friedemann. "Sorptionseffekte von Mn^{2+}-Ionen mit amorphem Aluminiumhydroxid," *Z. Anal Chem.* 268: 185-189 (1974).
201. Fricke, R., W. Neugebauer and H. Schäfer. "Beiträge zur anorganischen Chromatographie. I. Über die Einwirkung wässriger Kupferchlorid Lösungen auf γ-Aluminiumoxyd," *Z. Anorg. Allg. Chem.* 273: 215-226 (1953).
202. Williams, K. C., J. L. Daniel, W. J. Thomson, R. I. Kaplan and R. W. Maatman. "Reactions of Aqueous Salts with High Area Alumina," *J. Phys. Chem.* 69: 250-253 (1965).
203. Anderson, P. J. "On the Ion Adsorption Properties of Synthetic Magnetite," *Proc. 2nd Int. Cong. Surf. Activity* 3: 67-80 (1957).
204. Sing, K. S. W. "Adsorption at the Gas/Solid Interface," *Specialist Periodical Rep., Colloid Sci.* 1: 1-48 (1973).
205. Bowden, J. W., M. D. A. Bolland, A. M. Posner and J. P. Quirk. "Generalized Model for Anion and Cation Adsorption at Oxide Surfaces," *Nature Phys. Sci.* 245: 81-83 (1973).
206. Bowden, J. W., A. M. Posner and J. P. Quirk. "A Model for Ion Adsorption on Variable Charge Surfaces," *Trans. 10th Int. Cong. Soil Sci. (Moscow)* 2: 29-35 (1974).
207. Tanford, C. *Physical Chemistry of Macromolecules* (New York: John Wiley, 1961), pp. 526-586.
208. Levi, H. W., and E. Schiewer. "Austauschadsorption von Kationen an TiO_2.aq," *Radiochim. Acta* 5: 126-133 (1965).
209. Vol'khin, V. V., and G. V. Leont'eva. "Physicochemical Study of the Ion-Exchange Properties of Manganese Dioxide," *Inorg. Mat.* 5: 1041-1045 (1969).
210. Allen, L. H., E. Matijević and L. Meites. "Exchange of Na^+ for the Silanolic Protons of Silica," *J. Inorg. Nucl. Chem.* 33: 1293-1299 (1971).
211. Boistelle, R., M. Mathieu and B. Simon. "Isothermes d'Adsorption en Solution Aqueuse des Ions Cadmium sur les Formes (100) and (111) du Chlorure de Sodium," *Surf. Sci.* 42: 373-388 (1974).
212. Shimomura, S., Y. Nishihara, Y. Fukumoto and Y. Tanase. "Adsorption of Mercuric Ions by Ferric Hydroxide," *J. Hyg. Chem.* 15: 84-89 (1969).
213. French, C. M., and J. P. Howard. "Adsorption on Silica Gel," *Chem. Indy* 572 (1956).
214. Subuktagin, S. W. M., and R. Prasad. "Hydrous Thorium Oxide: A Study of its Amphoteric Behaviour," *J. Inorg. Nucl. Chem.* 34: 1053-1058 (1972).
215. Russell, J. D. "Infrared Spectroscopy of Ferrihydrite: Evidence for the Presence of Structural Hydroxyl Groups," *Clay Minerals* 14: 109-114 (1979).

216. Lahann, R. W. "Surface Charge Variation in Aging Ferric Hydroxide," *Clays Clay Minerals* 24: 320-326 (1976).
217. Shuman, L. M. "Adsorption of Zn by Fe and Al Hydrous Oxides as Influenced by Aging and pH," *Soil Sci. Soc. Am. J.* 41: 703-706 (1977).
218. Gadde, R. R., and H. A. Laitinen. "Study of the Sorption of Lead by Hydrous Ferric Oxide," *Environ. Lett.* 5: 223-235 (1973).
219. Kozawa, A. "Ion-Exchange Adsorption of Zinc and Copper Ions on Silica," *J. Inorg. Nucl. Chem.* 21: 315-324 (1961).
220. Vuceta, J., and J. J. Morgan. "Chemical Modeling of Trace Metals in Fresh Water: Role of Complexation and Adsorption," *Environ. Sci. Technol.* 12: 1302-1309 (1978).
221. Davis, J. A., and J. O. Leckie. "Effect of Adsorbed Complexing Ligands on Trace Metal Uptake by Hydrous Oxides," *Environ. Sci. Technol.* 12: 1309-1315 (1978).
222. Schulze, W. "Sorptionseffekte an Eisen(III)-Hydroxid-Fällungen III. Sorption von Co^{2+} Ionen bei Fällung mit Ammoniak," *Z. Anal. Chem.* 241: 207-211 (1968).
223. Upor, E., A'. Rónai and M. Görbicz. "Some Problems in the Separation by Precipitation, of Traces of Elements. V. Sorption of Ammine Complex Forming Cations on Metal Hydroxides. Determination of Zinc in Rock Samples, after Separation with Ammonium Hydroxide," *Acta Chim. Acad. Sci. Hung.* 61: 1-11 (1969).
224. Upor, E. "Separation of Trace Elements by Precipitation. VI. Sorption of Ammine Forming Cations on Metal Hydroxides. Investigation of the Mechanism of Adsorption and Desorption," *Acta Chim. Acad. Sci. Hung.* 64: 17-28 (1970).
225. Simon, J., W. Schulze and M. Völtz. "Sorptionseffekte an Metall(III)-Hydroxid-Fällungen. II. Mitfällung von Cd^{2+} Ionen mit Aluminiumhydroxid in Gegenwart von Ammoniak," *Z. Anal. Chem.* 25: 184-187 (1971).
226. Mac Naughton, M. G., and R. O. James. "Adsorption of Aqueous Mercury(II) Complexes at the Oxide/Water Interface," *J. Colloid Interface Sci.* 47: 431-440 (1974).
227. Kinniburgh, D. G., and M. L. Jackson. "Adsorption of Mercury(II) by Iron Hydrous Oxide Gel," *Soil Sci. Soc. Am. J.* 42: 45-47 (1978).
228. Kinniburgh, D. G. "Cation Adsorption by Hydrous Metal Oxides," Ph.D. Thesis, University of Wisconsin, Madison (1974).
229. Bourg, A. C. M., and P. W. Schindler. "Tenary Surface Complexes. I. Complex Formation in the System Silica-Cu(II)-Ethylenediamine," *Chimia* 32: 166-168 (1978).
230. Dixon, J. B., and S. B. Weed, Eds. *Minerals in Soil Environments* (Madison, Wisconsin: Soil Science Society of America, 1977).
231. van Olphen, H. *An Introduction to Clay Colloid Chemistry*, 2 ed. (New York: John Wiley & Sons, Inc., 1977).
232. Swartzen-Allen, S. L., and E. Matijević. "Surface and Colloid Chemistry of Clays," *Chem. Rev.* 74: 385-400 (1974).
233. Brown, G., A. C. D. Newman, J. H. Rayner and A. H. Weir. "The Structures and Chemistry of Soil Clay Minerals," in *The Chemistry of Soil Constituents*, D. J. Greenland and M. H. B. Hayes, Eds. (New York: John Wiley & Sons, Inc., 1978), pp. 29-178

234. Grim, R. E. *Clay Mineralogy*, 2nd ed. (New York: McGraw-Hill Book Co., 1968).
235. Jackson, M. L., and F. H. Abdel-Kader. "Kaolinite Intercalation Procedure for all Sizes and Types with X-ray Diffraction Spacing Distinctive from Other Phyllosilicates," *Clays Clay Minerals* 26: 81-87 (1978).
236. Lee, S. Y., M. L. Jackson and J. L. Brown. "Micaceous Vermiculite, Glauconite, and Mixed-Layered Kaolinite-Montmorillonite Examination by Ultramicrotomy and High Resolution Electron Microscopy," *Soil Sci. Am. Proc.* 39: 793-800 (1975).
237. Sridhar, K., M. L. Jackson and J. K. Syers. "Cation and Layer Charge Effects on Blister-Like Osmotic Swelling of Micaceous Vermiculite," *Am. Mineralogist* 57: 1832-1848 (1972).
238. Wada, K., and Y. Harada. "Effects of Salt Concentration and Cation Species on the Measured Cation-Exchange Capacity of Soils and Clays," *Proc. Int. Clay Conf., Tokyo* 1: 561-571 (1969).
239. Cradwick, P. D. G., V. C. Farmer, J. D. Russell, C. R. Masson, K. Wada and N. Yoshinaga. "Imogolite, a Hydrated Aluminium Silicate of Tubular Structure," *Nature Phys. Sci.* 240: 187-189 (1972).
240. de Villiers, J. M., and M. L. Jackson. "Cation Exchange Capacity Variations with pH in Soil Clays," *Soil Sci. Soc. Am. Proc.* 31: 473-476 (1967).
241. Roth, C. B., M. L. Jackson and J. K. Syers. "Deferration Effect on Structural Ferrous-Ferric Iron Ratio and CEC of Vermiculites and Soils," *Clays Clay Minerals* 17: 253-264 (1969).
242. Lagaly, G., M. Gonzalez and A. Weiss. "Problems in Layer-Charge Determination of Montmorillonites," *Clay Minerals* 11: 173-187 (1976).
243. de Villiers, J. M., and M. L. Jackson. "Aluminous Chlorite Origin of pH-Dependent Cation Exchange Capacity Variations," *Soil Sci. Soc. Am. Proc.* 31: 614-619 (1967).
244. Schwertmann, U., and M. L. Jackson. "Hydrogen-Aluminum Clays: A Third Buffer Range Appearing in Potentiometric Titration," *Science* 139: 1052-1054 (1963).
245. Jackson, M. L. "Aluminum Bonding in Soils: A Unifying Principle in Soil Science," *Soil Sci. Soc. Am. Proc.* 27: 1-10 (1963).
246. Dolcater, D. L., E. G. Lotse, J. K. Syers and M. L. Jackson. "Cation Exchange Selectivity of Some Clay-Sized Minerals and Soil Materials," *Soil Sci. Soc. Am. Proc.* 32: 795-798 (1968).
247. Jackson, M. L. *Soil Chemical Analysis Advanced Course*, 2nd ed. (Madison, Wisconsin: M. L. Jackson, 1979).
248. Jackson, M. L. "Interlayering of Expansible Layer Silicates in Soils by Chemical Weathering," *Clays Clay Minerals* 11: 29-46 (1963).
249. Jacobs, D. G., and T. Tamura. "The Mechanisms of Ion Fixation Using Radio-Isotope Techniques," *Trans. Int. Soc. Soil Sci. 7th Congr., Madison* 2: 206-214 (1960).
250. Sawhney, B. L. "Sorption and Fixation of Microquantities of Cesium by Clay Minerals: Effect of Saturating Cations," *Soil Sci. Soc. Am. Proc.* 28: 183-186 (1964).
251. Lim, C. H., M. L. Jackson and P. A. Helmke. "Kaolins: Sources of CEC Differences and Cs-Retention," *Clays Clay Minerals* (in press).
252. Jackson, M. L. "Weathering of Primary and Secondary Minerals in Soils," *Trans. Int. Soc. Soil Sci.* 4: 281-292 (1968).

253. Gaines, G. L., and H. C. Thomas. "Adsorption Studies on Clay Minerals II. A Formulation of the Thermodynamics of Exchange Adsorption," *J. Chem. Phys.* 21: 714-718 (1953).
254. Gast, R. G. "Alkali Metal Cation Exchange on Chambers Montmorillonite," *Soil Sci. Soc. Am. Proc.* 36: 14-19 (1972).
255. Laudelout, H., R. van Bladel and J. Robeyns. "Hydration of Cations Adsorbed on a Clay Surface from the Effect of Water Activity on Ion Exchange Selectivity," *Soil Sci. Soc. Am. Proc.* 36: 30-34 (1972).
256. Cremers, A., and H. C. Thomas. "The Thermodynamics of Sodium-Cesium Exchange on Camp Berteau Montmorillonite: an Almost Ideal Case," *Israel J. Chem.* 6: 949-957 (1968).
257. Eliason, J. R. "Montmorillonite Exchange Equilibria with Sodium-Strontium-Cesium," *Am. Mineralogist* 51: 324-335 (1966).
258. Carson, C. D., and J. B. Dixon. "Potassium Selectivity in Certain Montmorillonite Soil Clays," *Soil Sci. Soc. Am. Proc.* 36: 838-843 (1972).
259. Barrer, R. M., and D. L. Jones. "Chemistry of Soil Minerals: Part IX. Ion Exchange and Ion Fixation in Synthetic Fluorhectorites," *J. Chem. Soc.* (A): 503-508 (1971).
260. Robeyns, J., R. van Bladel and H. Laudelout. "Thermodynamics of Singly Charged Ion Exchanges in the Trace Regions on Camp Berteau Montmorillonite," *J. Soil Sci.* 22: 336-341 (1971).
261. Gast, R. G., and W. D. Klobe. "Sodium-Lithium Exchange Equilibria on Vermiculite at $25°$ and $50°C$," *Clays Clay Minerals* 19: 311-319 (1971).
262. Ferris, A. P., and W. B. Jepson. "The Exchange Capacities of Kaolinite and the Preparation of Homoionic Clays," *J. Colloid Interface Sci.* 51: 245-259 (1975).
263. Bolland, M. D. A., A. M. Posner and J. P. Quirk. "Surface Charge on Kaolinite," *Aust. J. Soil Res.* 14: 197-216 (1976).
264. Wear, A. H. "Potassium Retention in Montmorillonites," *Clay Minerals* 6: 17-22 (1965).
265. McBride, M. B., T. J. Pinnavia and M. M. Mortland. "Electron Spin Resonance Studies of Cation Orientation in Restricted Water Layers on Phyllosilicate (smectite) Surfaces," *J. Phys. Chem.* 79: 2430-2435 (1975).
266. Laudelout, H., R. van Bladel, G. H. Bolt and A. L. Page. "Thermodynamics of Heterovalent Cation Exchange Reactions in a Montmorillonitic Clay," *Trans. Faraday Soc.* 64: 1477-1488 (1968).
267. Levy, R., and I. Shainberg. "Calcium-Magnesium Exchange in Montmorillonite and Vermiculite," *Clays Clay Minerals* 20: 37-46 (1972).
268. Keay, J., and A. Wild. "The Kinetics of Cation Exchange in Vermiculite," *Soil Sci.* 92: 54-60 (1961).
269. Udo, E. J. "Thermodynamics of Potassium-Calcium and Magnesium-Calcium Exchange Reactions on a Kaolinitic Soil Clay," *Soil Sci. Soc. Am. J.* 42: 556-560 (1978).
270. Gilbert, M., and R. van Bladel. "Thermodynamics and Thermochemistry of the Exchange Reaction Between NH_4^+ and Mn^{2+} in a Montmorillonitic Clay," *J. Soil Sci.* 21: 38-49 (1970).
271. McBride, M. B. "Exchange and Hydration Properties of Cu^{2+} on Mixed-Ion Na^+-Cu^{2+} Smectities," *Soil Sci. Soc. Am. J.* 40: 452-456 (1976).
272. El-Sayed, M. H., R. G. Burau and K. L. Babcock. "Thermodynamics of Copper(II)-Calcium Exchange on Bentonite Clay," *Soil Sci. Soc. Am. Proc.* 34: 397-400 (1970).

273. Kishk, F. M., and M. N. Hassan. "Sorption and Desorption of Copper by and from Clay Minerals," *Plant Soil* 39: 497-505 (1973).
274. El-Sayed, M. H., R. G. Burau and K. L. Babcock. "Reaction of Copper Tetrammine with Bentonite Clay," *Soil Sci. Soc. Am. Proc.* 35: 571-574 (1971).
275. Bittel, J. E., and R. J. Miller. "Lead, Cadmium, and Calcium Selectivity Coefficients on a Montmorillonite, Illite, and Kaolinite," *J. Environ. Qual.* 3: 250-253 (1974).
276. Griffin, R. A., and A. K. Au. "Lead Adsorption by Montmorillonite Using a Competitive Langmuir Equation," *Soil Sci. Soc. Am. J.* 41: 880-882 (1977).
277. Bolt, G. H. "Cation-Exchange Equations used in Soil Science—a Review," *Neth. J. Agric. Sci.* 15: 81-103 (1967).
278. van Bladel, R., and H. Laudelout. "Apparent Irreversibility of Ion Exchange Reactions in Clay Suspensions," *Soil Sci.* 104: 134-137 (1967).
279. Hutcheon, A. T. "Thermodynamics of Cation Exchange on Clay: Ca-K-Montmorillonite," *J. Soil Sci.* 17: 339-355 (1966).
280. Fink, D. H., F. S. Nakayama and B. L. McNeal. "Demixing of Exchangeable Cations in Free-Swelling Bentonite Clay," *Soil Sci. Soc. Am. Proc.* 35: 552-555 (1971).
281. Coulter, B. S., and O. Talibudeen. "Calcium: Aluminium Exchange Equilibria in Clay Minerals and Acid Soils," *J. Soil Sci.* 19: 237-250 (1968).
282. Coulter, B. S. "The Chemistry of Hydrogen and Aluminium Ions in Soils, Clay Minerals and Resins," *Soils Fertilizers* 32: 215-223 (1969).
283. Nye, P., D. Craig, N. T. Coleman and J. L. Ragland. "Ion Exchange Equilibria Involving Aluminum," *Soil Sci. Soc. Am. Proc.* 25: 14-17 (1961).
284. Bloom, P. R., M. B. McBride and B. Chadbourne. "Adsorption of Aluminum by a Smectite: I. Surface Hydrolysis During Ca^{2+}-Al^{3+} Exchange," *Soil Sci. Soc. Am. J.* 41: 1068-1073 (1977).
285. Brown, G., and A. C. D. Newman. "The Reactions of Soluble Aluminium with Montmorillonite," *J. Soil Sci.* 24: 339-354 (1973).
286. Jackson, M. L. "Structural Role of Hydronium in Layer Silicates During Soil Genesis," *Trans. 7th Int. Cong. Soil Sci.* (Madison) 2: 445-455 (1960).
287. Colombera, P. M., A. M. Posner and J. P. Quirk. "The Adsorption of Aluminium from Hydroxy-Aluminium Solutions on to Fithian Illite," *J. Soil Sci.* 22: 118-128 (1971).
288. McBride, M. B., and P. R. Bloom. "Adsorption of Aluminum by a Smectite: II. An Al^{3+}-Ca^{2+} Exchange Model," *Soil Sci. Soc. Am. J.* 41: 1073-1077 (1977).
289. Sposito, G., and S. V. Mattigod. "On the Chemical Foundation of the Sodium Adsorption Ratio," *Soil Sci. Soc. Am. J.* 41: 323-329 (1977).
290. Veitch, F. P. "The Estimation of Soil Acidity and the Lime Requirement of Soils," *J. Am. Chem. Soc.* 24: 1120-1128 (1902).
291. Chernov, V. A. *The Nature of Soil Acidity*, Izd. Akad. Nauk, USSR. (Jenny, H., translator, Univ. of Calif., Berkeley; Madison, Wi: Soil Science Society of America, 1947).
292. Yariv. D., and H. Cross. *Geochemistry of Colloid Systems for Earth Scientists* (Berlin: Springer-Verlag, 1979).

293. Barshad, I. "The Effect of the Total Chemical Composition and Crystal Structure of Soil Minerals on the Nature of the Exchangeable Cations in Acidified Clays and in Naturally Occurring Acid Soils," *Trans. 7th Int. Cong. Soil Sci.* (Madison) 2: 435-444 (1960).
294. Kerr, G. T., R. H. Zimmerman, H. A. Fox, Jr. and F. H. Wells. "Degradation of Hectorite by Hydrogen Ion," *Clays Clay Minerals* 4: 322-329 (1956).
295. Paver, H., and C. E. Marshall. "The Role of Aluminum in the Reaction of the Clays," *J. Soc. Chem. Ind. (London)* 53: 750-760 (1934).
296. Barshad, I., and A. E. Foscolos. "Factors Affecting the Rate of the Interchange Reaction of Adsorbed H^+ on the 2:1 Clay Minerals," *Soil Sci.* 110: 52-60 (1970).
297. Bradfield, R. "The Nature of the Acidity of the Colloidal Clay of Acid Soils," *J. Am. Chem. Soc.* 45: 2669-2678 (1923).
298. Chernov, V. A. "Genesis of Exchangeable Aluminum in Soils," *Soviet Soil Sci.* 10: 1150-1156 (1959).
299. Coleman, N. T., and G. W. Thomas. "The Basic Chemistry of Soil Acidity," in *Soil Acidity and Liming*, Agron. No. 12, R. W. Pearson and F. Adams, Eds. (Madison, WI: American Society of Agronomy, 1967), pp. 1-41.
300. Jenny, H. "Reflections on the Soil Acidity Merry-go-round," *Soil Sci. Soc. Am. Proc.* 25: 428-432 (1961).
301. Truog, E. "Putting Soil Science to Work," *J. Am. Soc. Agron.* 30: 973-985 (1938).
302. Barshad, I. "Significance of the Presence of Exchangeable Magnesium Ions in Acidified Clays," *Science* 131: 988-990 (1960).
303. Wildman, W. E., M. L. Jackson and L. D. Whittig. "Iron-Rich Montmorillonite Formation in Soils Derived from Serpentinite," *Soil Sci. Soc. Am. Proc.* 32: 787-794 (1968).
304. Sridhar, K., and M. L. Jackson. "Layer Charge Decrease by Tetrahedral Cation Removal and Silicon Incorporation During Natural Weathering of Phlogopite to Saponite," *Soil Sci. Soc. Am. Proc.* 38: 847-850 (1974).
305. Jackson, M. L. "Clay Transformations in Soil Genesis During the Quaternary," *Soil Sci.* 99: 15-22 (1965).
306. Gruner, J. W. "The Structures of Vermiculites and Their Collapse by Dehydration," *Am. Mineralogist* 19: 557-575 (1934).
307. Jackson, M. L. "Chemical Composition of Soils," in *Chemistry of the Soil*, Amer. Chem. Soc. Monograph No. 160, 2nd ed. (New York: Van Nostrand Reinhold Co., 1964), pp. 71-141.
308. Norrish, K. "Factors in the Weathering of Mica to Vermiculite," *Proc. Int. Clay Conf. (Madrid)* (1973), pp. 417-432.
309. Veith, J. A., and M. L. Jackson. "Iron Oxidation and Reduction Effects on Structural Hydroxyl and Layer Charge in Aqueous Suspensions of Micaceous Vermiculites," *Clays Clay Minerals* 22: 345-353 (1974).
310. Scott, A. D., and S. J. Smith. "Susceptibility of Interlayer Potassium in Micas to Exchange with Sodium," *Clays Clay Minerals* 14: 69-81 (1966).
311. Newman, A. C. D. "Cation Exchange Properties of Micas I. The Relation between Mica Composition and Potassium Exchange in Solution at Different pH," *J. Soil Sci.* 20: 358-373 (1969).

312. Feigenbaum, S., and I. Shainberg. "Dissolution of Illite—a Possible Mechanism of Potassium Release," *Soil Sci. Soc. Am. Proc.* 39: 985-990 (1975).
313. Mehmel, M. "Ab- und Umbau am Biotit," *Chemie der Erde* 11: 307-332 (1937).
314. Roth, C. B., M. L. Jackson, J. M. de Villiers and V. V. Volk. "Surface Colloids on Micaceous Vermiculite," *Trans. Int. Soc. Soil Sci. (Aberdeen)* Comm. II / IV: 217-221 (1966).
315. Brindley, G. W., and R. F. Youell. "Ferrous Chamosite and Ferric Chamosite," *Mineral Mag.* 30: 57-70 (1953).
316. Farmer, V. C., J. D. Russell, W. J. McHardy, A. C. D. Newman, J. L. Ahlrichs and J. Y. H. Rimsaite. "Evidence for Loss of Protons and Octahedral Iron from Oxidized Biotites and Vermiculites," *Mineral Mag.* 38: 121-137 (1971).
317. Gilkes, R. J., R. C. Young and J. P. Quirk. "Oxidation of Ferrous Iron in Biotite," *Nature Phys. Sci.* 236: 89-91 (1972).
318. Gilkes, R. J., R. C. Young and J. P. Quirk. "Artificial Weathering of Oxidized Biotite: I. Potassium Removal by Sodium Chloride and Sodium Tetraphenylboron Solutions," *Soil Sci. Soc. Am. Proc.* 37: 25-28 (1973).
319. Raman, K. V., and M. L. Jackson. "Layer Charge Relations in Clay Minerals of Micaceous Soils and Sediments," *Clays Clay Minerals* 14: 53-68 (1966).
320. Scott, A. D., and M. G. Reed. "Expansion of Potassium-Depleted Muscovite," *Clays Clay Minerals* 13: 247-261 (1965).
321. Reichenbach, H. G. V., and C. I. Rich. "Preparation of Dioctahedral Vermiculites from Muscovite and Subsequent Exchange Properties," *Trans. 9th Int. Cong. Soil Sci. (Adelaide)* 1: 709-719 (1968).
322. Brown, G., and A. C. D. Newman. "Cation Exchange Properties of Micas III. Release of Potassium Sorbed by Potassium-Depleted Micas," *Clay Minerals* 8: 273-278 (1970).
323. Hodgson, J. F. "Cobalt Reactions with Montmorillonite," *Soil Sci. Soc. Am. Proc.* 24: 165-168 (1960).
324. Maes, A., and A. Cremers. "Cation-Exchange Hysteresis in Montmorillonite: a pH-Dependent Effect," *Soil Sci.* 119: 198-202 (1975).
325. O'Connor, T. P., and D. R. Kester. "Adsorption of Copper and Cobalt from Fresh and Marine Systems," *Geochim. Cosmochim. Acta* 39: 1531-1543 (1975).
326. Chester, R. "Adsorption of Zinc and Cobalt on Illite in Sea-Water," *Nature* 206: 884-886 (1965).
327. Garcia-Miragaya, J., and A. L. Page. "Influence of Ionic Strength and Inorganic Complex Formation on the Sorption of Trace Amounts of Cd by Montmorillonite," *Soil Sci. Soc. Am. J.* 40: 658-663 (1976).
328. Farrah, H., and W. F. Pickering. "Sorption of Copper Species by Clays, I. Kaolinite," *Aust. J. Chem.* 29: 1167-1176 (1976).
329. Farrah, H., and W. F. Pickering. "Sorption of Copper Species by Clays II. Illite and Montmorillonite," *Aust. J. Chem.* 29: 1177-1184 (1976).
330. Farrah, H., and W. F. Pickering. "The Sorption of Zinc Species by Clay Minerals," *Aust. J. Chem.* 29: 1649-1656 (1976).

331. Peigneur, P., A. Maes and A. Cremers. "Heterogeneity of Charge Density on Montmorillonite as Inferred from Cobalt Adsorption," *Clays Clay Minerals* 23: 71-75 (1975).
332. Tiller, K. G. "Stability of Hectorite in Weakly Acidic Solutions III. Adsorption of Heavy Metal Cations and Hectorite Solubility," *Clay Minerals* 7: 409-419 (1968).
333. Tiller, K. G. "The Interactions of some Heavy Metal Cations and Silicic Acid at Low Concentrations in the Presence of Clays," *Trans. 9th Int. Cong. of Soil Sci. (Adelaide)* 2: 567-575 (1968).
334. McBride, M. B. "Copper(II) Interactions with Kaolinite—Factors Controlling Adsorption," *Clays Clay Minerals* 26: 101-106 (1978).
335. Shainberg, I., P. F. Low and U. Kafkafi. "Electrochemistry of Sodium-Montmorillonite Suspensions: I. Chemical Stability of Montmorillonite," *Soil Sci. Soc. Am. Proc.* 38: 751-756 (1974).
336. Steger, H. F. "On the Mechanism of the Adsorption of Trace Copper by Bentonite," *Clays Clay Minerals* 21: 429-436 (1973).
337. Weaver, C. E., and K. C. Beck. *Clay Water Diagenesis during Burial: How Mud becomes Gneiss* (Boulder, CO: Geological Society of America, Inc., 1971), pp. 1-91.

CHAPTER 4

ADSORPTION VS PRECIPITATION

Richard B. Corey
    Department of Soil Science
    University of Wisconsin
    Madison, Wisconsin

INTRODUCTION

Specific adsorption and precipitation processes are similar in many respects; the major difference being that adsorption is a two-dimensional process while precipitation is three dimensional. In natural systems, adsorption generally occurs on the surfaces of minerals that are themselves precipitates. If an adsorbing species happens to be identical to one of the mineral components, adsorption will contribute to crystal growth, making the adsorption reaction an integral part of the precipitation process. If the adsorbing species is not a component of the mineral adsorbent, a number of reaction products are possible, depending on the chemical characteristics and concentration of the adsorbate. This chapter considers the following possible consequences of adding an adsorbate ion to a system containing mineral adsorbents:

1. Crystal growth. This occurs if the adsorbate is a component of the mineral adsorbent.
2. Crystal growth and/or diffusion into the solid phase. If the adsorbate is not a component of the adsorbent but can substitute isomorphously for a component of the adsorbent and form a stable, three-dimensional, solid solution, this will take place.
3. Formation of a stable surface compound (two-dimensional solid solution). This takes place if the adsorbate is not capable of forming a three-dimensional solid solution with the adsorbent.
4. Stabilization of metastable, polynuclear ions. This occurs by adsorption

onto oppositely charged surfaces of the adsorbent.
5. Heterogeneous nucleation of a new solid phase. This involves a new phase composed of the adsorbate and a component from the solution (hydroxides, carbonates, etc.).
6. Heterogeneous nucleation of a new solid phase. This refers to a new phase composed of the adsorbate and a component of the adsorbent resulting in dissolution of the adsorbent.

Generally, adsorption reactions are considered in terms of the formation of a stable surface compound in equilibrium with the solution (number 3 above). Failure to recognize the possibility of other reactions has sometimes led to errors in interpretation [1]. The conditions under which these other reactions can become significant are dealt with in the subsequent sections of this chapter.

THE PRECIPITATION PROCESS

In all of the reactions listed above, the initial reaction is adsorption. In four of the six, the ultimate result is the formation of a precipitate. The transition from adsorption to precipitation is not a simple process, but knowledge of the principles governing precipitation reactions can lead to a better understanding of the major factors involved.

The Solubility Product

For a solid electrolyte, $M_x A_y$, in equilibrium with its saturated solution at constant temperature and pressure, the reaction can be written as follows:

$$M_x A_y + (xb + yc)H_2O \rightleftharpoons x(M^{z+} \cdot bH_2O) + y(A^{z-} \cdot cH_2O) \quad (1)$$

where x = the number of moles of the cation, M, with charge, z+,
y = the number of moles of the anion, A, with charge, z−,
b = the number of water molecules associated with an M^{z+} ion in solution, and
c = the number of water molecules associated with an A^{z-} ion in solution.

The comprehensive thermodynamic solubility product, K_a, for this reaction is

$$K_a = \frac{(a_M)^x (a_A)^y}{(a_{M_xA_y})(a_{H_2O})^{(xb+yc)}} \qquad (2)$$

where a represents the activity of the designated chemical species. This equation can be derived by mass action [2], thermodynamic [2] or kinetic [3] approaches.

If one assumes that the activity of the solid phase is constant and that the hydration of the ions has an insignificant effect on the activity of the water, the standard states of these reactants can be selected so that their activities are equal to one, and a reduced thermodynamic solubility product, K_{so}, results in

$$K_{so} = (a_M)^x (a_A)^y \qquad (3)$$

If the concentrations in solution are very low (in the range of 10^{-5} M or below), concentrations can be substituted for activities and the classical concentration solubility product, K_c, of Nernst results in

$$K_c = [M]^x [A]^y \qquad (4)$$

where brackets denote concentrations.

Reduced thermodynamic solubility products can be calculated from solubility data, but care must be exercised to assure that true equilibrium has been attained. This can usually be checked by approaching the equilibrium concentration from both undersaturation and supersaturation. Reduced thermodynamic solubility products can also be calculated from standard free energies of formation [4], if suitably accurate values are available for a particular compound.

Activity of the Solid Phase

The comprehensive thermodynamic solutility product includes the activity of the solid phase as one of the variables. This activity depends on the interfacial tension of the solid and on the particle size. The interfacial tension may be affected by a number of factors, including lattice defects, impurities, surface heterogeneity and solvent properties.

Particle size is of particular importance in affecting the activity of the solid phase. During the course of a precipitation reaction, particle sizes may range from the very small particles forming the initial nuclei to very large crystals formed after a digestion period. A modified form of the Ostwald-Freundlich equation [5] relates the reduced thermodynamic solubility product, $K_{so(r)}$, of spherical crystals of radius, r, and that of very large crystals, K_{so}, to the interfacial tension, σ, in accord with the following equation:

$$\ln \frac{K_{so(r)}}{K_{so}} = \frac{2\sigma \bar{V}}{r\,RT} \qquad (5)$$

where $\bar{V}$ = the molar volume of the solid,
R = the gas constant, and
T = the absolute temperature.

This equation clearly points out the critical role that particle size plays in solubility and also the pronounced effect of the interfacial tension on this relationship.

Nucleation and Crystal Growth

The precipitation process requires, first of all, the formation of stable nuclei, after which crystal growth can take place. Nucleation may occur through chance collisions of reactants in solution (homogeneous nucleation) or from adsorption and subsequent nucleus formation on solid surfaces (heterogeneous nucleation).

Homogeneous Nucleation

The classical treatment of precipitation distinguishes three processes: nucleation, growth and aging. To initiate precipitation, a critical supersaturation must be attained. This degree of supersaturation is required because the solubility of the precipitate increases with decreasing particle size (Equation 5), and the size of the nucleus initially is very small. For the simplest case of a precipitate consisting of a cation and an anion of equal valence, the minimum size, r_n, of a particle that can exist at a given supersaturation ratio, $K_{so(r)}:K_{so}$, can be obtained by rearrangement of Equation (5):

$$r_n = \frac{2\sigma \overline{V}}{RT \ln(K_{so(r)}/K_{so})} \qquad (6)$$

At any specified supersaturation, all nuclei larger than r_n will grow and all smaller than r_n will dissolve. Supersaturation is generally high at the initiation of nucleation but decreases as the solute is incorporated into new nuclei and into growing crystals. As the supersaturation decreases, nuclei that have not attained the critical size for that condition dissolve, while those that have exceeded the critical size continue to grow. Thus, Kahlweit [6] considers precipitation to be a continuous race among particles to grow fast enough that they will remain stable throughout the precipitation process, deriving their sustenance from the dissolution of particles that could not maintain the required pace. If this process were to continue to its thermodynamic conclusion, only one large crystal would remain, and the supersaturation ratio would be equal to one. In practice, the rate of the dissolution-precipitation reaction becomes very slow as the supersaturation ratio approaches one, and numerous relatively small crystals may persist for very long time periods, particularly if the precipitate is of very low solubility.

According to conventional rate theory, the rate, J, at which nuclei form is given by the following equation:

$$J = A \exp(-\Delta G_a/RT) \qquad (7)$$

in which A is a factor related to the efficiency of collision of the molecules and ΔG_a is the activation energy. ΔG_a is related to the supersaturation ratio by

$$\Delta G_a = \frac{16\pi\sigma^3 \overline{V}^2}{3[RT \ln(IAP/K_{so})]^2} \qquad (8)$$

In this equation, the determined or assumed ion activity product (IAP) is used in the supersaturation term in place of the previously used $K_{so(r)}$, which referred to the solubility of a particle of a given size.

It is apparent from Equations 7 and 8 that the rate of nucleation is very dependent on the degree of supersaturation. For example, Stumm and Morgan [7], using estimated values for variables other than supersaturation, calculated that homogeneous nucleation should be almost instantaneous at $IAP/K_{so} = 100$, but should not occur even within geological time spans for $IAP/K_{so} = 10$.

Heterogeneous Nucleation

In natural systems, homogeneous nucleation is the exception rather than the rule. There are almost always foreign particles in the system that can catalyze the nucleation reaction by lowering the activation energy (and thus the supersaturation) required for nucleation. According to Walton [8], the action of solid substrates in promoting nucleation is primarily the result of an ordering process brought about by interactions across the interface. This ordering process is most pronounced when the lattice of the heterogeneous substrate closely matches the lattice of the precipitating solid. Stumm and Morgan [9] state that the catalytic effect of a solid surface is due to the interfacial energy between the two solids being lower than the interfacial energy between the crystal and the solution. In the extreme case of the substrate and the nucleus being identical, the interfacial energy between crystal and substrate is zero.

Steps, ledges and kinks on the surface of a matching substrate are particularly effective in nucleation because they allow more of the substrate surface to be in contact with the nucleus (Figure 1). Cavities in the surface can act as effective nucleation sites even without a matching lattice because a cavity allows maximum contact with the nucleus [10].

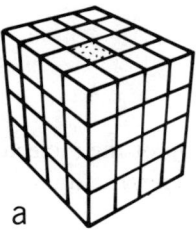
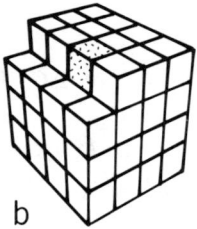

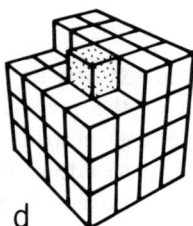
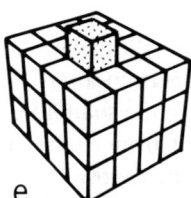

Figure 1. Pictorial representation of adsorption of an ion (the stippled cube) at different types of sites on the surface of a crystal. Strength of bonding should decrease in the following order: (a) a hole > (b) a notch > (c) a kink > (d) a ledge > (e) a face.

Even with relatively good matching of substrate and nucleus, a significant supersaturation will be required for nucleation. For example, an actual ice crystal is the only substance that has been found to have the capability of nucleating ice near 0°C [11]. Silver iodide has a crystal structure similar to ice and is used extensively in cloud-seeding operations, but it will nucleate ice only below −4°C in the case of 1-μ particles [12].

Mica, because of its clean cleavage and well-defined structure, has been used extensively to study the effect of lattice matching on heterogeneous nucleation [13]. Upreti and Walton [14] showed that the critical supersaturation for nucleation of alkali halides on mica generally decreased with the lattice mismatch. As would be expected, the critical supersaturation also decreased with increasing temperature.

Crystal Growth

The growth of crystals involves the following steps [15]: (1) transport of solute to the crystal-solution interface; (2) adsorption of the solute at the surface; (3) incorporation of the crystal constituents into the lattice. The rate-limiting step may be the rate of diffusion that limits transport to the interface (diffusion control) or it may be the rate of adsorption or incorporation (interface control).

The rate of diffusion-controlled growth, dc/dt, depends on the surface area of the crystal, S, a transport coefficient, k (which includes both diffusion and convection terms), for the limiting reactant in solution, and, particularly, the difference between the concentration of the reactant at the surface of the crystal, C_0, and in the bulk solution, C:

$$dc/dt = -kS(C - C_0) \qquad (9)$$

Growth rate slows as C approaches C_0. For precipitates of very low solubility, crystal sizes are usually small. If C is the same order of magnitude as C_0 for low-solubility precipitates, diffusion is very slow because of the small absolute concentration difference. If C were increased to a level that would support relatively rapid growth, the critical supersaturation for nucleation would be exceeded and much of the reactant would be used up in forming new small crystals rather than in increasing the size of the preexisting crystals.

From a mechanistic standpoint, the free energy of an ion adsorbed onto a crystal should be lowered by more contact with the surface. If the adsorbed ion is visualized as a cube (Figure 1), maximum stability should occur when it is a part of a three-dimensional lattice with all six sides in contact with other constituents in the crystal. Stability should decrease as the sides in contact decrease progressively to 5 (a "hole"), 4 (a "notch"), 3 (a "kink"), 2 (a "ledge") and 1 (a "face"). Therefore, the rate-limiting step in interface-controlled crystal growth should be the adsorption of the first ion on a flat face.

Gunn [16] pointed out that ions, particularly those with small radius and high charge, bind water molecules very tightly, and that the high energy requirement for dehydration may result in a high activation energy that slows the adsorption process. He postulated that formation of ion pairs would lower the energy required to remove the water molecules and would thus accelerate interface-controlled growth.

Effects of Impurities

If a substance that cannot be incorporated readily into the crystal is strongly adsorbed onto the surface, it may slow crystal growth drastically [17,18]. However, if the impurity is weakly adsorbed, it may only change the pattern of growth of preferential adsorption on specific crystal faces, or it might even increase the rate of growth by decreasing the "ledge" energy in the growth of a layer on a flat face [18].

If the impurity "fits" into the crystal structure, it may contribute to crystal growth by forming a solid solution. If it does not "fit," but the supersaturation is high, it may restrict crystal growth at the adsorption site, but become trapped (occluded) in voids within the rapidly growing precipitate [19]. These principles are often used in removing trace quantities of interfering ions by coprecipitation or occlusion.

Polymorphism

At any given temperature, except at a transition point, only one solid phase of a compound is thermodynamically stable. However, metastable polymorphic forms may exist if the thermodynamically stable form is not the kinetically favored form. Once the metastable product has formed, the transition to the stable form may be extremely slow, even effectively blocked. For instance, Berner [20] states that dolomite is thermodynamically more stable than calcite or aragonite in seawater, but its occurrence in fresh sediments is rare, whereas calcite and aragonite are commonly found.

As mentioned previously, impurities in the solution can have a marked effect on the rate of crystal growth. Impurities incorporated as solid solutions may also influence crystal form [21]. Some organisms produce or excrete specific solid phases, sometimes metastable forms, even though the surrounding solutions are undersaturated with respect to those solid phases [22]. The energy required for this feat is derived from exergonic reactions occurring within the organisms.

For polymorphic forms where K_{so} values do not differ greatly, Equation 5 indicates that large particles of the metastable form could be more stable than small particles of the thermodynamically stable form. If the metastable form were kinetically favored and quickly formed large particles, or if seed crystals of this form were produced by organisms, the particle-size effect on solubility could effectively limit nucleation and growth of the thermodynamically stable form. This could account for the persistence of some metastable forms in nature.

For some compounds, the nuclei tend to be amorphous, particularly if nucleation occurs at high supersaturation. Interfacial tensions of many oxide and hydroxide crystals are relatively large [23]. Therefore, according to Equation 6, for crystal growth to occur, either relatively large nuclei must be present at low supersaturations or high supersaturations must be employed to initiate nucleation. As supersaturation increases, the probability of nucleating metastable, amorphous forms increases, and, once formed, they may remain stable for a long time.

According to Sears [24], as cited by Griffin and Jurinak [25], a crystallite will grow in a random fashion with no well-defined crystal structure if the "critical cluster" required to form a stable nucleus is less than one unit cell. At high supersaturations, the size of a stable nucleus may be less than one unit cell for some compounds. In qualitative accord with this concept, Griffin and Jurinak [25] found that calcium phosphate nucleating on calcite surfaces showed equilibrium solution activities approaching the solubility product of hydroxyapatite at low P additions but showed considerable supersaturation with respect to either hydroxyapatite or octacalcium phosphate at high P additions. This suggests that hydroxyapatite was either the nucleating species or that the nucleating species was rapidly converted to hydroxyapatite at low supersaturation, but that a more stable form was nucleated and remained relatively stable at high supersaturations.

Precipitation from Homogeneous Solution

If the concentration of one or more of the reactants in a precipitation reaction can be raised slowly and uniformly throughout the solution, heterogeneous nucleation will usually start at a supersaturation well below that required for homogeneous nucleation, if suitable adsorbent surfaces are present. If the rate of crystal growth keeps pace with the production of the precipitant, no additional nuclei will form beyond the relatively few that formed at the start. Compared with conventional precipitation systems, the product contains fewer but larger crystals with fewer impurities. This is the principle behind precipitation from homogeneous solution [26]. An example of precipitation from homogeneous solution is the precipitation of calcium oxalate from an oxalic acid-urea solution. The solution is heated to hydrolyze the urea to NH_3 and CO_2 at the desired rate. This NH_3 then reacts with the oxalic acid resulting in a gradual production of oxalate ions that form a precipitate with the Ca.

If the precipitate formed during precipitation from homogeneous solution has a very low solubility, the rate of crystal growth may be so slow that it cannot keep pace with the rate of generation of the precipitant even though the precipitant is generated very slowly. In this case, additional nuclei will be formed, and the homogeneous-solution system will offer little advantage over the conventional system in terms of crystal size and purity of the precipitate.

Precipitation reactions in natural systems frequently occur under near homogeneous-solution conditions. Concentrations of reactants may be raised slowly and uniformly by such processes as evaporation, weathering, microbial decomposition of organic materials, reducing conditions, etc. On the other hand, rapid changes in concentration are also possible as in the cases of lake turnover and soil fertilization.

Solid Solutions

Perfect stoichiometric crystals do not exist in nature. Wagner and Schottky [27], as reported by Jagodzinski [28], proposed three types of defects that might appear in crystals: (1) substitution (interchange of ions), (2) interstitials (Frenkel defect), and (3) voids (Schottky defect). They pointed out that any defect will result in an unfavorable energy, Q, and that the probability, P, of finding such a defect would be

$$P = \exp(-Q/kT) \qquad (10)$$

If Q is large compared to kT, the defects will be rare. However, if Q is small, which could be the case for some substitution defects or for some combinations of defects, complete miscibility might result, particularly at high temperatures. This would give rise to a solid solution.

The AgCl-AgBr system is frequently used to demonstrate the properties of solid solutions [29]. The solubility product of AgBr is lower than that of AgCl. Therefore, if $AgNO_3$ were added to a solution containing equal concentrations of Br^- and Cl^-, one might expect that Br^- would be almost completely precipitated before any AgCl was formed. In practice, a solid solution results.

To develop a theoretical approach for this system, we can visualize the reaction as solid AgCl equilibrating with Br^- in solution:

$$AgCl_{(s)} + Br^- \rightleftharpoons AgBr_{(s)} + Cl^- \quad (11)$$

The equilibrium constant for this reaction, D, generally referred to as a distribution constant, is

$$D = \frac{a_{Cl^-}}{a_{Br^-}} \cdot \frac{a_{AgBr_{(s)}}}{a_{AgCl_{(s)}}} \quad (12)$$

If we use the comprehensive thermodynamic solubility products (Equation 2) for AgCl and AgBr, we find that

$$\frac{K_{a(AgCl)}}{K_{a(AgBr)}} = \frac{\frac{(a_{Ag^+})(a_{Cl^-})}{a_{AgCl_{(s)}}}}{\frac{(a_{Ag^+})(a_{Br^-})}{a_{AgBr_{(s)}}}} = \frac{a_{Cl^-}}{a_{Br^-}} \cdot \frac{a_{AgBr_{(s)}}}{a_{AgCl_{(s)}}} = D \quad (13)$$

If the activities of the solid phases were proportional to their mole fractions, the comprehensive solubility products could be replaced by the reduced solubility products (Equation 3), and D would be equal to the ratio of the reduced solubility products. Actually, Vaslow and Boyd [29] found D to be about two-thirds that ratio. Bodine et al. [30] found much greater deviation for this relationship with solid solutions of $MnCO_3$ or $SrCO_3$ in $CaCO_3$, with D being highly temperature dependent. Large deviations from the theoretical D value indicate unfavorable energy relationships for the solid solution, that is, the minor component does not mesh very well with the major component, and the extent of solid solution formation is limited.

Solid solutions are ubiquitous in natural systems. Given the complex chemical composition of those systems, the probability of forming solid solutions is very great, which leads to the opening statement of this section that stoichiometric crystals do not exist in nature. Incorporation of an "impurity" into a crystal would be limited to a very small amount if the defect formation energy, Q, were large or the temperature very low. On the other hand, low defect formation energy and high temperature would favor unrestricted mixing. Voids or interstitial additions may accompany solid solution formation to maintain charge balance if the "impurity" has a charge different from the ion it replaces.

Most silicate minerals are solid solutions of a sort. For instance, albite feldspar is essentially a quartz crystal with one of four Si^{4+} ions replaced by Al^{3+} and the charge compensated by an interstitial addition of Na^+ to give the formula $NaAlSi_3O_8$. Albite, in turn, forms a complete solid solution series with anorthite, $CaAl_2Si_2O_8$ (the plagioclase series). As another example, the surface charge on the faces of expanding 2:1 layer silicates usually arises from isomorphic substitution (solid solution) of Al^{3+} for Si^{4+} in the outer tetrahedral sheets or a divalent cation for Al^{3+} in the inner octahedral sheet.

Many common carbonate and oxide minerals are actually solid solutions. Chave [31] reported single-phase, biogenic calcite containing up to 29% by weight $MgCO_3$. Taylor and McKenzie [32] found that most of the soil Co was associated with the Mn minerals birnessite and lithiophorite, when these minerals were present. That these minerals can be a reservoir for transition elements is supported by the following formula that Glover [33] determined for a marine birnessite:

$$(Na,Ca,K)_{0.83} Mg_{1.04} Mn^{IV}_{5.96} (Fe,Co,Ni,Cu)_{0.16} O_{13.6} (H_2O)_{3.8}$$

The apatite minerals are another example of a natural solid solution. According to McClellan and Lehr [34], carbonate apatite is a substituted fluorapatite with this general formula:

$$Ca_{(10-0.42x)} Na_{0.3x} Mg_{0.12x} (PO_4)_{6-x} (CO_3)_x F_{(2+0.4x)}$$

Chien and Black [35] claim that the free energy of formation of these species decreases and the phosphate solubility increases with carbonate substitution. Solid solutions of hydroxyapatite and fluorapatite have also been postulated [36].

Solid solutions also exist in the $AlPO_4 \cdot 2H_2O$ – $FePO_4 \cdot 2H_2O$ system. Arlidge et al. [37] prepared phosphates with various Al:Fe ratios by digesting mixtures of the pure phosphates. One of the products they obtained was a highly crystalline barrandite with an Al:Fe ratio of 1.04. Infrared absorption bands were generally intermediate between those of variscite, $AlPO_4 \cdot 2H_2O$, and strengite, $FePO_4 \cdot 2H_2O$.

The examples of solid solutions cited above illustrate the many such systems found in nature. Interpretation of solubility data from these systems is fraught with problems. If the solid solution involves a trace constituent and a major constituent, i.e., $MnCO_3$ in calcite, the solubility relationship of the minor constituent is impossible to predict accurately [38]. The problem is frequently accentuated by inhomogeneity of the solid solution, i.e., the proportions of the solid-solution components vary within a single crystal. This can result from a changing solution composition during growth of the crystal and insufficient time for homogenization to occur by diffusion within the solid. Another disturbing characteristic of some solid solutions (and some "pure" minerals) is a tendency toward incongruent dissolution in which ions of the solid phase do not accumulate in stoichiometric quantities on dissolution because of the formation of a second solid phase. Particular difficulties are encountered in attempts to estimate free energies of formation of silicates from solubility data [39]. For many solid solutions, true equilibrium is possible only in a solution identical in composition to the solution in contact with the mineral surface at the time of precipitation. Dissolution of the crystal will almost invariably result in ratios of cations in solution that differ from that original solution, so that a truly reversible reaction is impossible.

Precipitates Involving Polynuclear Complexes

Some substances, particularly hydrolysis products of polyvalent cations, form soluble polynuclear complexes in solutions supersaturated with respect to the hydroxide or oxide. An example often quoted is the formation of $Al_{13}O_4(OH)_{24}(H_2O)_{12}^{7+}$ by adding the proportionate amount of base to an Al^{3+} solution [40]. Coagulation of these metastable polynuclear cations can be achieved if the surface charge is neutralized by addition of further base, specific adsorption of an anion, or interaction with a negatively charged colloid. Hsu [41] showed that polynuclear cations of hydroxy-Al were precipitated when enough phosphate was added to neutralize the net positive charge. In solutions containing both Al^{3+} and polynuclear cations, each form reacted separately with added phosphate because of a slow attainment of equilibrium between the two forms of Al.

Coagulation of a negatively charged silica dispersion by polynuclear hydroxy-Al cations was reported by Hahn and Stumm [42]. They concluded that the primary mechanism for coagulation was reduction of the negative surface potential of the silica colloids by adsorption of the hydroxy-Al cations. Precipitates formed by the mechanisms discussed above are not thermodynamically stable and usually do not have a well-defined chemical composition. Both the composition of the polynuclear complex and its surface charge are generally related to the solution composition so that the precipitate composition is variable. Some of these precipitates may persist in metastable form for relatively long periods of time. They may occur in nature, particularly in acid systems undergoing rapid changes in pH or solute concentrations, but they are more frequently creatures of the laboratory.

THE ADSORPTION-PRECIPITATION BOUNDARY

Specific adsorption of ions on mineral surfaces involves the formation of relatively strong chemical bonds similar to those occurring in precipitates. The strength of bonding may be enhanced by linking the adsorbed ion to more than one surface ion. Atkinson et al. [43] proposed that an adsorbed phosphate ion displaces hydroxide ions from an iron oxide surface and bonds to two Fe^{3+} ions forming a binuclear bond between those ions. This hypothesis was supported by the work of Parfitt et al. [44,45], but their investigations suggested that coordinated water molecules as well as hydroxide ions are displaced by the phosphate. Parfitt [46] proposed that binuclear bonding of the adsorbed phosphate also occurs on surfaces of aluminum hydroxides and kaolinite.

Adsorption as a Two-dimensional Solid Solution

If the adsorption mechanism proposed above is valid, the adsorption of phosphate on mineral surfaces can be looked on as a two-dimensional solid solution in which phosphate and hydroxide ions and water molecules are all competing for adsorption sites, much as the Cl^- and Br^- compete for sites on the surface of the AgBr-AgCl solid solution. Scholten [47] attempted to use such a solid-solution approach to describe phosphate adsorption on goethite, but he considered only the phosphate-hydroxide exchange. Even without considering displacement of coordinated water molecules, the solid-solution model described phosphate adsorption reasonably well over a pH range from 4 to 7. If the solid-solution equilibrium is written (in oversimplified form for purposes of clarity)

$$\text{Fe-OH}_{(s)} + H_2PO_4^- \rightleftharpoons \text{Fe-}H_2PO_{4(s)} + OH^- \quad (14)$$

the distribution constant, D, is as follows:

$$D = \frac{a_{OH^-}}{a_{H_2PO_4^-}} \cdot \frac{a_{\text{Fe-}H_2PO_{4(s)}}}{a_{\text{Fe-OH}_{(s)}}} \quad (15)$$

or

$$\frac{a_{\text{Fe-H}_2\text{PO}_4(s)}}{a_{\text{Fe-OH}(s)}} = \frac{D}{a_{\text{OH}^-}} \cdot a_{\text{H}_2\text{PO}_4^-} \qquad (16)$$

Equation 6 has the form of a Langmuir equation in which the Langmuir binding energy term, k, is equal to D/a_{OH^-}. If the model is correct, the adsorption should follow a Langmuir plot of $a_{\text{Fe-H}_2\text{PO}_4}$ versus $a_{\text{H}_2\text{PO}_4}$ if pH is held constant. Displacement of water molecules can also be incorporated into this approach, but the Langmuir k then becomes a more complicated function of pH [48].

The concept of specific adsorption as a two-dimensional solid solution was presented here in terms of phosphate adsorption on hydrous oxide surfaces. The concept can be applied generally to adsorption systems if the competitive reactions occurring at the surface are known. Whether such an approach will yield accurate quantitative results is still open to question.

Nucleation of a Second Solid Phase

The fact that an ion is specifically adsorbed on the surface of a substrate suggests that it has a tendency to form an insoluble compound or stable complex with the ion of opposite charge in the substrate. Indeed, Grimme [49] has shown that selectivity of cation adsorption on oxide surfaces correlates reasonably well with the solubilities of the corresponding metal hydroxides. As more and more adsorbate ion is added, nucleation of a new solid phase will occur at some point, and the solubility of the adsorbate ion would then be controlled by the solubility of the new solid phase, rather than by an adsorption reaction.

Nucleation of Precipitates Involving Solution Components

If a metal ion that forms an insoluble hydroxide is added incrementally to a solution buffered at a pH high enough to induce precipitation at some critical supersaturation of the metal ion, nucleation will occur when that critical concentration is exceeded. For example, if the solution contains particles of hematite, the metal ion will initially be adsorbed on the hematite surfaces. If enough metal ion is added to exceed the critical supersaturation for heterogeneous nucleation, the metal hydroxide will form on the surface of the hematite. The critical supersaturation will be lower for heterogeneous nucleation than for homogeneous nucleation, but the solubility product of any resulting crystalline precipitate of a given particle size should be the same regardless of the nucleation mechanism. A similar situation could exist for precipitation of carbonates in systems exposed to atmospheric CO_2, and for many other species of low solubility.

The major problem in this type of system is to determine when the reaction shifts from adsorption control to solubility product control. Healy et al. [50] studied the adsorption of Co^{2+} on silica surfaces as a function of pH and concluded that below pH 6 and at less than $10^{-5} M$, Co^{2+} acted as a counterion in the diffuse layer. Above pH 6.5, Co^{2+} was specifically adsorbed and charge reversal accompanied this process. At combinations of pH and Co concentrations just below the K_{so} of $Co(OH)_2$, adsorption of a polynuclear $Co_x(OH)_{2x-y}^{y+}$ cation was postulated, while at higher pH values, nucleation of $Co(OH)_2$ occurred. Under the latter conditions the surface behavior was similar to $Co(OH)_2$ rather than silica.

The extent of formation and adsorption of polynuclear cations is often difficult to predict. These species are not stable and form under supersaturated conditions, often in localized supersaturated regions associated with reagent additions. Once formed under these conditions they may persist for some time, even if the bulk solution is undersaturated with respect to the stable solid phase. Polynuclear cations are strongly adsorbed onto negatively charged surfaces and could be responsible for changes in adsorption characteristics, which some investigators have interpreted as nucleation under conditions of undersaturation with respect to the precipitate [51]. Clearly, nucleation cannot occur under undersaturated conditions because even large crystals would dissolve rather than grow. However, polynuclear cations produced as described above could be stabilized sufficiently by adsorption to persist for relatively long time periods.

Polynuclear cations can be formed directly on surfaces as well as in solution. Robarge and Corey [52] saturated a strong-acid resin with Al^{3+} ions and added incremental amounts of $Ca(HCO_3)_2$. The concentration of exchangeable Al^{3+} ions decreased linearly with base addition, and the slope of the $[Al^{3+}]$ vs $[OH^-]$ plot indicated an OH:Al ratio between 2.0 and 2.1 for the adsorbed polynuclear cation that was formed. As base was added to give OH:Al ratios between 2:1 and 2.7:1, the Al species remained soluble in 0.1 N HCl. Above an OH:Al ratio of 2.7:1, a part of the Al found to be insoluble in 0.1 N HCl was presumed to be crystalline gibbsite in accord with the findings of Hsu and Bates [53].

Nucleation Involving Dissolution of the Adsorbent

If the adsorbate ion forms an insoluble precipitate with one of the components of the adsorbent, and if there are no other ions in solution that will form precipitates with the adsorbate, incremental additions of the adsorbate ion will eventually result in nucleation of a second solid phase made up of adsorbate ions and ions derived from dissolution of the adsorbent. An example is the adsorption of phosphate on calcite surfaces. On addition of phosphate to calcite suspensions, Griffin and Jurinak [25,54] found a rapid initial adsorption reaction, followed by a nucleation induction period that was inversely related to the amount of P added (and thus to the supersaturation), and finally surface nucleation and growth of a calcium phosphate solid phase. On the basis of solubility data, the new solid phase appeared to be hydroxyapatite at low initial P additions and an amorphous form at higher additions.

Stumm and Leckie [55] postulated a three-step precipitation process of adsorption, heterogeneous nucleation and crystal growth. In a study similar to that of Griffin and Jurinak, Stumm and Leckie confirmed the presence of apatite on the surface of the calcite by grazing angle electron diffraction and suggested that surface diffusion of phosphate over the calcite surface limited the growth rate of the crystals. In both these studies, the Ca in the phosphate phase was derived from dissolution of the calcite.

A dissolution-precipitation mechanism has also been shown for the reaction of phosphate with kaolinite, $Al_2Si_2O_5(OH)_4$, and greenalite (thought to be the Fe analogue of kaolinite) in 1 M solutions of sodium or potassium phosphates [56]. At high temperatures, crystalline phosphates formed rapidly as Al and Fe were dissolved from the silicate minerals and reprecipitated as crystalline phosphates (usually forms resembling taranakite or minyulite). While the concentration of P in this study was very high compared with most natural systems, such concentrations, accompanied by very low pH, do exist temporarily around fertilizer granules containing monobasic calcium phosphate (MCP). For example, Lindsay and Stephenson [57] found that the "metastable triple point solution" in a saturated solution of MCP contained 3.98 M P and 1.44 M Ca and had a pH of 1.48. When MCP dissolved in a soil system, large quantities of Al, Fe and Mn were dissolved from the soil minerals and precipitated as phosphates.

Veith and Sposito [58] worked at much lower initial phosphate concentrations ($10^{-3}M$ sodium phosphate) in studies with synthetic allophanic aluminosilicates and hydrous aluminum oxides. The P:Al ratios in these systems were 1:1. Two amorphous Al phosphates were formed with proposed formulas of $Al(OH)_2H_2PO_4$ and $AlOHNaPO_4$. The pH stability range of the proposed $Al(OH)_2H_2PO_4$ was from 2.5 to 9 and the calculated pK_{so} was 28.1±0.1.

Robarge and Corey [52] and Robarge [59] studied P adsorption onto hydroxy-Al polynuclear cations adsorbed on a cation exchange resin. Their results suggested adsorption of phosphate by the adsorbed polynuclear cations until the P:Al ratio of the nonexchangeable (polynuclear) fraction approached 1:2. At this point the $H_2PO_4^-$ concentration was approximately 5 × 10^{-6} M at a pH of 4.6. When the P:Al ratio exceeded 1:2 the pH decreased, whereas it had not changed significantly during the adsorption phase because phosphate probably was displacing mostly coordinated water molecules. The decrease in pH suggested that exchangeable Al^{3+} was starting to react with the phosphate to form a precipitate of the type proposed by Veith and Sposito [58], according to the following reaction:

$$Al^{3+} + H_2PO_4^- + 2H_2O \rightleftharpoons Al(OH)_2H_2PO_4 + 2H^+ \qquad (17)$$

The negative log of the ion activity product at an overall P:Al ratio of 1:1.2 for the nonexchangeable Al compound $[Al(OH)_2H_2PO_4]$ was 28.1 to 28.3 in a 1.1×10^{-3} M $Ca(ClO_4)_2$ solution. This is very close to the pK_{so} of 28.1 found by Veith and Sposito in a Na system. The agreement is probably fortuitous as the apparent pK_{so} was decreasing steadily with increasing P:Al ratio to that point. The increasing P:Al ratio of the Al solid phase was probably caused partly by Reaction 17 and partly by dissolution and reprecipitation of the adsorbed polynuclear complex, whose proposed formula at a P:Al ratio of 1:2 was $Al_6(OH)_{12}(H_2PO_4)_3(H_2O)_6^{3+}$.

Adsorption-Precipitation Boundary Problems

Predicting the critical solution concentration marking the boundary between simple adsorption and an adsorption-precipitation reaction on theoretical grounds would appear to be a formidable task at present, even for the simplest systems. Such a calculation would require an accurate knowledge of the supersaturation required for heterogeneous nucleation in the particular system used. It would also require careful control over the formation of polynuclear cations in the case of hydrolyzable cations, and it might involve activation energies for dissolution of the adsorbent in cases of dissolution-precipitation reactions or surface diffusion rates, if these were rate limiting.

Predicting rates of crystal growth once nucleation has occurred would be even more difficult. Crystal growth might be either diffusion controlled or interface controlled, and surface "poisons" are often plentiful in natural systems.

Dissolution-precipitation reactions could also be either diffusion or interface controlled at either precipitation or dissolution steps. Walton [60] states that dissolution is generally diffusion controlled. However, if dissolution were interface controlled, as discussed by Lieser et al. [61] for recrystallization of ionic crystals, the rate of dissolution would be related exponentially to an activation energy of dissolution. Thus, a relatively high supersaturation of the newly formed solid phase might be required to surmount the activation energy barrier and initiate dissolution of the adsorbent.

Although it may not be possible, given the present state of knowledge, to spell out the precise conditions that mark the boundary between adsorption and precipitation, a lower limit can be set. Nucleation of a new solid phase will not occur if the solution is undersaturated with respect to that solid phase. Nucleation under conditions of *apparent* undersaturation would be possible if the K_{so} were based on a solid phase that was more soluble than the one actually formed or if the precipitate were a solid solution and one or more of the components were not considered in the ion activity product.

Any time an investigator does exceed the K_{so} of a potential solid phase in an adsorption experiment, even if only in localized areas for a brief time period, there is a risk of forming a precipitate or a polynuclear species. The risk increases exponentially with increased supersaturation. Many direct determinations of maximum adsorption capacities are suspect because of the extreme supersaturations employed.

CONCLUSIONS

1. Specific adsorption and precipitation reactions differ primarily in geometry, adsorption being two-dimensional and precipitation three-dimensional. Bonding mechanisms are similar.

2. Ions specifically adsorbed on surfaces form solid solutions similar to crystalline solid solutions, except that adsorption is a two-dimensional process that has fewer restrictions on size and bonding mechanism.

3. Concentration of ions that are major constituents in natural systems are usually controlled by conventional solubility relationships of reasonably well-defined solid phases. Relatively small deviations from theoretical concentrations may occur due to particle-size effects, presence of polymorphic forms or incorporation of impurities into the crystals.

4. Concentrations of ions that are minor constituents in natural systems are usually controlled by adsorption reactions or solid-solution equilibria involving poorly defined systems. Concentrations of these elements calculated by conventional solution chemistry approaches are likely to be in error.

5. Heterogeneous nucleation of a new solid phase involving ions adsorbed on a mineral surface occurs when a critical supersaturation of adsorbate is exceeded. Below that critical concentration, adsorption controls the activity of the adsorbate.

6. Although it is difficult to determine the exact boundary between adsorption and precipitation, interpretations of many adsorption experiments are suspect because they were carried out in systems supersaturated with respect to a potential new solid phase involving the adsorbate ion.

REFERENCES

1. Veith, J. A., and G. Sposito. "On the Use of the Langmuir Equation in the Interpretation of Adsorption Pheomena," *Soil Sci. Soc. Am. J.* 41: 697-702 (1977).
2. Lewin, S. *The Solubility Product Principle* (New York: Wiley-Interscience, 1960), p. 12.
3. Stumm, W., and J. J. Morgan. *Aquatic Chemistry* (New York: Wiley-Interscience, 1970), p. 12.
4. Garrels, R. M., and C. L. Christ. *Solution, Minerals, and Equilibria* (New York: Harper and Row Publishers, Inc., 1965), p. 12.
5. Freundlich, H. *Colloid and Capillary Chemistry* (London: Methuen, 1926), p. 155.
6. Kahlweit, M. In: *Reactivity of Solids,* J. W. Mitchell, R. C. DeVries, R. W. Roberts and P. Cannon, Eds. (New York: Wiley-Interscience, 1969), p. 93.
7. Stumm, W., and J. J. Morgan. *Aquatic Chemistry* (New York: Wiley-Interscience, 1970), p. 230.
8. Walton, A. G. In: *Nucleation,* A. C. Zettlemoyer, Ed. (New York: Marcel Dekker, Inc., 1969), p. 265.
9. Stumm, W., and J. J. Morgan. *Aquatic Chemistry* (New York: Wiley-Interscience, 1970), p. 232.
10. Walton, A. G. In: *Nucleation,* A. C. Zettlemoyer, Ed. (New York: Marcel Dekker, Inc., 1969), p. 274.
11. Wylie, R. B. "The Freezing of Supercooled Water in Glass," *Proc. Phys. Soc.* (London) B66: 241-254 (1953).
12. Vonnegut, B. "Nucleation of Ice Formation by Silver Iodide," *J. Appl. Phys.* 18: 593-595 (1947).
13. Walton, A. G. In: *Nucleation,* A. C. Zettlemoyer, Ed. (New York: Marcel Dekker, Inc., 1969), p. 300.
14. Upreti, M. C., and A. G. Walton. "Heterogeneous Nucleation of Alkali Halide Crystals from Solution," *J. Chem. Phys.* 44: 1936-1939 (1966).
15. Stumm, W., and J. J. Morgan. *Aquatic Chemistry* (New York: Wiley-Interscience, 1970), p. 233.
16. Gunn, D. J. "Mechanism for the Formation and Growth of Ionic Precipitates from Aqueous Solution," *Faraday Disc. Chem. Soc.* 61: 133-140 (1976).
17. Davies, C. W., and G. H. Nancolles. "The Precipitation of Silver Chloride from Aqueous Solutions. IV. Temperature Coefficients of Growth and Solution," *Trans. Faraday Soc.* 51: 818-823 (1955).
18. Hartman, P. In: *Crystal Growth: An Introduction,* P. Hartman, Ed. (New York: Elsevier North-Holland, Inc., 1973), p. 390.
19. Walton, A. G. In: *Nucleation,* A. C. Zettlemoyer, Ed. (New York: Marcel Dekker, Inc., 1969), p. 87.
20. Berner, R. A. *Principles of Sedimentology* (New York: McGraw-Hill Book Co., 1971), p. 87.
21. Stumm, W., and J. J. Morgan. *Aquatic Chemistry* (New York: Wiley-Interscience, 1970), p. 204.
22. Kitano, Y., and D. W. Hood. "Influence of Organic Material on the Polymorphic Crystallization of Calcium Carbonate," *Geochim. Cosmochim. Acta* 29: 29-41 (1965).

23. Stumm, W., and J. J. Morgan. *Aquatic Chemistry* (New York: Wiley-Interscience, 1970), p. 214.
24. Sears, G. W. In: *Physics and Chemistry of Ceramics*, C. Klingsberg, Ed. (New York: Gordon and Breach, 1963).
25. Griffin, R. A., and J. J. Jurinak. "The Interaction of Phosphate with Calcite," *Soil Sci. Soc. Am. Proc.* 37: 847-850 (1973).
26. Gordon, L., M. L. Salutsky and H. H. Willard. *Precipitation from Homogenous Solution* (New York: John Wiley & Sons, Inc., 1969), p. 52.
27. Wagner, C., and W. Schottky. "The Theory of Arranged Mixed Phases," *Z. Physik. Chem.* (Leipzig) B11: 163-210 (1938).
28. Jagodzinsky, H. In: *Problems of Nonstoichiometry*, A. Rabenau, Ed. (New York: Elsevier North-Holland, Inc., 1970), p. 131.
29. Vaslow, F., and G. E. Boyd. "Thermodynamics of Coprecipitation: Dilute Solid Solutions of AgBr in AgCl," *J. Am. Chem. Soc.* 74: 4691-4695 (1952).
30. Bodine, M. W., H. D. Holland and M. Borcsik. "Coprecipitation of Manganese and Strontium with Calcite," *Symposium on Problems of Postmagnetic Ore Deposition*, Prague, Vol. II (1965) pp. 401-406.
31. Chave, K. E. "A Solid Solution Between Calcite and Dolomite," *J. Geol.* 60: 190-192 (1952).
32. Taylor, R. M., and R. M. McKenzie. "The Association of Trace Elements with Manganese Minerals in Australian Soils," *Aust. J. Soil Res.* 4: 29-39 (1966).
33. Glover, E. D. "Characterization of a Marine Birnessite," *Am. Mineralogist* 62: 278-285 (1977).
34. McClellan, G. H., and J. R. Lehr. "Crystal Chemical Investigation of Natural Apatites," *Am. Mineralogist* 54: 1374-1391 (1969).
35. Chien, S. H., and C. A. Black. "Free Energy of Formation of Carbonate Apatites in Some Phosphate Rocks," *Soil Sci. Soc. Am. J.* 40: 234-239 (1976).
36. Chaverri, J. G., and C. A. Black. "Theory of the Solubility of Phosphate Rock," *Iowa State J. Sci.* 41: 77-95 (1966).
37. Arlidge, E. Z., V. C. Farmer, B. D. Mitchell and W. A. Mitchell. "Infrared, X-ray and Thermal Analysis of Some Aluminum and Ferric Phosphates," *J. Appl. Chem.* 13: 17-27 (1963).
38. Garrels, R. M., and C. L. Christ. *Solutions, Minerals and Equilibria* (New York: Harper and Row Publisher, Inc., 1965), p. 42.
39. Helgeson, H. C. "Evaluations of Irreversible Reactions in Geochemical Processes Involving Minerals and Aqueous Solutions. I. Thermodynamic Relations," *Geochim. Cosmochim. Acta* 32: 853-857 (1968).
40. Baes, C. F., Jr., and R. E. Mesmer. *The Hydrolysis of Cations* (New York: Wiley-Interscience, 1976), p. 118.
41. Hsu, P. H. In: *Trace Inorganics in Water*, R. F. Gould, Ed.,, Advances in Chemistry Series, No. 73 (Washington, DC: American Chemical Society, 1968), p. 115.
42. Hahn, H. H., and W. Stumm. In: *Adsorption from Aqueous Solution*, R. F. Gould, Ed., Advances in Chemistry Series No. 79 (Washington, DC: American Chemical Society, 1968), p. 91.
43. Atkinson, R. J., A. M. Posner and J. P. Quirk. "Kinetics of Isotopic Exchange of Phosphate at the α-FeOOH-Aqueous Solution Interface," *J. Inorg. Nucl. Chem.* 34: 2201-2211 (1972).

44. Parfitt, R. L., R. J. Atkinson and R. St. C. Smart. "The Mechanism of Phosphate Fixation on Iron Oxides," *Soil Sci. Soc. Am. Proc.* 39: 837-841 (1975).
45. Parfitt, R. L., and R. J. Atkinson. "Adsorption of Phosphate on Goethite," *Nature* 264: 740-741 (1976).
46. Parfitt, R. L. "Phosphate Adsorption on an Oxisol," *Soil Sci. Soc. Am. J.* 41: 1064-1067 (1977).
47. Scholten, A. G. *The Reaction of Phosphate with Mineral Surfaces and Iron Oxide Gels,* Ph.D. Thesis, University of Wisconsin, Madison (1965), p. 46.
48. Corey, R. B. Unpublished results.
49. Grimme, H. "Die Adsorption von Mn, Co, Cu, and Zn durch Goethite aus verdünnten Lösungen," *Z. Pflanzenernahr. Dung. Bodenkunde* 121: 58-65 (1969).
50. Healy, T. W., R. O. James and R. Cooper. In: *Adsorption from Aqueous Solution,* R. F. Gould, Ed., Advances in Chemistry Series, No. 79 (Washington, DC: American Chemical Society, 1968), p. 62.
51. James, R. O., and T. W. Healy. "Adsorption of Hydrolyzable Metal Ions at the Oxide-Water Interface. II. Charge Reversal of SiO_2 and TiO_2 Colloids by Co(II), La(III) and Th(IV) as Model Systems," *J. Interface Sci.* 40: 53-64 (1972).
52. Robarge, W. P., and R. B. Corey. "Adsorption of Phosphate by Hydroxy-Aluminum Species on a Cation Exchange Resin," *Soil Sci. Soc. Am. J.* 43: 481-487 (1979).
53. Hsu, P. H., and T. F. Bates. "Fixation of Hydroxy-Aluminum Polymers by Vermiculite," *Soil Sci. Soc. Am. Proc.* 28: 763-768.
54. Griffin, R. A., and J. J. Jurinak. "Kinetics of the Phosphate Interaction with Calcite," *Soil Sci. Soc. Am. Proc.* 38: 75-79 (1974).
55. Stumm, W., and J. O. Leckie. In: *Advances in Water Pollution Research,* S. H. Jenkins, Ed. (Elmsford, NY: Pergamon Press, Inc., 1970), III-26/1-16.
56. Kittrick, J. A., and M. L. Jackson. "Electron-Microscope Observations of the Reaction of Phosphate with Minerals Leading to a Unified Theory of Phosphate Fixation in Soils," *J. Soil Sci.* 7: 81-89 (1956).
57. Lindsay, W. L., and H. F. Stephenson. "Nature of the Reactions of Monocalcium Phosphate Monohydrate in Soils: IV. Repeated Reactions with Metastable Triple-Point Solution," *Soil Sci. Soc. Am. Proc.* 23: 440-445 (1959).
58. Veith, J. A., and G. Sposito. "Reactions of Aluminosilicates, Aluminum Hydrous Oxides, and Aluminum Oxide with o-Phosphate: The Formation of X-ray Amorphous Analogs of Variscite and Montebrasite," *Soil Sci. Soc. Am. J.* 41: 870-876 (1977).
59. Robarge, W. P. *Characterization of Sorption Sites for Phosphate,* Ph.D. Thesis, University of Wisconsin, Madison (1975).
60. Walton, A. G. *The Formation and Properties of Precipitates* (New York: Wiley-Interscience, 1967), p. 70.
61. Lieser, K. H., H. Mager and G. Pallikaris. "Mechanism of Recrystallization of Ionic Crystals Under Solution," *Z. Phys. Chem.* (Wiesbaden) 105: 35-46 (1977).

CHAPTER 5

THE SURFACE ACIDITY OF HYDROUS SOLIDS

C. P. Huang
Environmental Engineering Program
Department of Civil Engineering
University of Delaware
Newark, Delaware

INTRODUCTION

Surface acidity is an important property of hydrous solids to which are closely related the mode and extent of interfacial reactions such as adsorption and coagulation. Much has been reported on the determination of surface acidity of solid catalysts [1]. Forni [2] has made a comprehensive review of the various methods for its determination. In dealing with surface acidity, three variables must be considered: (1) the acid strength of the surface sites; (2) their density (number of acid centers per unit surface area of the solid); and (3) their nature (Brönsted or Lewis type).

For most hydrous solids, specifically the oxides and hydroxides, Brönsted acid-base sites are readily developed on their introduction to an aqueous solution. For instance, Hindin and Weller [3] pictured the development of surface acid and base sites of alumina as follows:

The presence of Brönsted acid and base sites on hydrous solids expedites the quantitative characterization of hydrous solid-solution interfaces by conventional alkalimetric titration.

Many studies have been reported on the titration of hydrous solids. Of note are those on $Fe_2O_3(s)$, by Atkinson et al. [4] and Parks and de Bruyn [5]; on FeOOH(s), by Atkinson et al. [4]; on $SiO_2(s)$, by Bolt [6], Abendroth [7], Breeuwsma and Lyklema [8] and Schindler and Kamber [9]; γ-$Al_2O_3(s)$, by Huang and Stumm [10], Hohl and Stumm [11]; on $TiO_2(s)$, by Berube and de Bruyn [12] and Yates [13]; on $ZrO_2(s)$ and $ThO_2(s)$ by Ahmead [14]; on $Fe_3O_4(s)$ by Ahmead and Maksimov [15]; on $CaCO_3(s)$ by Huang [16]; and on $Ca_{10}(OH)(PO_4)_6(s)$ and $Ca_{10}F_2(PO_4)_6(s)$ by Bell et al. [17]. However, most of the above work has focused on the description of double-layer structure and related phenomena such as adsorption.

Jacob has attempted to calculate the surface acidity of Al_2O_3 using Belot's data [18]. However, he visualized the surface as only a "monoprotic" solid acid. Huang and Stumm [10], using a modified Langmuir plot method and treating the oxide as a "diprotic" solid acid, have calculated the acidity of hydrous γ-Al_2O_3. Hohl and Stumm [11] determined the surface acidity of γ-$Al_2O_3(s)$ by plotting the surface microscopic equilibrium constant, against the degree of deprotonation with predetermined total numbers of surface functional centers. Recently, Davis et al. [19] by applying the charge-balance concept, have calculated the surface acidity constants of some solid oxides. Details of this technique are seen in Chapter 6.

The objective of this chapter is to demonstrate the determination of surface acidity of hydrous solids using the modified Langmuir plot method developed by Huang and Stumm [10]. Special emphasis will be placed on the titration technique, its set-up, limitations and improvement. No attempt will be made to compare the various methods of surface acidity determination. An extension of the method described here to systems containing specifically adsorbed multivalent metal ions is also presented.

POTENTIOMETRIC TITRATION

Titration Apparatus

It has been reported that glass membranes doped with different quantities of aluminum oxide exhibit different electromagnetic force (EMF) responses with pH [20]. Most oxides have been found to be strongly adsorbed on most laboratory glassware, including the glass electrode. This may possibly cause defects in the proper functioning of a glass electrode together with the salt effect error present with a double-bridge electrode. Under such circumstances, the use of a standard hydrogen electrode against a Ag/AgCl reference electrode may be beneficial. Figure 1 demonstrates the correlation of responses between a glass electrode and a hydrogen electrode. The pH values from glass electrodes were obtained with a Beckman special E2 combined glass electrode together with a Beckman Model G7 pH-meter.

The hydrogen electrode is constructed by bubbling extra pure hydrogen gas over a platinum foil, the surface of which is able to catalyze the half cell reaction:

$$H^+ + e^- \rightleftharpoons \tfrac{1}{2} H_2 \text{ (gas)} \tag{1}$$

and hence establish the equilibrium between hydrogen molecules (or atoms) and the hydronium ions in the solution in which the platinum metal electrode is immersed. The platinum surface has been precoated with platinum black to ensure proper catalytic behavior. See Table I for a summary of procedures for the preparation of hydrogen and reference electrodes.

The silver-silver chloride electrode is one of the family of silver-silver halide electrodes, so-called electrodes of the second kind, by contrast to those of

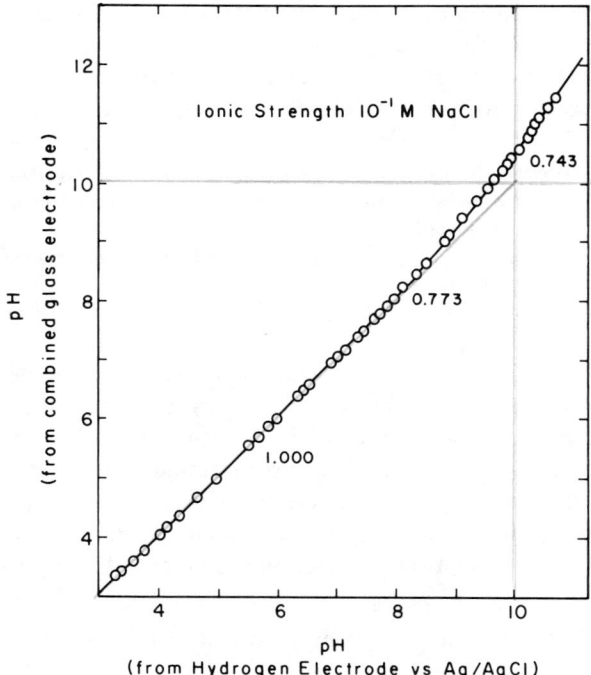

Figure 1. Correlation of response between glass and hydrogen electrodes.

Table I. Preparation of Hydrogen Electrodes—Procedures

To clean the platinum surface:
1. Dip the electrode in 1 M H_2SO_4 solution.
2. Connect the electrodes to a 4.5-V mercury cell.
3. Reverse the polarity every 15 seconds. Continue cleaning for 2 minutes.
4. Rinse the electrode with distilled water.

To replatinize the electrode:
1. Dip the electrodes in a platinizing solution containing 3 g chloroplatinic acid and 0.02 g lead acetate per 100 ml distilled water.
2. Connect the electrodes to the same dc cell. Continue platinizing until the electrode surface becomes black.
3. Rinse the electrode with distilled water. Immerse the electrode in distilled water when not in use.

---

the first kind, which are directly reversible to the ions of the metal phase. The electrode has a solid phase in the form of a sparingly soluble salt (AgCl(s), $pK_{so} = 9.764$ at 25°C) in equilibrium with a saturated solution of this salt participating in the electrode reaction.

To achieve stability, reproducibility and sensitivity of this electrode, four main types of electrode construction have been developed [3]:

1. *electrolytic,* the electrolytic deposition of both silver and silver halide;
2. *thermal,* the decomposition of a paste of silver, silver halide and water to form silver-silver halide by thermal process;
3. *thermal-electrolytic,* the electrolytic formation of silver halide on thermally reduced silver oxide paste; and
4. *miscellaneous,* such as the use of precipitated silver halide on silver.

In our experiments, a Ag-AgCl electrode was prepared by direct coating of silver chloride onto a silver billet surface from saturated potassium chloride solution at a polarizing potential of 1.5 V, as described in Table II. To test the working condition of the silver-silver electrode, the electrode was constantly checked by immersing the electrode in a series of NaCl or KCl solutions of given concentration against a standard calomel reference electrode. The overall system for a titration setup can be described by the cell:

$$Pt|H_2(g)|HCl(m_1),NaOH(m_2),NaCl(m_3), Solid, H_2O\ AgCl|Ag$$

A Kiethley Model 660 differential voltmeter was used to record the EMF readings. This instrument has a stable, reproducible resolution of 100 μV. The

SURFACE ACIDITY OF HYDROUS SOLIDS

Table II. Preparation of Ag-AgCl Electrodes—Procedures

1. Clean metallic tips of a pair of silver electrodes with mild detergent and scouring powder.
2. Rinse tips in distilled water and immerse in saturated KCl solution.
3. Connect the electrodes to a 1.5-V dry cell. Reverse current polarity to alternatively clean and recoat the electrodes.
4. When adequately coated, the electrode will turn violet in color. Rinse the electrode with distilled water.

titration cell was a flat-bottomed 300 ml Teflon®* beaker with a rubber stopper through which passed the electrodes and the microburet. Mixing was provided by a Teflon-covered magnetic bar. Room temperature was controlled to 25±0.5°C.

Calibration of Hydrogen Electrode

To express measured EMF in terms of the conventional pH scale, the hydrogen electrode was standardized against solutions of given hydrogen activity [21]. To standardize the electrode, standard HCl was titrated with standard NaOH at a constant ionic strength, adjusted with NaCl. The observed EMF can be expressed by:

$$E = E^{0\prime} + E' = E^{0\prime} - \frac{2.303 \, RT}{F} \log [H^+] \qquad (2)$$

where E and $E^{0\prime}$ = the observed and standard EMF values at Ionic strength, I, and

F, R and T = Faraday's Constant, the universal gas constant and the absolute temperature, respectively.

$[H^+]$ is the proton concentration in mol/l, which can be found by the following:

$$[H^+] = C_a - \frac{v}{V_0 + v} C_b$$

where C_a and C_b = the molar concentrations of total acid and NaOH titrant, respectively, and

*Registered trademark of E. I. du Pont de Nemours and Co., Inc., Wilmington, Delaware.

V_0 and v = the volume of the total solution titrated and the titrant, ml.

Therefore, at equilibrium and 25°C for aqueous solutions:

$$E^{0'} = E + 0.05915 \log [H^+] \qquad (3)$$

The $E^{0'}$ values for various ionic strengths are given in Table III. To calculate pH from emf measurement, the following approach was used:

Rearrangement of (3) yields the following:

$$\log [H^+] = \frac{E^{0'} - E}{\frac{RT}{F}} = \frac{E^{0'} - E}{0.05915} \qquad (4)$$

By definition,

$$\{H^+\} = [H^+](\gamma_\pm) \qquad \text{wrong!} \qquad (5)$$

$$-\log \gamma_\pm = \frac{S_f \sqrt{I}}{1 + A\sqrt{I}} \qquad (6)$$

$$S_f = \frac{1}{\nu}\left(\sum_{j=1}^{p} \nu_j z_j\right)^{3/2} \frac{1.283 \times 10^6}{(\epsilon T)^{3/2}} \qquad (7)$$

where $\gamma_\pm$ = the mean activity coefficient,
ν = $(\nu_+ + \nu_-)$,
ν_j = the numbers of ion of the jth type produced by dissociation of one molecule of electrolyte,
z_j = the valence of ion of the jth type with sign,
ϵ = the dielectric constant, and
A = a constant depending on the characteristics of the solvent and the size of the ion.

At 25°C, for 1:1 electrolyte

S_f = 0.5056
A = 1.529 for NaCl in aqueous solution

Hence:

$$-\log \gamma_\pm = \frac{0.5056\sqrt{I}}{1 + 1.529\sqrt{I}}$$

pH can thus be calculated with the aid of Table IV.

$$pH = -\log \{H^+\} = -\log [H^+] - \log \gamma_\pm \quad (8)$$

$$= \frac{E - E^{0'}}{0.05915} - \frac{0.5056\sqrt{I}}{1 + 1.529\sqrt{I}} \quad (9)$$

Titration Procedure and Evaluation of Isotherms

To better illustrate the modified Langmuir plot method, Alon (trade name for $\gamma\text{-Al}_2\text{O}_3$) was used as an example due to its slightly alkaline pH_{zpc}.
A weighted quantity of Alon powder was suspended in 10^{-2} M standard NaOH solution and stirred for 30 minutes. The solid was then separated by centrifuging for another 30 minutes at 19,600 g. The base-washed Alon was further cleaned with triple-distilled water by following the previous procedure until constant conductivity of the supernatant was reached. Three to five rinses with conductivity water were found satisfactory. Loss of solid due to washing was shown to be insignificant. Base-treated Alon was redispersed in a measured volume of triple-distilled water for the stock suspension and kept underwater for two weeks before being used for any titration or adsorption experiments.

Purity of sample contributes greatly to the reproducibility of titration. It is recommended that some kind of acid- or base-washing, followed by rinsing with conductivity water, be performed. The example, ($\gamma\text{-Al}_2\text{O}_3$) illustrated

Table III. Measured $E^{0'}$ Values

Ionic Strength (NaCl)	$E^{0'}$ (mV)
1 x 10^{-3}	0.38964
3 x 10^{-3}	0.36957
1 x 10^{-2}	0.34353
3 x 10^{-2}	0.31858
1 x 10^{-1}	0.29295

Table IV. $-\log \gamma_\pm$ vs NaCl (M)

NaCl (M)	$-\log \gamma_\pm$
1 x 10^{-3}	0.0152
3 x 10^{-3}	0.0254
1 x 10^{-2}	0.0438
3 x 10^{-2}	0.0693
1 x 10^{-1}	0.1077

in this chapter used Mallinckrodt AR-grade salts, NaCl and KCl, without further purification. Anachimia standard acid 10^{-1} M HCl and base 10^{-1} M NaOH were used to adjust the pH. Medtech pure N_2 gas was used as anoxic agent.

A 100-ml sample of stock Alon solution, weight concentration 6.23 g/l, was pipetted into a Teflon beaker. To the beaker was added triple-distilled water to a total volume of 200 ml at preselected ionic strength.

The emf (mV) or pH readings were taken every two minutes after each measured addition of acid or base. Base-washed Alon, after immersion in water for two weeks, displayed fast equilibration in alkimetric-acidimetric titration.

The adsorption isotherm is readily obtainable by a simple mass balance equation:

$$\Delta\Gamma_\pm = \frac{\Delta(C_A - C_B) - [V_1(C_{H^+} - C_{OH^-}) - V_2(C_{H^+} - C_{OH^-})]}{S} \tag{10}$$

where $\Delta\Gamma_\pm = \Delta(\Gamma_{H^+} - \Gamma_{OH^-})$ is the change in relative surface excess of hydrogen over hydroxide ions due to each increment of acid or base added, $\Delta(C_A - C_B)$,

S = the total surface area of the solid in suspension, m²/g,
V = the total volume of solution, ml and
C_{H^+} and C_{OH^-} = the concentration of hydrogen ions and hydroxide ions derived from pH, mol/l.

Subscripts "1" and "2" refer to physical quantities before and after each titrant addition.

While this equation may sound simple, one can avoid laborious calculation and exclude impurities from being counted as surface consumption of H^+ or OH^- by doing sample and blank titrations as follows:

Four identical samples are prepared at each ionic strength. Two are directly titrated, one with acid and the other with base (sample titration) and the other two samples are completely stirred for 30 minutes, then centrifuged for another 30 minutes at 19,600 g. The clear supernatant solutions later receive the same titration procedure as the previous two samples that are titrated in the presence of colloidal Alon (blank titration). Figures 2a and 2b are examples of this operation.

Recording pH (or mV) during the course of titration of a suspension provides data that can be interpreted in terms of adsorption of hydrogen or hydroxide ions or their complexes. Operationally, the quantity of "adsorption"

SURFACE ACIDITY OF HYDROUS SOLIDS

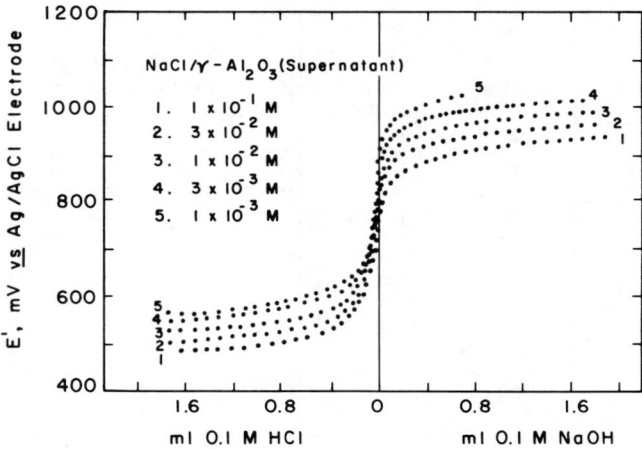

Figure 2a. Titration curves for the supernatants of γ-Al_2O_3 suspensions in the presence of NaCl at various concentrations.

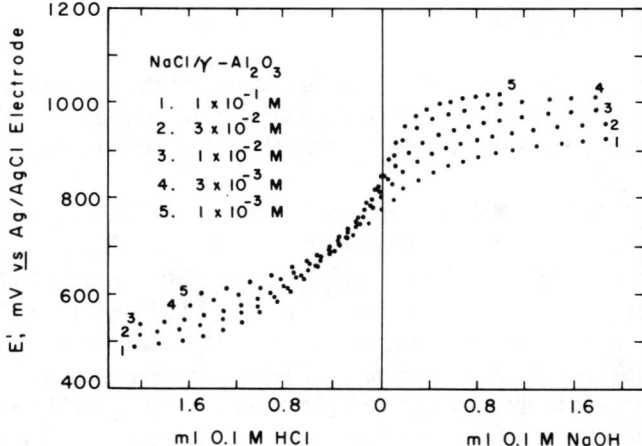

Figure 2b. Titration curves for γ-Al_2O_3 suspensions in the presence of NaCl at various concentrations.

is the number of appropriate ions that cannot be accounted for in a mass balance between the titration curve of suspension and that of the supernatant. The difference that cannot be balanced in the solution part, yet is present in the system, must be ascribed to the surface. Figure 3 is the result of such a calculation from Figure 2.

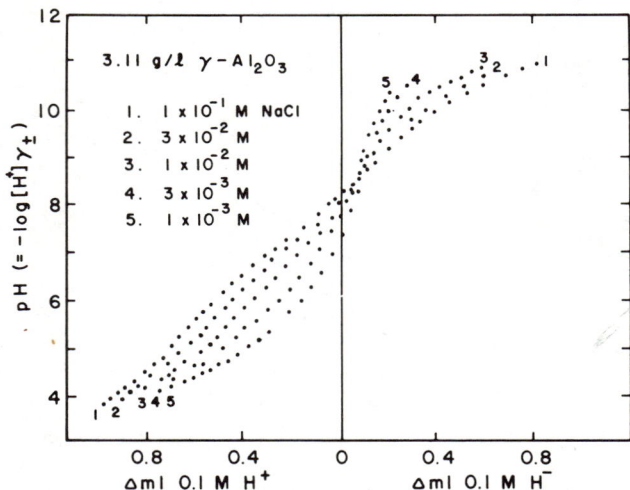

Figure 3. Net titration curves for γ-Al_2O_3 suspension in the presence of NaCl at various concentrations.

The Point of Zero Charge

The pH_{pzc}, the point of zero charge, is one of the key points of information for the determination of surface acidity by titration, even though no hydrous oxide has a unique pH_{pzc} [22,23], and the exact value depends on sample origin.

Because H^+ and OH^- are presumably the potential-determining species for hydrous oxides, it is convenient to determine the pH_{pzc} by alkimetric-acidimetric titration [1]. Ideally, pH_{pzc} can be distinguished by pH at zero net adsorption of protons and hydroxide ions, i.e., $\Gamma_{H^+} - \Gamma_{OH^-} \equiv 0$. Following this argument, the ideal case is that pH_{pzc} will be independent of electrolyte concentration (Figure 4a). However, since specific adsorption of anions tends to shift pH_{pzc} to a higher value and pH_{pzc} will be lowered on specific adsorption of cations (Figures 4b and 4c), for most real situations pH_{pzc} is quantitatively determined at an intersection point of the "net" titration curves of lower ionic strength when specific adsorption is usually absent or insignificant.

The pH_{pzc} of hydrous oxides is a function of many variables, including its previous history, aging, doping by impurities, nature of crystallinity, synthetic processes, temperature and sorbability of electrolytes, and degree of hydration. Oxides of the same chemical composition usually display different pH_{pzc} because of dissimilar crystallinity, synthetic processes, length of hydration period and batch reproducibility.

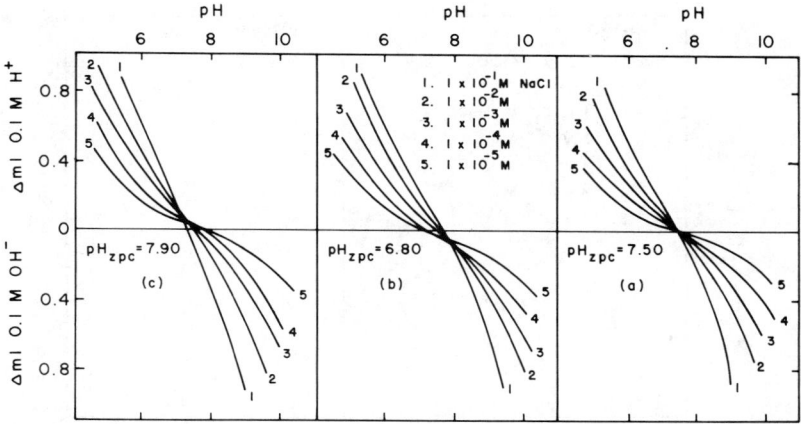

Figure 4. Schematic representation of the determination of pH_{pzc} from alkalimetric titration: (a) in the absence of specific adsorption of electrolytes; (b) in the presence of specific cationic adsorption; and (c) in the presence of specific anionic adsorption.

Surface Charge

Essentially, the method consists of measuring the proton or hydroxide ion consumption by the solid phase, by comparing the titration curve of a suspension of the inorganic oxide (Figure 2b) with that of the medium alone (Figure 2a). (Instead of titrating the ionic medium alone, in our experiment the supernatant of the suspension was titrated.) A net adsorption curve for H^+ and OH^- (in unit of ml strong base and acid) is presented in Figure 3. These results are obtained by taking the difference in ml of strong base and/or acid (0.1 M) taken up in titration of solution containing the solids and the supernatant alone.

Surface charge density can be calculated from the data of Figure 3, with Equation 13. Figure 5 shows the change of surface charge with solution pH found in this way:

$$\sigma_0 = F(\Gamma_{H^+} - \Gamma_{OH^-}) \tag{11}$$

Surface Potential

The surface potential, ψ_0, must be known to extract K_{int} from the experimental data by means of Equations 25 and 26. In the simplest model, ψ_0 may be computed from the surface charge using the elemental double-layer

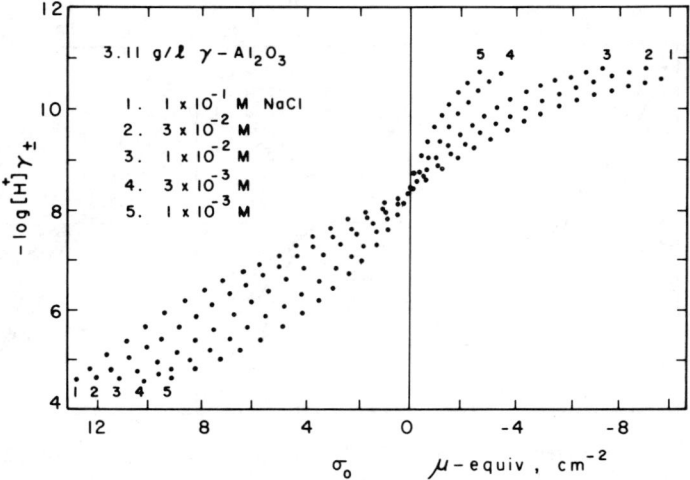

Figure 5. Surface charge distribution as computed from alkalimetric titration.

theory, i.e., setting $\psi_0 = \psi_\delta$, where ψ_δ is the potential at the plane of closest approach. From the following equation:

$$\sigma_\delta = 2 \left(\frac{RT\epsilon I}{2\pi}\right)^{1/2} \sinh\left(\frac{zF}{2RT}\psi_\delta\right) \tag{12}$$

or at 25°C in aqueous solution:

$$\psi_\delta = 0.05 \sinh^{-1}\left[\sigma_\delta/(11.74\sqrt{I})\right] \tag{13}$$

where F is Faraday's Constant, I refers to the electrolyte concentration in M and ψ_δ is expressed in volts. The charge at the diffuse double layer in $\mu C/cm^2$ is represented by σ_δ, which, ideally, may be set equal to σ_0.

THE DETERMINATION OF SURFACE ACIDITY

The Amphoteric Properties of Hydrous Oxides

The amphoteric behavior of hydrous oxides can be compared, at least operationally, with amphoteric polyelectrolytes [24,25]. The potential-determining role of H^+ and OH^- is most readily accounted for by realizing

SURFACE ACIDITY OF HYDROUS SOLIDS

that the pH-dependent charge of an oxide results from acid-base reactions at the surface:

$$M\text{-}OH_2^+ \underset{}{\overset{K_{int}^1}{\rightleftharpoons}} M\text{-}OH \underset{}{\overset{K_{int}^2}{\rightleftharpoons}} M\text{-}O^- \qquad (14)$$

where M stands for a metal or metalloid central ion. For an oxide, solvation (hydration) of the surface has to precede proton transfer [26]. Proposed models for the oxide-water interaction are presented in Chapters 1, 2 and 6 of this book. It is convenient to visualize the surface sites as converted on hydration to surface hydroxo groups of M-OH groups that can then dissociate protons:

$$M\text{-}OH + H_2O \rightleftharpoons M\text{-}O^- + H_3O^+$$

or

$$M\text{-}OH + OH^- \rightleftharpoons M\text{-}(OH)_2^- \qquad (15)$$

or accept protons:

$$M\text{-}OH + H_3O^+ \rightleftharpoons M\text{-}OH_2^+ + H_2O$$

or

$$M\text{-}OH \rightleftharpoons M^+ + OH^- \qquad (16)$$

One cannot distinguish by conventional analytical means between dissociation of protons from the surface or the binding or adsorption of OH^- ions to the surface. Near the oxide surface the water must become structurally well ordered. It is also possible that the surface charge may arise within a few layers of the surface, i.e., by penetration of hydroxide ions into the structurally ordered water layers [27-29].

The Acidity of the M-OH Group

The acidity of a surface M-OH group depends on the acidity of the central ion and is influenced by electrostatic field strength and induction effects of the solid. Also, it is modified by the structural ordering of the water layers immediately adjacent to the solid surface. In particular, hydrogen bonding tends to facilitate proton transfer and hence increase the acidity. The acidity and basicity of the surface M-OH group characterize the amphoteric properties of the hydrous oxide.

Microscopic Acidity Constant

The microscopic acidity constant of the polyacid or surface hydroxyl group is given by the following equation:

$$HX \underset{}{\overset{K'}{\rightleftharpoons}} H^+ + X^-$$

and

$$K' = [H^+] \frac{\{X^-\}}{\{HX\}} = [H^+] \frac{f}{1-f} \qquad (17)$$

or

$$pK' = pH - \log \frac{f}{1-f} \qquad (18)$$

where $\{i\}$ = surface concentration of ith species,
$[i]$ = the concentration of the ith species in the bulk phase, and
f = the degree of deprotonation.

If all the acid groups were identical and ionized independently of one another, then the titration curve of a n-protic acid would be indistinguishable from that of nmoles of monoprotic acid with a pK_a value equal to that of the microscopic constant pK'. However, each loss of a proton reduces the charge on the polyacid, thus affecting the acidity of the neighboring groups. Eventually, with progressively reduced charge of the polyacid, it becomes increasingly difficult to remove a proton. Thus, pK' becomes larger with increasing degree of titration.

Intrinsic Constant

The free energy of deprotonation, as expressed by K', is a summation of free energy of dissociation of H^+, as measured by K_{int} (intrinsic constant), and a free energy of removal of the proton from the site of dissociation into the bulk of the solution, as expressed by the Boltzman (or electrostatic) factor [24,25]. Thus:

$$K' = K_{int} \exp(F\psi_0/RT) \qquad (19)$$

The intrinsic constant is the acidity constant of the acid group in a completely chargeless environment.* "Chargeless" is distinguished from zero net charge, because a zero net charge surface still contains positive and negative sites, although in equal number, and these might influence the acidity constant.

Equation 19 can also be derived from Stern's treatment of the double layer. If n_c is the number of OH$^-$ ions "sorbed" per cm^2 from a bulk concentration of mole fraction, X_{OH}, of OH$^-$ ions, n_s is the number of possible sorption sites in the surface and Φ is the chemical energy of adsorption, the **ratio of occupied (with OH$^-$)** to unoccupied sites may be related to the corresponding ratio in the solution as follows:

$$\frac{n_c}{n_s - n_c} = \frac{f}{1-f} = X_{OH} \exp(F\psi_\delta - \Phi)/RT \tag{20}$$

By substituting

$$X_{OH} \text{ with } [OH^-]/55 = \frac{K_w}{55\,[H^+]}$$

one obtains the following expression:

$$pH = \log \frac{f}{1-f} - pK_{int} - \frac{F\psi_\delta}{2.3\,RT} \tag{21}$$

with $pK_{int} = -\log K_w - \log 55 - \Phi/2.3\,RT$.

Sum of Functional Groups (Acidity Density) and K_{int}, K'

If for hydrous oxide surfaces the model of functional M-OH groups is adopted, a surface acidity density ($[X_t^-] = \{MOH\} + \{M\text{-}O^-\}$, or $[X_t^+] = \{MOH_2^+\} + \{MOH\}$), can be defined for such surfaces. Operationally, this density may be determined by adding known excess of base or acid, which, after separation of the solid phase, is back-titrated with acid or base [11]. In general,

$$K_{int} = K'\, e^{-\frac{F\psi_0}{RT}} \tag{22}$$

*A "completely chargeless environment" is often considered, due to experimental limitations, as the absence of a potential gradient. That is, when $\psi_0 = \psi_{bulk}$.

For positively charged surfaces with the aid of equations from Table V, this equation becomes

$$K_{int}^1 = \frac{[X_t^+] - [X_+]}{[X_+]} [H^+]_{bulk} \, e^{-\frac{F\psi_0}{RT}} \qquad (23)$$

For a negatively charged surface equation it is

$$K_{int}^2 = \frac{[X_-]}{[X_t^-] - [X_-]} [H^+]_{bulk} \, e^{-\frac{F\psi_0}{RT}} \qquad (24)$$

Rearrangement of Equations 23 and 24 yields

$$\frac{e^{\frac{F\psi_0}{RT}}}{[H^+]} = \frac{[X_t^+]}{K_{int}^1 [X^+]} - \frac{1}{K_{int}^1} \qquad (25)$$

$$[H^+] \, e^{-\frac{F\psi_0}{RT}} = -K_{int}^2 + \frac{K_{int}^2 [X_t^-]}{[X_-]} \qquad (26)$$

Then, a plot of $1/[X_+]$ versus $\exp F\psi_0/RT/[H^+]$ for Equation 25 and of $1/[X_-]$ versus $[H^+] \exp - F\psi_0/RT$ for Equation 26 gives K_{int}^1 and K_{int}^2; $[X_t^+]$ and $[X_t^-]$ are obtained from the abscissa intercepts of the plots.

Figure 6 shows the plots of Equations 25 and 26 for γ-Al_2O_3 based on the data presented in Figure 4.

Microscopic Acidity Constant

A constant, K', like that defined by Equation 19, can also be obtained. For surfaces that are positively charged, K_1' can be calculated by the following equation:

SURFACE ACIDITY OF HYDROUS SOLIDS

Table V. Evaluation of Intrinsic Constants and Total Exchange Capacity

$AlOH^+ \rightleftharpoons AlOH + H^+; K^1_{int}$	$AlOH \rightleftharpoons AlO^- + H^+; K^2_{int}$
$K^1_{int} = \dfrac{\{AlOH\}\{H^+\}}{\{AlOH_2^+\}}$	$K^2_{int} = \dfrac{\{AlO^-\}\{H^+\}}{\{AlOH\}}$
$[X_t^+] = \{AlOH_2^+\} + \{AlOH\}$	$[X_t^-] = \{AlOH\} + \{AlO^-\}$
$[X_+] = \{AlOH_2^+\}$	$[X_-] = \{AlO^-\}$
$K^1_{int} = \dfrac{[X_t^+ - X_+]\{H^+\}}{[X_+^+]}$	$K^2_{int} = \dfrac{[X_-]\{H^+\}}{[X_t^- - X_-]}$
$\dfrac{K^1_{int}}{[X_+]\{H^+\}} = -\dfrac{1}{[X_t^+]} + \dfrac{1}{[X_+]}$	$K^2_{int}[X_t^-] - K^2_{int}[X_-] = [X_-]\{H^+\}$
$\{H^+\} = [H^+]_{bulk}\, e^{-(F\psi/RT)}$	$\{H^+\} = [H^+]_{bulk}\, e^{-(F\psi_0/RT)}$
$\dfrac{e^{+(F\psi_0/RT)}}{[H^+]_{bulk}} = \dfrac{[X_t^+]}{K^1_{int}}\dfrac{1}{[X_+]} - \dfrac{1}{K^1_{int}}$	$e^{-(F\psi_0/RT)}[H^+]_{bulk} = \dfrac{1}{[X_-]}K^2_{int}[X_t^-]$ $- K^2_{int}$
plot $\dfrac{e^{-(F\psi_0/RT)}}{[H^+]_{bulk}}$ vs $\dfrac{1}{[X_+]}$	plot $e^{-(F\psi_0/RT)}[H^+]_{bulk}$ vs $\dfrac{1}{[X_-]}$
at $\dfrac{1}{[X_+]} = 0$; $\dfrac{1}{K^1_{int}} = \dfrac{e^{(F\psi_0/RT)}}{[H^+]_{bulk}}$	at $\dfrac{1}{[X_-]} = 0$; $K^2_{int} = -e^{-(F\psi_0/RT)}[H^+]_{bulk}$
at $\dfrac{e^{-(F\psi_0/RT)}}{[H^+]_{bulk}} = 0$; $\dfrac{1}{[X_+]} = \dfrac{1}{[X_t^+]}$	at $e^{-(F\psi_0/RT)}[H^+]_{bulk} = 0$; $\dfrac{1}{[X_-]} = \dfrac{1}{[X_t^-]}$

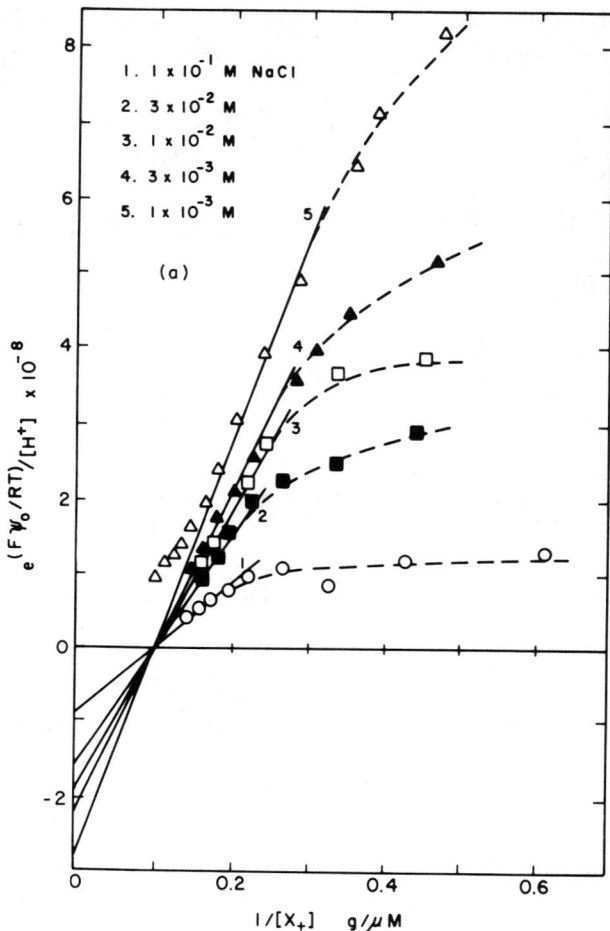

Figure 6. The determination of surface acidity by graphic procedure: (a) for positive surface; and (b) for negative surface.

$$K'_1 = \frac{(1-f_1)[H^+]}{f_1} \qquad (27)$$

while for a negative surface, K'_2 is given by

$$K'_2 = \frac{f_2[H^+]}{(1-f_2)} \qquad (28)$$

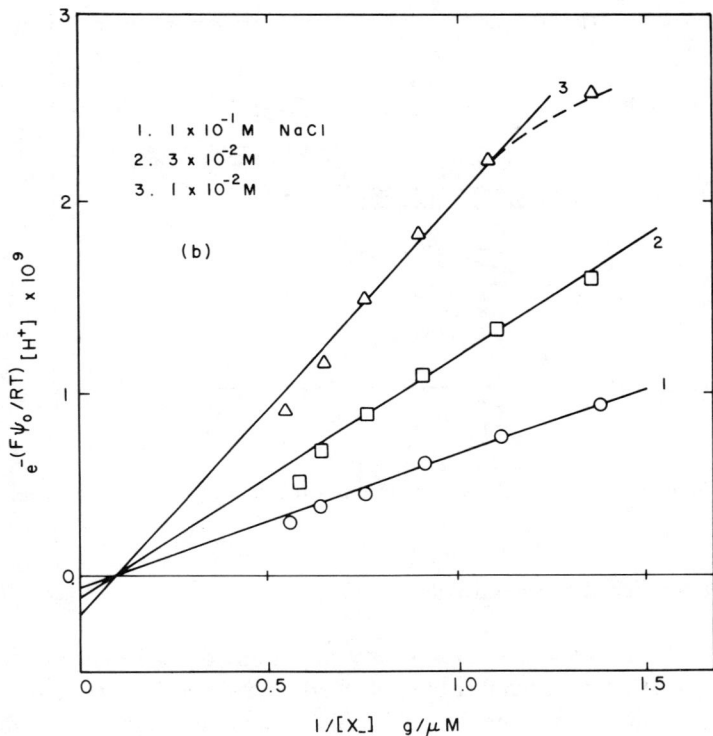

the f's being the fractions of acidity density in ionized form.

These equations imply that the hydrous oxides are like diprotic acids with two dissociable protons. Figure 7 shows the results of K' calculation for γ-Al$_2$O$_3$. It is further noted that in certain cases, such as γ-Al$_2$O$_3$, α-TiO$_2$ and α-FeOOH, which show two distinguishable (proton dissociation) regions in the titration curve, the pH$_{pzc}$ can be calculated analytically by pH where

$$[M-OH_2^+] = [M-O^-]$$

or

$$pH_{pzc} = \tfrac{1}{2}\,(pK_{int}^1 + pK_{int}^2) \qquad (29)$$

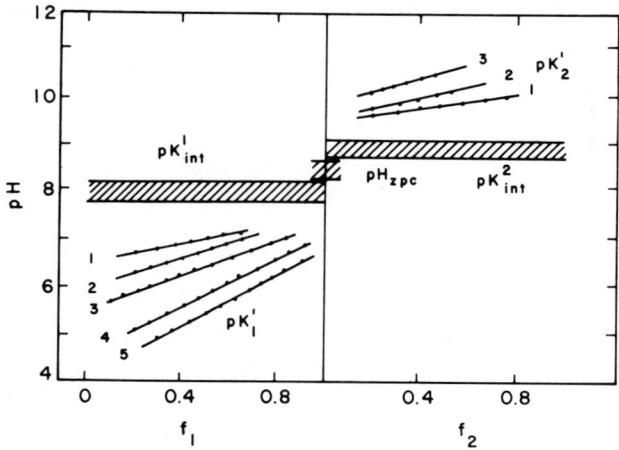

Figure 7. Plot of surface acidity constants as a function of the degree of titration (f_1 and f_2) [10].

Surface Acidity of Various Solids

Table VI shows the surface acidity of various solids based on titration curves published by a number of researchers. The surface acidity constant for $SiO_2(s)$, as determined by the modified Langmuir plot method, is close to what was determined by Schindler and Kamber [9]. The total number of surface acid sites of $SiO_2(s)$, however, is only available from the modified Langmuir plot method.

The surface acidity constants of γ-Al_2O_3, calculated by Huang and Stumm [10] and Hohl and Stumm [11], are identical. However, the total number of acid sites as obtained by Hohl and Stumm [11], who first added large amounts of strong base to the γ-Al_2O_3 suspension then back-titrated the excess base with strong acid, is slightly larger than that determined by Huang and Stumm [10]. Peri [32] reported a total surface acid sites of 80×10^{13} sites/cm^2 for gas-dried samples. The acidity constants calculated by Davis et al. [19], based on Huang's [33] titration data, are 5.70 and 11.50, respectively, for pK_{int}^1 and pK_{int}^2. These values are at variance with what was calculated by Huang and Stumm [10] and Hohl and Stumm [11]. Jacob [18] calculated the second surface acidity constant for α-Al_2O_3 and reported a value of 9.70. This value is in good agreement with what was calculated by the modified Langmuir plot method based on the data of Yorpps and Fuerstenau [31].

No comparison of surface acidity calculated by the modified Langmuir plot method can be made against other techniques on other hydrous solids listed in Table V.

Table VI. Surface Acidity of Various Solids

Solids	pK_{int}^1	pK_{int}^2	X_t^{+a}	X_t^{-a}	s^b	I	pH_{pzc}(expt)	Source
SiO_2	-2.77^c	6.77	—	5.82	264	0.1 M NaCl	−(2.0)	16
		5.80						9
	-0.30^c	6.30	—	0.64	40	0.1 M KCl	−(3.0)	30
	—	7.20	—	—	—	—	—	19
α-FeOOH	6.99	8.40	5.60	5.60	71	0.1 M KCl	7.70 (7.30)	4
	4.20	10.80	—	—	—	—	—	19
α-Fe$_2$O$_3$	8.85	9.75^c	—	10.98	45	0.1 M KCl	−(9.30)	4
	4.20	8.95^c	—	7.70	18	0.1 M KCl	−(8.51)	8
Fe_3O_4	6.28	7.66	—	21.50	—	0.1 M KCl	6.97 (7.00)	15
γ-Al$_2$O$_3$	7.70	9.30	6.43	6.43	117	0.1 M NaCl	8.50 (8.50)	10
	7.20	9.50	—	12.86	117	0.1 M NaClO$_4$	8.35 (3.50)	11
	8.37	8.70	—	—	—	—	—	16
	5.70	11.50	—	68.47	117	0.1 M NaCl	—	19
α-Al$_2$O$_3$	8.50	9.70	—	27.10	15	0.1 M NaCl	−(9.10)	31
	—	9.70	—	—	—	—	—	18
ThO_2	5.68	7.49	18.91	18.91	0.019	1 M KNO$_3$	6.59 (7.00)	14
ZrO_2	5.67	7.91	24.03	24.03	—	1 M KNO$_3$	6.79 (6.20)	14
TiO_2	5.72	—	5.78	—	125	—	—	12
	5.40	6.40	12.48	15.60	43	0.1 M KCl	5.90 (6.00)	12
Fluorapatite	6.35^c	7.05	—	8.00	35.3	0.1 M KCl	−(6.70)	17
Hydroxylapatite	8.8^c	8.40	—	18.36	26.6	0.1 M KCl	−(8.60)	17

^{a} In 10^{13} sites/cm^2; ^{b} In m^2/g; ^{c} Calculated from pK_{int}^1 and pH_{pzc}. Note that under most circumstances, X_t^+ is identical to X_t^-.

The surface acidity of several oxides, based on intrinsic constants, has the following series:

$$SiO_2 > ThO_2 \approx TiO_2 \approx ZrO_2 > Fe_3O_4 \approx \text{fluorapatite} > \alpha\text{-FeOOH} >$$

$$\gamma\text{-}Al_2O_3 > \alpha\text{-}Al_2O_3 \approx \alpha\text{-}Fe_2O_3 \approx \text{hydroxylapatite}$$

The total number of acid sites follows a different order, but generally are on the order of 10^3 sites/cm^2.

It is further noted that the method developed by Davis et al. [19] gives a larger K_{int}^1 (or smaller pK_{int}^1) and smaller K_{int}^2 (or larger pK_{int}^2) than those obtained by the modified Langmuir plot method, or methods used by other researchers. This can be attributed to the consideration of inert electrolytes in the charge balance calculation, an important factor for consideration when specific adsorption of inert electrolyte ions is present. However, specific adsorption of simple electrolyte ions usually does not take place at dilute concentrations. This can be further verified from the titration curve, as demonstrated in Figure 4.

Difficulties

In principle, the modified Langmuir plot method is simple, although several operational problems have been experienced in the determination of surface acidity by this method. Most of the following difficulties are also related to the titration technique *per se.*

Precise Information on Specific Surface Area

To calculate the surface charge, precise measurement of specific surface area is needed. Generally, the BET method should give a reasonable specific surface area measurement. While conducting BET experiments, it is advantageous to complete the adsorption and the desorption isotherm over rather broad relative pressure (P/P_0) ranges to allow analyses of surface porosity by the method of de Boer et al. [34] or the t-method [35].

Should the surface show evidence of high porosity, then van den Hull's method of negative adsorption might be used to determine the specific area of a porous solid in solution [36]. Huang and Stumm [37] have determined the specific surface area of $\gamma\text{-}Al_2O_3$ by BET measurement, negative adsorption, and electrochemical methods.

SURFACE ACIDITY OF HYDROUS SOLIDS

Dynamics of Titration

Many researchers have experienced erratic pH responses in titrating colloidal suspension. Bérubé [38] reported that a slow pH drifting was observed for five months in a rutile-sodium chloride aqueous system. Similar observations of slow pH drifting was reported by Blok [39] on zinc oxide, Onada and de Bruyn [40] on hematite and Korpi [41] on alumina. Slow pH response can be attributed to contamination of the system, such as impurities introduced to the sample during preparation or titration, CO_2 expulsion, leaking from reference electrode or malfunction of the glass indicator electrode.

According to our own experience, slow pH drifting can be partially prevented by thorough purification followed by hydrating the solids for at least two weeks under conductivity water. In cases where slow pH drifting prevails, a "fast" titration can still be accomplished by taking pH readings at a reasonable time period, i.e., 2-5 minutes after the addition of titrant. The use of a hydrogen indicator electrode and silver-silver chloride reference electrode can also improve the situation (less drifting). It would also be useful to continuously record the pH response with a pH-recorder. The "fast" titration is then done by reading the appropriate equilibrium pH values from the recording chart.

Determination of Surface Charge

Surface charge is calculated from the adsorption isotherm $\Gamma_{H^+} - \Gamma_{OH^-}$, with reference to pH_{pzc}. In principle, this is obtained by the mass balance equation (Equation 10).

Operationally, $\Gamma_{H^+} - \Gamma_{OH^-}$, is obtained by subtracting the blank titration curve from the sample titration curve. The only setback in this operation is that $\Gamma_{H^+} - \Gamma_{OH^-}$ may be overestimated when the solid becomes substantially soluble. To overcome this difficulty, it is suggested that one use a modified form of Equation 10:

$$\Delta\Gamma_{\pm} = \frac{\Delta(C_A - C_B) - [V_2(C_{H^+} - C_{OH^-})_2 - V_1(C_{H^+} - C_{OH^-})_1] - (V_2 \Sigma M_{t2} - V_1 \Sigma M_{t1})}{S}$$

where ΣM_t represents total soluble metal species from the solids. For an oxide, $\Sigma M_t = M^{+z} + MOH^{z-1} + MOH_2^{z-2} + \ldots - MOH_{2+1}^{-1} - MOH_{2+2}^{-2} - \ldots$ Under such circumstances, the amount of solid solubilized must be determined and incorporated into the isotherm calculation. Operationally, this is done by making background titration at various critical pH values. Ideally, the background titration curves, beginning at various conditioning pH values, should converge at certain pH values when solubility of the solid is insignificant.

Linearity of Plot

Another difficulty frequently encountered in the determination of acidity constants by the method proposed by Huang and Stumm [10] is the observation of a straight line by plotting $1/\{H^+\}$ vs $1/X_+$ or $\{H^+\}$ vs $1/X_-$. Linearity is better obtained with widespreading titration curves; i.e., data points must be at least one to two pH units away from pH_{pzc}. However, since one can always have one successful linear plot at either $pH < pH_{pzc}$ or $pH > pH_{pzc}$, and theoretically X_t^+ should be identical to X_t^-, and $pH_{pzc} = (pK_{int}^1 + pK_{int}^2)/2$, it is possible to calculate the pK_{int}s for those curves where data points do not fall within the linear portion of the plot; i.e., Equations 25 and 26. It is noted that data within one pH unit on both sides of pH_{pzc} always yield a flat horizontal line, thereby rendering the method impractical. Data that are far away from the pH_{pzc} or that are beyond X_t^+ or X_t^- tend to give less steep a tail. Once again this region of data points must be ignored. A successful linear plot is warranted at higher ionic strength, i.e., $10^{-1}\ M$ ~ $1\ M$, when a widespreading titration curve is readily attainable.*

ALKALIMETRIC TITRATION OF HYDROUS SOLIDS IN THE PRESENCE OF SPECIFICALLY ADSORBED METAL IONS

In the above treatment it was assumed, and later proved, that the species of NaCl, namely Na^+ and Cl^-, do not become specifically adsorbed onto the γ-Al_2O_3 surface at concentrations less than $10^{-1}\ M$. Specific adsorption of metal ions, such as the alkaline-earth and the heavy metals, tends to shift the titration curve toward the alkaline pH region, a clear indication of releasing H^+ as the result of specific adsorption of multivalent metal ions. The purpose of this section is to illustrate the determination of the stability constants of surface complexes within the context of surface acidity, as determined by the modified Langmuir plot method.

Coordination with Alkaline-Earth Metal Ions

From laboratory observation, the adsorption of cations; e.g., Mg^{+2}, Ca^{+2}, Sr^{+2} and Ba^{+2}, onto hydrous solids such as γ-Al_2O_3 can be interpreted as surface complex formation or as ion exchange:

---

*Davis et al. [19] have shown the adsorption of "inert" electrolytes is significant at $10^{-1}\ M$ and cannot be omitted from the charge balance equation.

SURFACE ACIDITY OF HYDROUS SOLIDS

$$J(AlOH) + M^{+2} \rightleftharpoons M(AlO^-)_j^{2-j} + jH^+; \quad *\beta_j^S(\text{int})$$

where $*\beta_j^S(\text{int})$ is the intrinsic stability constant of the jth order surface complex and may be expressed as follows:

$$*\beta_j^S(\text{int}) = \frac{\{M(AlO)_j^{2-j}\}\{H^+\}^j}{\{AlOH\}\{M^{+2}\}} \tag{30a}$$

$$= \frac{\{M(AlO)_j^{2-j}\}[H^+]^{-jF\psi_0/RT}}{AlOH \; [M^{+2}]^{-2F\psi_0/RT}} \tag{30b}$$

$$= *\beta_j' \, e^{(2-j)F\psi_0/RT} \tag{30c}$$

where $*\beta_j'$ is the apparent stability constant and can be calculated from given bulk concentrations of $[H^+]$ and $[M^{+2}]$ and surface concentrations of $\{M(Al)_j^{2-j}\}$ and $\{AlOH\}$.*

By applying a proton balance equation at a given pH, one has, in the absence of complexing metal ion,

$$C_b' + [H^+] + \{AlOH_2^+\} = [OH^-] + \{AlO^-\} \tag{31}$$

and on introduction of metal ions to the suspension,

$$C_b' + [H^+] + \{AlOH_2^+\} = [OH^-] + \{AlO^-\} + \sum_j \{M(AlO)_j^{2-j}\} \tag{32}$$

By combining Equations 31 and 32 one has

$$\{M(AlO)_j^{2-j}\} = (C_b' - C_b)/(j) \tag{33}$$

Equation 33 depicts exactly the extent of metal ion coordination that results in the shift in titration curves (Figure 8). At constant pH, $\Delta C_b = C_b' - C_b$, is proportional to the amount of metal ion association with surface AlO^- groups. The quantities of $[M^{+2}]$ and $\{AlOH\}$ are calculated from mass balance equations as follows:

---

*A more complete discussion of the distinction between the concentration and activity of ions and surface complexes is given in Chapter 1.

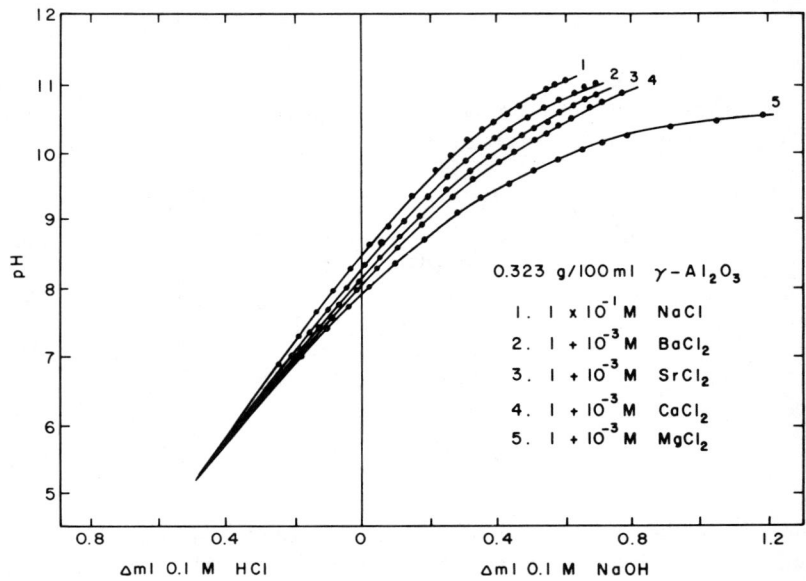

Figure 8a. Net titration curves for $\gamma\text{-}Al_2O_3$ in the presence of alkaline-earth metal ions.

$$[M^{+2}] = M^{+2}_{Total} - \Delta C_b/(j)$$

$$\{AlOH\} = [X_t^- - \Delta C_b](1 + f_1 - f_2) \qquad (34)$$

or

$$\{AlOH\} \cong [X_t^- - \Delta C_b]\left(\frac{K^1_{int}\{H^+\}}{K^1_{int} K^2_{int} + K^1_{int} H^+ + {H^+}^2}\right)$$

By assuming the Nerstian relationship,

$$\psi_0 = \frac{2.303\ RT}{F}(pH_{pzc} - pH) \qquad (35)$$

and from Equation 30c,

$$*\beta_j^S(int) = *\beta_j' \exp(2-j)(pH_{pzc} - pH) \qquad (36)$$

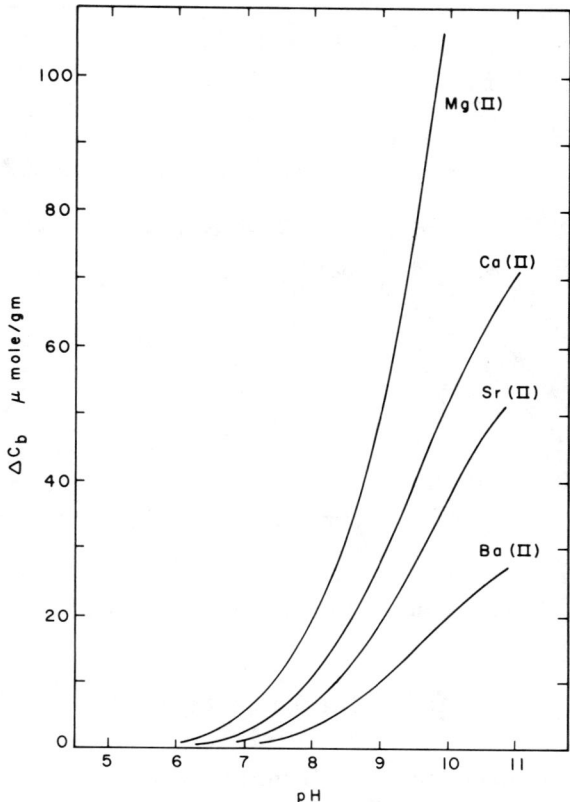

Figure 8b. Plot of $\Delta C_b'$ vs pH for $\gamma\text{-}Al_2O_3$ suspensions in the presence of alkaline-earth metal ions.

a plot of log $^*\beta_j$ vs pH should yield a straight line with slope equal to $(2-j)$.

Figure 8a shows the titration of $\gamma\text{-}Al_2O_3$ in the presence of multivalent metal ions, specifically alkaline-earth. The degree of surface complex formation, ΔC_b, as a function of pH is shown in Figure 8b. The stability constants calculated with Equation 36 from data presented above are shown in Figure 9. The results indicated that alkaline-earth metal ions form the following surface complexes with $\gamma\text{-}Al_2O_3$. At low pH (i.e., pH $<$ pH$_{pzc}$) and especially for Mg^{+2} and Ca^{+2} ions,

$$M^{+2} + AlOH = (AlO)M^+ + H^+$$

At high pH (i.e., pH $>$ pH$_{pzc}$) and for all metals,

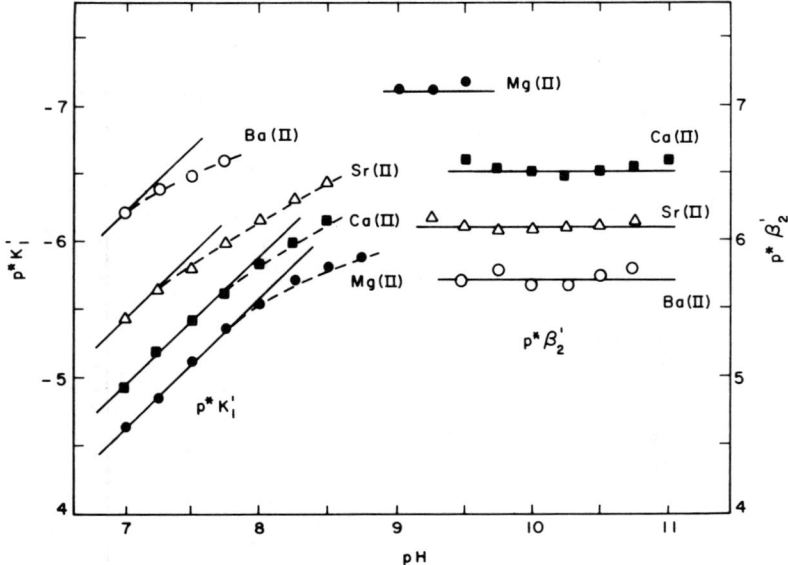

Figure 9. Plot of apparent surface stability constants vs pH for γ-Al_2O_3/alkaline-earth metal systems.

$$M^{+2} + 2(AlO^-) \rightleftharpoons (AlO)_2 M$$

The results demonstrated that Nernstian relation (Equation 22) is operative in these systems (Figure 9).

Coordination with Heavy Metal Ions

The results shown in Figure 10 again indicate that without direct adsorption measurement it is possible to quantitatively evaluate the association between transition metals, e.g., Cu(II), Zn(II), Co(II), and Ni(II) and hydrous solids, i.e., γ-Al_2O_3. The titration procedures are exactly the same as described previously, except that a glass electrode with a calomel reference electrode were used for pH measurement. In comparison with the alkaline-earth system, the titration curves for transition metal systems shift to the more alkaline side than the alkaline-earth system, indicating a stronger surface association at more acidic levels. The ability to form aqueous hydroxyl species of the heavy metal ions renders the calculation of stability constants more complicated than those of the alkaline-earth metal ions. For this reason, only the first (more acidic pH) portions of the titration curves were used to

SURFACE ACIDITY OF HYDROUS SOLIDS

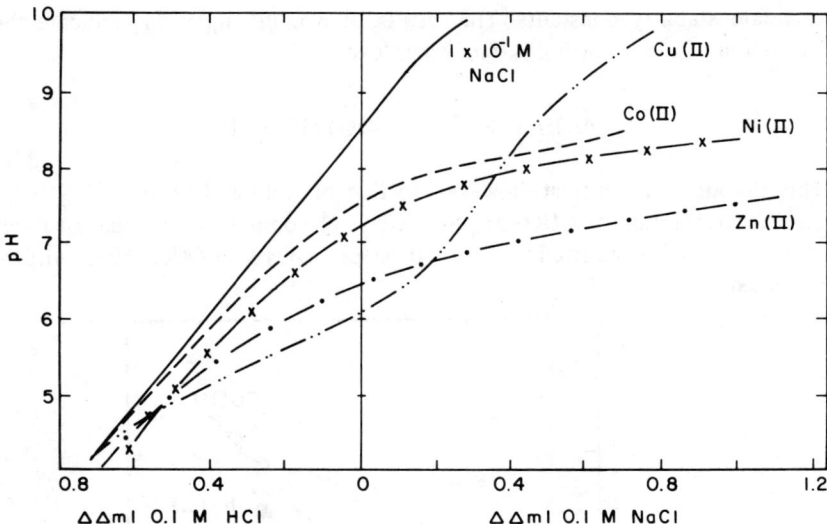

Figure 10a. Net titration curves for $\gamma\text{-Al}_2\text{O}_3$ suspensions in the presence of heavy metal ions.

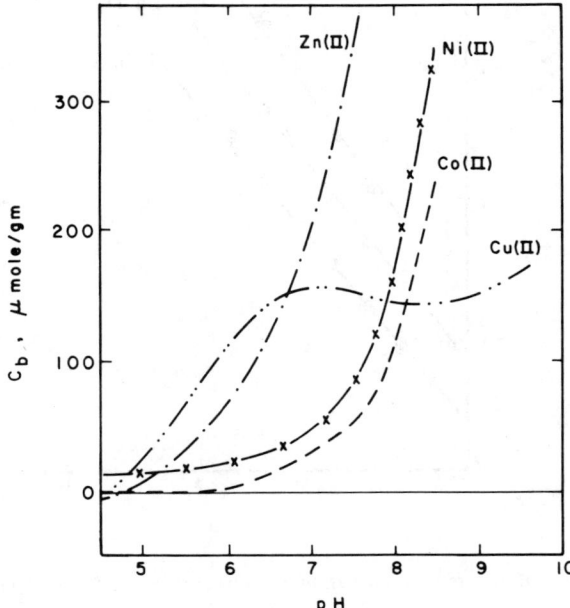

Figure 10b. Plot of $\Delta C_b'$ vs pH for $\gamma\text{-Al}_2\text{O}_3$ suspensions in the presence of heavy metal ions.

calculate stability constants. The results, shown in Figure 11, indicate the formation of the following surface complexes:

$$AlOH + M^{+2} \rightleftharpoons (AlO)M^+ + H^+$$

This finding is in contrast, however, to that of Hohl and Stumm [11], who reported formation of (AlO) M plus $(AlO)_2$ M complexes, and that of Davis et al. [19], who reported the formation of (AlO) M + (AlO) MOH surface complexes.*

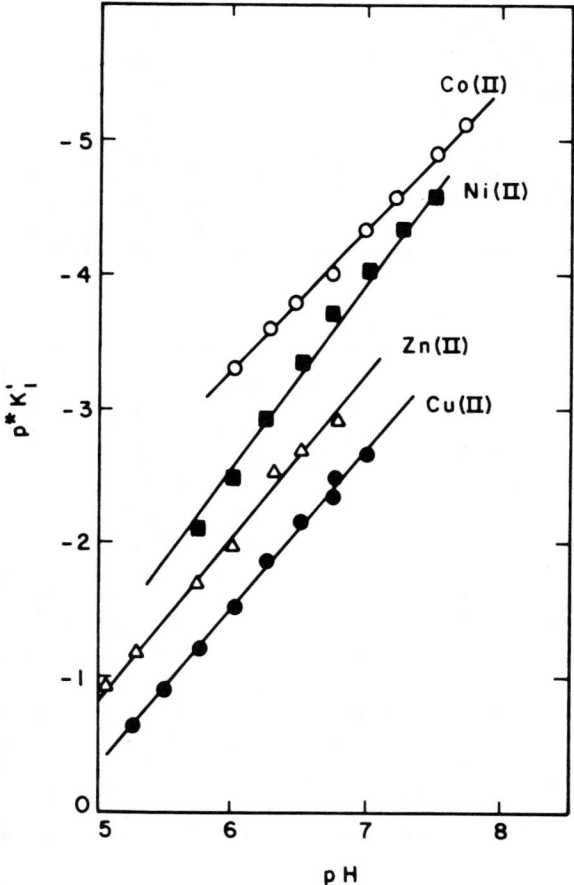

Figure 11. Plot of apparent surface stability constants vs pH for $\gamma\text{-}Al_2O_3$/heavy metal systems.

*In Chapter 7, Morel emphasizes the distinction between this type of reported complex and a more complex that has been verified chemically.

SURFACE ACIDITY OF HYDROUS SOLIDS

The applicability of the Nernstian Equation has been discussed by several authors [42,43]. However, neither Schindler et al. [44] nor Hohl and Stumm [10] observed any pH effect on the apparent surface stability constants. While the slope of the plot, log $*\beta$ vs pH, may not exactly fall to the predicted slope of 2-j, the linear relationship observed in Figures 9 and 11 can only be interpreted as pH effect on surface stability constants.

The results shown in Table VII give the intrinsic stability constants of various systems. It clearly demonstrates that the stability constants are of the same order of magnitude for equivalent systems.

Table VII. Comparison of Surface Stability Constants

Solids/Electrolyte	Metal Ions	$pK_1^S(\text{int})$	$p\beta_2^S(\text{int})$	References
$\gamma\text{-Al}_2\text{O}_3/0.1\ M\ \text{NaCl}$	Mg^{2+}	6.1	11.4	32
	Ca^{2+}	6.4	12.1	
	Sr^{2+}	6.9	12.5	
	Ba^{2+}	7.7	12.9	
	Cu^{2+}	4.6		
	Zn^{2+}	5.2		
	Co^{2+}	5.9		
	Ni^{2+}	6.1		
$\gamma\text{-Al}_2\text{O}_3/0.1\ M\ \text{NaClO}_4$	Pb^{2+}	2.2	8.1	11
$\text{TiO}_2/1\ M\ \text{NaClO}_4$	Cu^{2+}	1.5	5.0	45
	Cd^{2+}	3.2	10.5	
	Pb^{2+}	-0.2	2.0	
$\text{SiO}_2/1\ M\ \text{NaClO}_4$	Fe^{3+}	1.8	4.2	45
	Cu^{2+}	5.5	11.2	
	Cd^{2+}	6.1	14.2	
	Pb^{2+}	5.1	10.7	
	Mg^{2+}	8.1	16.7	
	Ca^{2+}	7.3	14.7	
$\text{Fe(OH)(am)}/0.1\ M\ \text{NaNO}_3$	Cu^{2+}	4.2		42
	CuOH^+	8.9		
	Zn^{2+}			
	ZnOH^+	8.4		
	Pb^{2+}	3.8		
	PbOH^+	7.4		
	Cd^{2+}	4.9		
	CdOH^+	8.7		
	Ag^+	5.0		
	AgOH	8.1		

CONCLUSIONS

The surface acidity of hydrous solids can be characterized in terms of its strength (or acidity constants) and density (or total number of acid sites per area). Surface hydration tends to develop Brönsted acid sites and facilitates the characterization of surface acidity by alkalimetric titration.

A modified Langmuir plot is developed to quantify the surface acidity of hydrous solids. According to this method, the solid acid is treated as a diprotic acid, which can dissociate two surface protons. The complete dissociation of surface protons involves two steps of action: breaking the covalently bound proton from the solid surface ("intrinsic" or chemical energy requirement), followed by transporting the untied proton from the surface to the bulk phase (coulombic energy requirement). Within the context of this method, the inert electrolytes must not become specifically adsorbed by the surface and are therefore not potential-determining ions. The surface specificity of inert electrolytes can be observed from titration curves. Specific adsorption of ions tends to shift the pH_{pzc} to a more alkaline value when anions are adsorbed, or to a more acidic value when cations are specifically associated.

There are circumstances in which the modified Langmuir plot might not yield a straight line for certain titration curves or certain portions of a titration curve. Data that are close to pH_{pzc} usually are found at the plateau region and should not be used to obtain the surface acidity information. Most useful data are those 2 to 3 pH units away from the pH_{pzc}.

In the presence of specifically adsorbed metal ions, the titration technique can still be used to obtain information on the mode and extent of metal adsorption. For all cases illustrated, it seems clear that the Nernstian relationship is applicable. Therefore, the apparent surface stability constant is a function of pH when surface complex formation results in the release of protons. The results indicate that for γ-Al_2O_3 and alkaline-earth systems, the formation of $(\equiv AlO)_2$ M surface complexes are present over pH > pH_{pzc}, while at pH < pH_{pzc}, only Mg^{+2} and Ca^{+2} ions form $(\equiv AlO)$ M^+ complexes. For the heavy metal ions, Cu(II), Zn(II), Co(II) and Ni(II), at pH < pH_{pzc}, only $(\equiv AlO)$ M^+ surface complexes were observed. Although this finding contrasts with what was reported by Davis [45], who found the formation of $(\equiv FeO)$ M^+ and $(\equiv FeO)$ MOH surface complexes at least for amorphous $Fe(OH)_3(s)$, the surface stability constants for all equivalent systems are of the same order of magnitude.

ACKNOWLEDGMENT

This research work was supported in part by grant No. ENG 75-07176 from the National Science Foundation. The author wishes to thank Prof.

Werner Stumm for many of the ideas suggested in this work. Major portion of the data were taken from Huang [33].

REFERENCES

1. Tanabe, K. *Solid Acids and Bases* (New York: Academic Press, 1970).
2. Forni, L. "Comparison of the Methods for the Determination of Surface Acidity of Solid Catalysts," *Catal. Rev.* 8: 65-115 (1973).
3. Hindin, S. G., and S. W. Weller. "The Effect of Pretreatment on the Activity of γ-Alumina I. Ethylene Hydrogenation," *J. Phys. Chem.* 60: 1501 (1956).
4. Atkinson, R. J., A. M. Posner, and J. P. Quirk. "Adsorption of Potential-Determining Ions at the Ferric Oxide Aqueous Electrolyte Interface," *J. Phys. Chem.* 71: 550 (1967).
5. Parks, G. A., and P. L. de Bruyn. "The Zero Point of Charge of Oxides," *J. Phys. Chem.* 66: 967 (1962).
6. Bolt, G. H. "Determination of the Charge Density of Silica Sols," *J. Phys. Chem.* 61: 166 (1957).
7. Abendroth, R. J. "Behavior of a Pyrogenic Silica in Simple Electrolytes," *J. Colloid Interface Sci.* 34: 591 (1970).
8. Breeuwsma, A., and J. Lyklema. "Interfacial Electrochemistry of Hematite (α-Fe_2O_3)," *Disc. Faraday Soc.* 52: 324 (1971).
9. Schindler, P. W., and H. R. Kamber. "The Acidity of Silano Groups," *Helv. Chim. Acta* 51: 1781 (1968).
10. Huang, C. P., and W. Stumm. "Specific Adsorption of Cations on Hydrous γ-Al_2O_3," *J. Colloid Interface Sci.* 43: 409 (1973).
11. Hohl, H., and W. Stumm. "Interaction of Pb^{2+} with Hydrous γ-Al_2O_3," *J. Colloid Interface Sci.* 55: 281 (1976).
12. Bérubé, Y. G., and P. L. de Bruyn. "Adsorption at Rutile-Solution Interface II Model of the Electrochemical Double Layer," *J. Colloid Interface Sci.* 27(92): 305 (1968).
13. Yates, D. E. "On the Structure of Oxide/Water Interface," Ph.D. Thesis, University of Melbourne (1975).
14. Ahmead, S. M. "Studies of the Double Layer at Oxide-Solution Interface," *Can. J. Chem.* 44: 1663 (1966).
15. Ahmead, S. M., and D. Maksimov. "Studies of the Oxide Surfaces at the Liquid-Solid Interface, Part II. Fe Oxide," *Can. J. Chem.* 46: 7841 (1968).
16. Huang, C. P. "The Determination of Surface Acidity by Alkalimetric Titration," Annual NSF Report (1977).
17. Bell, L. C., A. M. Posner, and J. P. Quirk. "The Point of Zero Charge of Hydroxyapatite and Fluorapatite in Aqueous Solution," *J. Colloid Interface Sci.* 42: 250 (1973).
18. Jacob, D. G. "An Interpretation of Cation Exchange by Alumina," *Health Phys.* 12: 1565 (1966).
19. Davis, J. A., R. O. James, and J. O. Leckie. "Surface Ionization and Complexation at the Oxide/Water Interface," *J. Colloid Interface Sci.* 63: 480 (1978).
20. Eisenman, G. "The Origin of Glass-Electrode Potential," in *Glass Electrodes for Hydrogen and Other Cations,* G. Eisenman, Ed. (New York: Marcel Dekker, Inc., 1967), p. 133.

21. Bates, R. G. *Determination of pH: Theory and Practice* (New York: Wiley & Sons, Inc., 1965), p. 25.
22. Parks, G. A. "The Aqueous Surface Chemistry of Oxides and Complex Oxide Minerals," *Advances in Chemistry Series No. 67* (Washington, DC: American Chemical Society, 1967), p. 121.
23. Parks, G. A. "The Isoelectric Points of Solid Oxides, Solid Hydroxides, and Aqueous Hydroxo Complex Systems," *Chem. Rev.* 65: 177 (1965).
24. King, E. J. "Polyprotic Acids," in *Acid-Base Equilibria* (New York: Macmillan Publishing Co., Inc., 1965), p. 218.
25. Tanford, C. "Multiple Equilibria," in *Physical Chemistry of Macromolecules* (New York: John Wiley & Sons, Inc., 1961), p. 526.
26. Stumm, W., C. P. Huang, and R. S. Jenkins. "Specific Chemical Interaction Affecting the Stability of Dispersed Systems," *Croat. Chim. Acta* 42: 223 (1970).
27. Murray, D. J., T. W. Healy, and D. W. Fuerstenau. "The Adsorption of Metal on Colloidal Hydrous Manganese Oxide," in *Adsorption from Aqueous Solution*, Advances in Chemistry Series No. 79 (Washington, DC: American Chemical Society, 1968), p. 74.
28. Bérubé, Y. G., G. Y. Onoda and P. L. de Bruyn. "Proton Adsorption at the Ferric Oxide-Aqueous Solution Interface," *Surface Sci.* 8: 448 (1967).
29. Bérubé, Y. G., and P. L. de Bruyn. "Adsorption at Rutile-Solution Interface: I. Thermodynamics," *J. Colloid Interface Sci.* 28: 92 (1968).
30. Lyklema, J., and T. F. Tadros. "Adsorption of Potential Determining Ions at the Silica-Aqueous Electrolyte Interface and the Role of Some Cations,"
31. Yorpps, J. A., and D. W. Fuerstenau. "The Zero Point of Charge of α-Al_2O_3," *J. Colloid Sci.* 19: 61 (1964).
32. Peri, J. B. "Infrared and Gravimetric Study of the Surface Hydration of γ-Alumina," *J. Phys. Chem.* 69: 211 (1965).
33. Huang, C. P. "The Chemistry of the Aluminum Oxide-Electrolyte Interface," Ph.D. Thesis, Harvard University, Cambridge, MA (1971).
34. de Boer, J. H., B. D. Lippens, B. G. Linsen, J. C. P. Brookhoff, A. van den Heuvel, and T. J. Osinga. "The t-Curve of Multimolecular N_2 Adsorption," *J. Colloid Interface Sci.* 21: 405 (1966).
35. Shüll, C. G. "The Determination of Pore Size Distribution from Gas Adsorption Data," *J. Am. Chem. Soc.* 70: 1405 (1948).
36. van den Hull, H. J., and J. Lyklema. "Determination of Specific Surface Areas of Dispersed Materials by Negative Adsorption," *J. Colloid Interface Sci.* 23: 500 (1972).
37. Huang, C. P., and W. Stumm. "The Specific Surface Area of γ-Al_2O_3," *Surface Sci.* 32: 287 (1972).
38. Bérubé, Y. G. "Adsorption of Inorganic Ions at the Titanium Oxide-Solution Interface," Ph.D. Thesis, Massachusetts Institute of Technology, Cambridge, MA (1967).
39. Blok, L., and P. L. de Bruyn. "The Ionic Double Layer at the ZnO/Solution Interface," *J. Colloid Interface Sci.* 32: 527 (1970).
40. Onoda, G. Y., Jr., and P. L. de Bruyn. "Proton Adsorption at the Ferric Oxide/Aqueous Solution Interface: I. A Kinetic Study of Adsorption," *Surface Sci.* 4: 48 (1966).
41. Korpi, G. S. "The Interfacial Chemistry of Aluminum Oxides," Ph.D. Thesis, Stanford University, Stanford, CA (1965).

42. Wright, H. J. L., and R. J. Hunter. "Adsorption at Solid/Liquid Interfaces: I. Thermodynamics and the Adsorption Potential," *Aust. J. Chem.* 26: 1183 (1973).
43. Levine, S., and A. L. Smith. "Theory of the Differential Capacity of the Oxide/Aqueous Electrolyte Interface," *Disc. Faraday Soc.* 52: 290 (1971).
44. Schindler, P. W., B. Furst, R. Dick, and P. U. Wolf. "Ligand Properties of Surface Silanol Groups: I. Surface Complex Formation with Fe^{+2}, Cu^{+2}, Cd^{+2} and Pb^{+2}," *J. Colloid Interface Sci.* 55(2): 469 (1976).
45. Davis, J. A. "Adsorption of Trace Metals and Complexing Ligands at the Oxide/Water Interface," Ph.D. Thesis, Stanford University, Stanford, CA (1977).

CHAPTER 6

SURFACE IONIZATION AND COMPLEXATION AT THE COLLOID/AQUEOUS ELECTROLYTE INTERFACE

R. O. James
    Department of Applied Earth Sciences
    Stanford University
    Stanford, California

INTRODUCTION

This chapter was originally intended to cover some of our earlier work on the adsorption of hydrolyzable metal ions and complex forms at the oxide-water interface. However, because of some recent advances in describing the properties of aqueous ionizable colloids, such as surface charge, adsorption density of solutes, electrokinetic potential and conductivity of colloid suspensions, these newer models will be the focus of this review. It is hoped that this will stimulate further experimental investigations and eventually lead to more complete understanding of the behavior of colloids in solutions containing complex ionic and molecular species.

The recent advances in electrical double layer (edl) models for simpler electrolyte systems have essentially resulted from the formulation of the adsorption and charge development reactions in terms of the reacting species, i.e., surface sites and solution species, and estimates of the "chemical" and electrostatic interactions that are subject to the charge-balance and charge-potential relationships in edl models. Improved methods for obtaining intrinsic ionization and binding constants have also been an important factor. It will be shown that even for simple systems there is a scarcity of pertinent data to conduct varied tests on the newer models.

Before proceeding, it is appropriate to point out where the newer models may have application for complex electrolyte adsorption (e.g., hydrolyzable metal ion uptake) and what challenges remain for these systems.

Concerning the adsorption of hydrolyzable metal ions, there are several models that can describe the adsorption of various ionic and molecular forms onto surfaces in terms of the H^+ released from the surface or the metal ion or the pH of solution [1-6]. The range of pH in which large increases in adsorption or percentage removed from solution occur depends largely on the metal ion, while the presence of complexing soluble ligands, variable amounts of surface sites, ionic strength and type of solid may also have some effect on uptake. Adsorption from dilute hydrolyzable metal solutions usually occurs at pH values below the hydrolysis pH region and may be considered as either ion exchange or surface complexation, in which metals replace surface protons or as the preferential adsorption of hydrolyzed species [6]. However, none of the models described so far can simultaneously give an accurate description of other measurable properties, e.g., zeta potential, adsorption density and surface charge [7]. It is particularly important to understand the effect of adsorbed solutes on the zeta potential of aqueous colloid dispersions because many processes involving dispersion or coagulation require control or regulation of this double layer potential.

During adsorption experiments at high coverages, the zeta potential may change sign, reversing the tendency for adsorption of counter- and coions to be adsorbed or repelled in the diffuse double layer. Experiments have shown that this is often due to the formation of a coating on the substrate by the adsorbing metal hydroxide [7-12]. One of the challenging problems that remains to be answered is whether the coating forms by surface condensation of simple ionic species, surface condensation of hydrolyzed species, including polymers, or whether it is due to adhesion of colloidal hydroxides formed in the solution (i.e., heterocoagulation)[12-16]. Again, more experimental work needs to be devised and performed before models will simultaneously predict adsorption densities, surface charge and zeta potential.

Due to the complexities of the behavior in such systems, it is necessary to be able to describe the effect of relatively simple electrolytes on colloid properties before returning to these systems. Thus, the remainder of this chapter will be devoted to the surface charge development of oxide or polymer colloids, their electrokinetic potential and specific conductance in simple strong electrolytes.

MODELS FOR OXIDE-WATER INTERFACES BY ANALOGY

The EDL at Hg-H_2O and AgI-H_2O Interfaces

Until recently, many of the models that were used to account for the behavior of mineral-water interfaces were directly analogous to the models developed from studies of the edl at the mercury-water interface and also the reversible AgI-water interface. In both of these systems, the potential difference between the phases could be controlled or measured by an external circuit. The surface charge density could be measured from the surface tension of the Hg-water interface or by the difference in adsorption density of potential-determining ions (pdi) at the AgI-water interface. The potential-determining ions were those ions that were present in both the colloid and aqueous electrode and participated in rapidly established equilibrium reactions. At equilibrium, the potential difference between the phases fixed the activity of the pdi in bulk solution, and vice versa. This subject has been reviewed by Grahame [17], de Bruyn and Agar [18-21], Lyklema and Overbeek [22], Wiese et al. [23] and Healy and White [24].

The models that have emerged represent the charge distribution at the interface according to the electroneutrality condition

$$\sigma_0 + \sigma_\beta + \sigma_d = 0 \tag{1}$$

where σ_0 = the charge density at the colloid surface,
σ_β = the Stern layer charge of "fixed" ions at their distance of closest approach to the surface, where there may be some specific chemical interaction with the surface, and
σ_d = the net charge of counterions and coions in the diffuse charge layer, where thermal motion diffuses the ordering of ionic distribution in the outer regions of the electric field.

Depending on the electrolyte composition and concentration, the Stern layer charge may or may not represent a significant contribution to the above equation when applied to Hg-H_2O and AgI-H_2O surfaces. It will be shown later that for oxides, σ_β is often a large contribution to the surface charge balance.

Due to the absence of charge between the surface, σ_0, and the Stern layer, σ_β, the potential decay in this molecular condenser may be written as follows:

$$\psi_0 = \psi_\beta + \frac{\sigma_0}{K_1} \tag{2}$$

where ψ_0 = the mean surface potential,
ψ_β = the mean Stern layer potential, and
K_1 = the inner layer capacitance.

The potential decay in the Stern layer and the outer layer or start of diffuse layer is similarly given by the following equation:

$$\psi_\beta = \psi_d - \frac{\sigma_d}{K_2} \qquad (3)$$

where K_2 = the capacitance of the outer layer molecular condenser.

These relationships, together with the equations of the diffuse layer theory, which relate ψ_d and σ_d, constitute the model proposed by Grahame [17] for ionic adsorptions at the mercury-water interface. Similar models have been found to apply reasonably well for the AgI-water interface in simple electrolytes and moderately low ionic strength by Lyklema and Overbeek [22].

Extension of Reversible Electrode EDL Models

Many of the oxides and other colloids, e.g., polymer latexes, are insulators and cannot be included in the class of conducting, reversible electrode behavior of the classical AgI colloids. Therefore, by way of convenience, experimental similarities and extended analogy to the potential-determining role of Ag^+ and I^- for AgI colloids, H^+ and OH^- have been called pdi for oxides and protolyzable colloids.

There is considerable experimental support for this definition. The sign and magnitude of the zeta potential of oxides depends on a_{H^+} or pH. Simple electrolytes usually only affect the magnitude of the ζ potential, so that the ζ-pH curves at different ionic strengths have a mutual intersection at the isoelectric point, pH_{iep}, where $\zeta = 0$. If the surface charge of oxides and other such colloids is determined from the net uptake of H^+ and OH^- at constant swamping electrolyte concentration, then for simple electrolytes, the titration curves for various ionic strengths also mutually intersect at the point of zero charge ($\Gamma_{H^+} - \Gamma_{OH} = 0$), pH_{pzc}. This pH_{pzc} usually closely matches the isoelectric point, pH_{iep}, from electrokinetic experiments.

So it has been established that H^+ and OH^- help determine the sign and magnitude of the surface charge and zeta potential. The question that remains is "what is the surface potential and how does the activity of pdi affect it?"

Thermodynamic arguments have been used [18,19] to show that H^+ and OH^- may be considered as potential-determining ions if these ions react with the metal and oxonium ions of the solid to form dissolved metal and metal hydroxo complexes, which are at equilibrium with the solid.

These reactions have been used to lead to the Nernstian relationship between the inner potential difference, $\Delta\phi^{\alpha\beta}$, between phases α (e.g., oxide) and β (e.g., aqueous electrolyte) by balancing the electrochemical potential of the components in both phases [19]. If the activity ratio of pdi species in the solid phase remains constant, then

$$\Delta\phi^{\alpha\beta} = \frac{RT}{nF} \ln \frac{a_i^\beta}{a_i^\alpha} + \frac{\Delta\mu_i^{\beta\alpha}}{z_i F} \tag{4}$$

Since only the outer or Volta potential difference is measurable in AgI-H_2O and Hg-H_2O systems, the inner potential is assumed to be represented by

$$\phi^\alpha = \psi^\alpha + \chi^\alpha \tag{5}$$

where ψ^α = the outer potential of phase α,
 χ^α = the surface jump potential,
 ψ^α = the contribution due to free charges, and
 χ^α = the contribution due to dipoles at the surface.

By suitable choice of reference conditions, e.g., $\Delta\psi^{\alpha\beta}_{stan} = 0$ at $a_i^\beta = a_i^{o\beta}$, $a_i^\alpha = a_i^{o\alpha}$, and assuming $\chi^{\alpha\beta}_{stan} = \chi^{\alpha\beta}$ in

$$\Delta\psi^{\alpha\beta} = \chi_s^{\alpha\beta} - \chi^{\alpha\beta} + \frac{RT}{z_i F} \ln \frac{a_i^{o\alpha} \cdot a_i^\beta}{a_i^{o\beta} \cdot a_i^\alpha} \tag{6}$$

one may obtain the simple form

$$\psi_0 = \frac{RT}{z_i F} \ln \frac{a_i^\beta}{a_i^{o\beta}} \tag{7}$$

by making two more assumptions, i.e., $\psi_0 = \Delta\psi^{\alpha\beta}$, and, if α is the solid phase, $a_i^\alpha = a_i^{o\alpha}$.

It has often been assumed that the surface potential in the edl calculations can be estimated (at 25°C) from

$$\psi_0 = 2.203 \frac{RT}{\gamma} \log_{10} \frac{a_{H^+}}{a_{H^+pzc}} = 0.0592 \, (\text{pH}_{pzc} - \text{pH}) \qquad (8)$$

However, several authors, including Hunter and Wright [25,26] and Bérubé and de Bruyn [19-21] began to show that Nernstian potentials are incompatible with the Stern-Grahame models applied to experimental observations of properties of oxide systems.

When attempts to test these models (the surface potential model and the edl model) fail, then one or both are inadequate. The following discussion summarizes recent studies in which the surface ionization of the solid together with the diffuse double layer or the Stern-Grahame model have been used to account for the surface charge (σ_0)-surface potential (ψ_0) relationships, zeta potential, and conductometric and potentiometric titration properties of oxide or polymer colloids.

SURFACE POTENTIAL-CHARGE RELATIONSHIP

Since the work of Hunter and Wright [25,26] and Levine and Smith [27] on the formulation of the surface potential-surface charge relationships for the Stern-Grahame edl model and also the work of Stumm et al. [2,3], a series of papers by Healy et al. [24,28] have led to very useful methods of describing the behavior of oxides and other colloids in aqueous electrolytes. These methods have been used to give qualitative and semiquantitative accounts of the ionization of surfaces when the surface charge is balanced only by a diffuse charge layer as a function of pH (or pdi) and ionic strength. They have used similar approaches to investigate the regulation of surface charge and potential as a function of interparticle separation during the "equilibrium" interaction of like and unlike colloidal particles [29,30]. Recently, Ruckenstein and Prieve [31-33] have used a similar approach to estimate the effect of solution conditions on the surface ionization of colloids to predict the rate of adhesion of colloidal particles to various solids as a theoretical background to the technique of "potential barrier chromatography."

The types of particles considered by Healy and co-workers [28] include:

1. amphoteric surfaces, e.g., oxides [34],
2. zwitterionic surfaces, e.g., cells, proteins and synthetic polymer colloids [35,36], and
3. monofunctional surfaces, e.g., carboxylated polystyrene latexes [37-39].

In each of these model colloids, simple ionization of the surface sites involving the uptake and release of H^+ from and to the solution phase was considered in the formulation of the surface charge, σ_0. The effect of indifferent electrolytes was estimated through the concentration dependence of the counterbalancing diffuse layer charge (at 25°C):

$$\sigma_d = -11.74 \sqrt{c_{salt}} \sinh \frac{ze\psi}{2kT} \; \mu C/cm^2 \tag{9}$$

In these models it is assumed that the surface charge and potential arise due to specific reactions occurring at the surface; that the colloid is essentially insoluble; and that there is no equilibrium established between components in the bulk solid and solution that might give rise to the Nernstian-type behavior of silver halides. The polystyrene latices should be excellent models for such conditions, but certain of the oxides are less appropriate models. For example, mercuric oxide has relatively high solubility and fairly rapid equilibrium reactions and, in fact, forms the reversible oxide electrode in alkaline solutions, $Hg/HgO/OH^-$. On the other hand, solids like crystalline TiO_2 and similar oxides should be good model colloids for testing the surface ionization hypothesis.

Amphoteric Surface Model

The surface ionization reactions of the amphoteric colloid with AH surface groups in noncomplexing 1:1 electrolyte solutions are given by [2,24,28]

$$AH_2^+ \rightleftharpoons AH + H_s^+ \quad K_{a_1} = \frac{[AH][H_s^+]}{[AH_2^+]} \tag{10}$$

and

$$AH \rightleftharpoons A^- + H_s^+ \quad K_{a_2} = \frac{[A^-][H_s^+]}{[AH]} \tag{11}$$

where H_s^+ = the concentration of protons at the surface, and
K_{a_1} and K_{a_2} = intrinsic surface acidity constants.

The surface charge may then be defined in terms of the ionized sites:

$$\sigma_0 = \frac{eN_s \left([AH_2^+] - [A^-] \right)}{\left([AH_2^+] + [AH] + [A^-] \right)} \tag{12}$$

where N_s is the total number of AH sites per unit area of colloid. To relate the surface charge to the solution conditions, e.g., pH, it is then assumed that the effective surface concentration of hydrogen ions is given by the Boltzmann equation:

$$[H^+]_s = [H^+] \exp\left(-\frac{e\psi_0}{kT}\right) \qquad (13)$$

where ψ_0 is the mean surface potential relative to the bulk solution. Thus we obtain the following:

$$\frac{\sigma_0}{eN_s} = \frac{\dfrac{[H^+]}{K_{a_1}} \exp\left(-\dfrac{e\psi_0}{kT}\right) - \dfrac{K_{a_2}}{[H^+]} \exp\left(\dfrac{e\psi_0}{kT}\right)}{1 + \dfrac{[H^+]}{K_{a_1}} \exp\left(-\dfrac{e\psi_0}{kT}\right) - \dfrac{K_{a_2}}{[H^+]} \exp\left(\dfrac{e\psi_0}{kT}\right)} \qquad (14)$$

It can be shown that this becomes [24,27]

$$\frac{\sigma_0}{eN_s} = \alpha = \frac{\delta \sinh(y_N - y_0)}{1 + \delta \cosh(y_N - y_0)} \qquad (15)$$

If $y_0 = e\psi_0/kT$, then

$$y_N = \frac{e\psi_N}{kT} = 2.303(pH_{pzc} - pH) = 2.303 \Delta pH \qquad (16)$$

where $pH_{pzc} = 1/2\,(pK_{a_1} + pK_{a_2})$ and

$$\delta = 2 \times 10^{-(\Delta pK/2)}$$

where

$$\Delta pK = pK_{a_2} - pK_{a_1}$$

ψ_N is the Nernst potential and ψ_0 is the equilibrium surface potential.

SURFACE IONIZATION AND COMPLEXATION

The surface charge must be balanced by adsorbed charges to maintain electroneutrality. In this case, the surface charge is assumed to be balanced entirely by the diffuse charge layer due to electrolyte counter- and coions of bulk concentration $C(mol/dm^3)$, i.e.,

$$\sigma_0 + \sigma_d = 0 \quad (17)$$

and

$$\sigma_d = -11.74 \ C \ \sinh\left(\frac{ze\psi_0}{2kT}\right) = -11.74\sqrt{C} \ \sinh\frac{y_0}{2} \quad (18)$$

where σ_0 and σ_d are in $\mu Coul/cm^2$.

It should be noted that this assumption using ψ_0 as the potential for the start of the diffuse layer is only a good approximation at low electrolyte concentrations. Since the ionic radii of H^+ and OH^- are different from other cations and anions, it might be expected that the electrolyte ions need not experience the surface potential, ψ_0, defined at the locus of adsorbed pdi.

Combination of Equations 15, 17 and 18 (at 25°C) leads to the following:

$$\left[\frac{C}{1.362 \times 10^{-14} \ N_s}\right] \sinh\left(\frac{y_0}{2}\right) = \alpha = \frac{\delta \sinh(y_N - y_0)}{1 + \delta \cosh(y_N - y_0)} \quad (19)$$

This equation may be solved for y_0 and then α. Numerical values must be known or given to K_{a_1}, K_{a_2}, $[H^+]$, N_s and C; this allows evaluation of the parameters ΔpH, y_N, ΔpK and δ. Note that for a given combination of K_{a_1}, K_{a_2} (or pH_{pzc}) and pH, y_N is fixed. δ and ΔpK are fixed by K_{a_1} and K_{a_2}. The electrolyte concentration, C, determines the effect of ionic strength on the surface charge, σ_0, and surface potential, ψ_0.

The solutions for y_0 and α at different values of pK_{a_1}, pK_{a_2}, pH, N_s and C may be obtained graphically by the method of intersecting curves. Both the left- and right-hand sides of Equation 19 are plotted as a function of y_0 (or ψ_0) for various parameter values. The solution for y_0 and α is given at the intersection of the matching curves.

Calculation of Surface Charge and Potential

These calculations are for stable colloid dispersions, i.e., particles at large separation distances. The surface charges and potentials can be obtained in a

similar manner for unstable or coagulating colloids; however, there are more restrictions and boundary conditions that must be considered [29,30].
Graphs of the functions

$$\alpha = \frac{\delta \sinh(y_N - y_0)}{1 + \delta \cosh(y_N - y_0)} \qquad (20)$$

for the fractional surface charge and

$$\alpha = \frac{1}{\gamma} \sinh \frac{y_0}{2} \qquad (21)$$

for the equivalent diffuse layer charge as functions of y_0 or $y_0 - y_N$ are shown in Figures 1 and 2 for specified values of $\gamma (= f(N_s \sqrt{C}, T), \delta (= 2 \times 10^{-(\Delta pK/2)})$ and $y_N (= 0.0592(pH_{pzc} - pH))$. The range of values of the parameter (at 25°C)

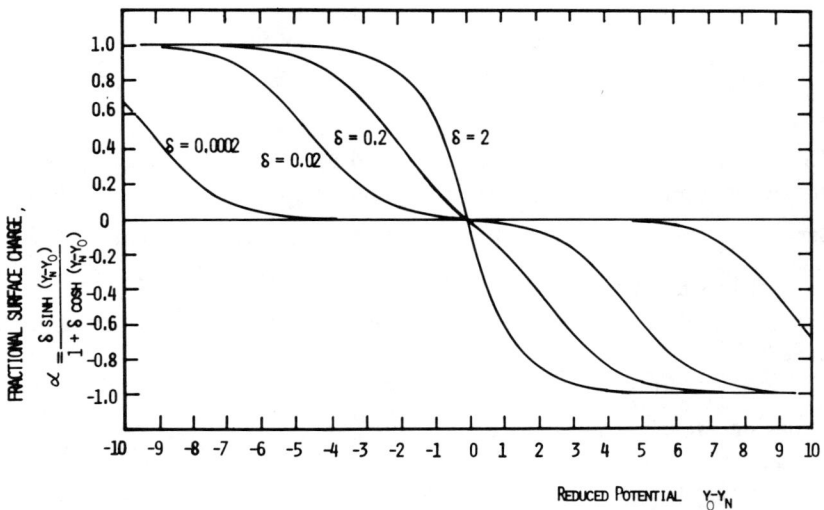

Figure 1. Variation of the fractional surface charge function arising from simple acid-base dissociation equilibria as a function of the reduced potential parameters $(y - y)$, where $y = e\psi_0/kT$ and $y_N = 2.303(pH_{pzc} - pH)$, for various values of $\delta = 2 \times 10^{-(\Delta pK/2)}$.

SURFACE IONIZATION AND COMPLEXATION

$$\gamma = 1.362 \times 10^{-14} \frac{N_s}{\sqrt{C}} \qquad (22)$$

are given in Table I. For most conditions, γ has values less than 2000 but greater than 1.

To obtain solutions for y_0 and α at particular pH (or y_N) values, the surface charge curve (Figure 1, Equation 15) is transposed parallel to the potential (y_0) axis of the diffuse layer charge curve (Figure 2) until the sigmodial curve cuts the $\alpha = 0$ axis at, say, $y_0 = y_{N_1}$. The intersection of the diffuse layer charge curve ($\sigma_0 = -\sigma_d = 1/\gamma \sinh y_0/2$) with the surface charge curve gives the coordinate $[y_0(Y_{N_1}), \alpha(y_{N_1})]$. Variation of y_{N_x} corresponding to $2.303(pH_{pzc} - pH_x)$ may be obtained by sliding the surface charge curve along the y_0-axis until $\alpha = 0$ at $y_0 = y_{N_x}$; this allows graphic estimation of the variation of $\psi_0 = 2.303(kT/e)y_0$ and $\sigma_0 = N_s e\alpha$ as a function of the pH for fixed ionic strength $I = 1/2 \; \Sigma \; C_i Z_i^2$. In addition, the effect of ionic strength on ψ_0 and σ_0 at fixed pH may be made similarly by allowing for the variation of $1/\gamma$ in the diffuse layer charge term with the square root of electrolyte concentration, while y_N or the surface charge curve (α) is fixed.

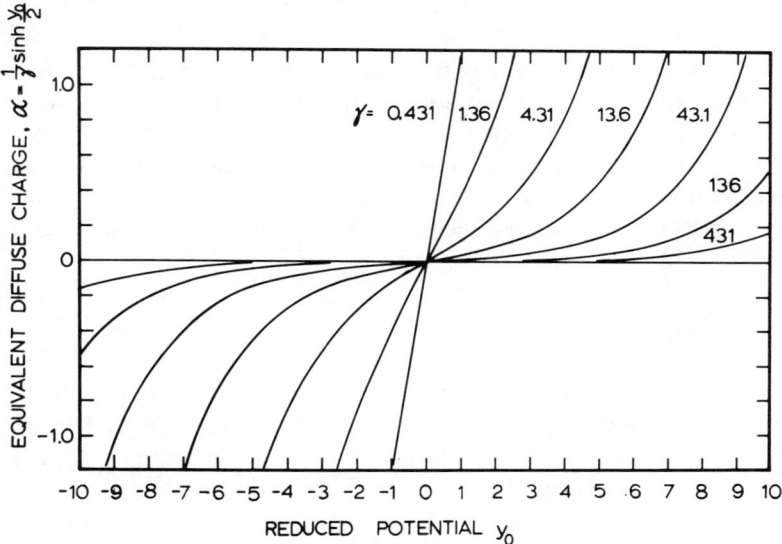

Figure 2. Variation of the equivalent diffuse layer charge as a function of the potential $y_0 = e\psi_0/kT$ for various ionic strength values, $1/\gamma = 1/(1.362 \times 10^{-14} \times N_s)$ at 25°C.

Table I. Variation in the Magnitude of Parameter γ^a with Changes in Surface Site Density (N_s) and Electrolyte Concentration (C)

C (mol/dm^3)	N (sites/cm^2)				
	1×10^{15}	5×10^{14}	1×10^{14}	5×10^{13}	1×10^{13}
	γ $(1.362 \times 10^{-13} [N_s/C])$				
10^{-5}	4307	2153.5	430.7	215.4	43.07
10^{-4}	1362	681	136.2	68.1	13.62
10^{-3}	430.7	215.35	43.07	21.54	4.307
10^{-2}	136.2	68.1	13.62	6.81	1.362
10^{-1}	43.07	21.535	4.307	2.154	0.4307

Using data thought to be pertinent to some oxides and also polystyrene latices, Healy et al. [24,28] have been able to obtain qualitative agreement with the observed effects of ionic strength and pH on the surface charge density and double layer potential for these surfaces. The results of graphic and computer solution of these equations show that the surface potential (ψ_0) - pH relationship is generally not given by the Nernst equation, except when the potential is low in magnitude and the difference in the surface ionization constants, $\Delta pK = pK_{a_2} - pK_{a_1}$, is small. This corresponds to the physical situation where positive and negative surface sites dominate the neutral surface sites, as considered for such ionic solids as AgI. For surfaces where neutral sites are always a large fraction of the total sites, i.e., ΔpK is large, the calculated surface potential is never that obtained from the Nernst equation, assuming H$^+$ and OH$^-$ as pdi. The results of graphic solution of the diffuse layer model for an amphoteric colloid with $pH_{pzc}6$, $\Delta pK_a = 2$ and 10^{14} sites/cm^2 are shown in Figure 3.

Zwitterionic Surfaces

Certain colloids may have more than one kind of surface site when there are different positive and negative sites such as -RCOOH and R_3N^+H on proteins or synthetic polymer latices [35,36] and possibly certain clays. They are zwitterionic and may be represented by the following:

$$AH \rightleftharpoons A + H_s^+ \quad K_{a_1} = \frac{[A^-][H_s^+]}{[HA]} \quad (23)$$

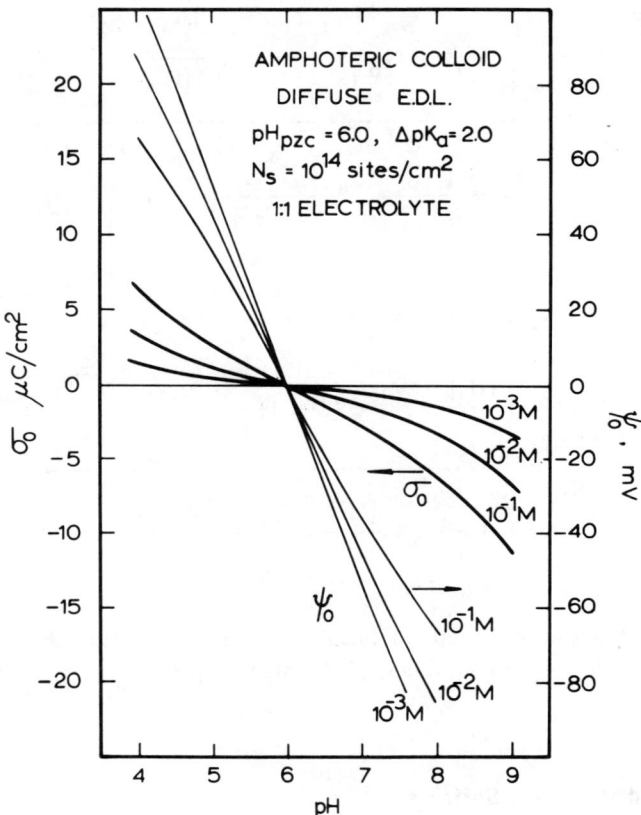

Figure 3. The variation of surface charge, σ_0, and surface potential, ψ_0, as a function of pH and electrolyte concentration derived from method of intersecting curves using Figures 1 and 2.

$$BH^+ \rightleftharpoons B + H_s^+ \qquad K_{a_2} = \frac{[B][H_s^+]}{[BH^+]} \qquad (24)$$

where N_A and N_B are the site densities of each type of site. Expressing $[BH^+]$ and $[A^-]$ in terms of $[H_s^+]$, K_{a_1}, K_{a_2} and N_A, N_B gives the surface charge density as

$$\sigma_0 = \frac{\left[e N_B \dfrac{[H^+]}{[K_{a_2}]} e^{y_0} + N_B \dfrac{[K_{a_1}]}{[K_{a_2}]} - N_A \dfrac{[K_{a_1}]}{[H^+]} e^{-y_0} - N_A \dfrac{[K_{a_1}]}{[K_{a_2}]}\right]}{1 + \dfrac{[H^+]}{[K_{a_2}]} e^{-y_0} + \dfrac{[K_{a_1}]}{[H^+]} e^{y_0} + \dfrac{[K_{a_1}]}{[K_{a_2}]}} \qquad (25)$$

or if $f = N_B/N_S$ and $N_S = N_A + N_B$,

$$\alpha = \frac{\sigma_0}{eN_S} = \frac{f \dfrac{[H^+]}{K_{a_2}} e^{-y_0} + (2f-1)\dfrac{K_{a_1}}{K_{a_2}} - (1-f)\dfrac{K_{a_1}}{K_{a_2}} e^{y_0}}{1 + \dfrac{K_{a_1}}{K_{a_2}} + \dfrac{[H^+]}{K_{a_2}} e^{-y_0} + \dfrac{K_{a_1}}{[H^+]} e^{y_0}} \qquad (26)$$

Again, this can be solved graphically for y_0 and α by using the electroneutrality equation, $\sigma_0 + \sigma_d = 0$, and given values of pH, pK_{a_1}, pK_{a_2}, C, N_A and N_B or f.

Single and Mixed Site Surfaces

Single-site surfaces are not particularly common in natural systems. Perhaps one approximation would be the surface of silica, although there has been suggestion of two types of sites from ion exchange studies on colloidal silica [40,41]. Carefully prepared synthetic polymer lattices, e.g., polystyrene, may contain only one type of surface group, however. Examples of such studies are given in the literature [37-42]. In addition, surface films of fatty acids and other amphipathic molecules or ions at the air-water interface are in this category.

These surface functional groups may behave as strong or weak acids according to $AH \rightleftharpoons A^- + H^+$ (or $BH^+ \rightleftharpoons B + H^+$).

$$K_a = \frac{[A^-][H_s^+]}{[HA]} \qquad (27)$$

and the fractional surface charge is represented in the simple anionic system as

SURFACE IONIZATION AND COMPLEXATION

$$\frac{\sigma_0}{eN_s} = \alpha_- = \frac{-\frac{K_a}{H}e^{y_0}}{1 + \frac{K_a}{H}e^{y_0}} \tag{28}$$

Although such surfaces do not have an iep or pzc as seen for amphoteric or zwitterionic surfaces, y_0 may again be obtained by the method of intersecting curves using $\sigma_0 + \sigma_d = 0$ by plotting the fractional surface charge, e.g.,

$$\alpha = \frac{\frac{K_a}{H}e^{y_0}}{1 + \frac{K_a}{H}e^{y_0}} \tag{29}$$

and the equivalent diffuse charge $\alpha = (1/\gamma) \sinh(y_0/2)$ as a function of y_0 for given values of H^+, K_a, N_s and $\sqrt{C}$.

In the case of mixed anionic sites, which may be present on latex surfaces, the model may be developed in a similar way. If the surface ionization reactions are

$$HA \rightleftharpoons A^- + H^+ \quad K_A = \frac{[A^-][H_s^+]}{[HA]} \tag{30}$$

and

$$HC \rightleftharpoons C^- + H_s^+ \quad K_B = \frac{[C^-][H_s^+]}{[HC]} \tag{31}$$

and total number of sites A and C are NA and N_C with $N_s = N_A + N_C$, then the surface charge is given by

$$\frac{\sigma_0}{eN_s} = \frac{-\frac{K_A[HA]}{[H^+]}e^{y_0} - \frac{K_C[HC]}{[H^+]}e^{y_0}}{[HA] + \frac{K_A[HA]}{[H^+]} + [HC] + \frac{K_C[HC]}{[H^+]}} \tag{32}$$

$$\alpha = \frac{-(K_A/[H^+])(1-g)e^{y_0}}{1+(K_A/[H^+])e^{y_0}} - \frac{(K_C/[H^+])(g)e^{y_0}}{1+(K_C/[H^+])e^{y_0}} \tag{33}$$

where $g = N_c/N_s$.

With given values for K_A, K_C, H^+, N_A and N_c or g, Equation 33 can be combined with the diffuse layer charge, $\sigma_d = (1/\gamma)\sinh(y_0/2)$, to give graphic or computer solution for y_0 and α as a function of pH and electrolyte concentration.

Surfaces for which this latter model may find application include polymer latexes with mixed $-SO_3H$ and $-COOH$ groups or clay mineral surfaces, which may possibly have different functional groups on faces and edges.

The methods reviewed here lay the foundation for a chemical approach to the interpretation of the edl properties of colloids in aqueous electrolytes. The maximum charge of the colloid is limited by the number of surface sites present; this, in turn, limits the value of surface charge density in the diffuse double layer to plausible values, regardless of the electrolyte concentration. The surface potential is adjusted until it satisfies the equilibrium relationship of pdi in the bulk solution and at the surface, and also the electroneutrality of the interfacial region.

In earlier models for electrochemically reversible colloids, in which the surface potential, ψ_0, was estimated from the Nernst equation, the surface charge density calculated from $\sigma_d = -11.74\sqrt{C}\sinh ze\psi_0/2kT$ could reach impossibly high values.

The surface ionization model shows the experimental trends of dependence of surface charge and zeta potential on the solution properties pH and ionic strength. In principle this simple model could be extended to account for specific interaction of electrolyte anions and cations with surfaces, but no reports have yet been published.

In many of their calculations, Healy et al. [27] assigned values to the binding constants that were thought to be appropriate to the properties of the particular colloid or they used previously determined surface acidity "constant." However, there is reason to suspect that some of the experimental ionization "constants" were not the intrinsic constants specified in this version of the theory [2,3,43-46]. The experimental ionization constants vary considerably with ionic strength, even though the electrostatic component had been eliminated by extrapolation to zero charge conditions. Hence, these simple calculations using one ionization "constant" cannot be expected to be quantitative over a very wide range of electrolyte concentrations.

The work summarized here features a new, simple model that can be very useful in qualitative discussion of ionization of many different types of colloids that are not part of the realm of classical colloids of reversible electrode systems, e.g., silver halides, whose study led to the development of models of the structure in the interfacial region. This model can be used with a book of tables, slide rule, or calculator and graph paper. It also serves as an introduction to our next subject—the surface ionization and complexation of oxide or polymer colloids in aqueous electrolytes using the structure of the Stern-Grahame model of the edl. It is interesting to note, however, that these types of models were used by Davies [47] and Payens [48] to describe surfactant monolayers.

SURFACE IONIZATION AND COMPLEXATION FOR THE EDL OF COMPACT AND DIFFUSE LAYERS

It has been seen that by simply considering the surface charge formed by the ionization or dissociation of surface functional groups and a balancing diffuse layer, one can offer a good qualitative description of the colloid properties of oxide or other dispersions in water. The magnitude of the surface charge increases and the surface potential decreases with ionic strength at fixed pH in agreement with experimental observations. Since H^+_{aq} and OH^- play a key role as reactants in the ionization of the surface sites and of the solvent, water, these ions have a major effect on the sign and magnitude of the surface charge and potential at fixed ionic strength. The model shows why Nernstian potentials cannot be expected in these electrochemically irreversible systems.

The electrolyte ions also have an effect on ψ_0 in this model through the dependence of the diffuse layer charge on the electrolyte concentration [28].

To account for more specific interactions of electrolyte ions with colloid surface sites, it will be shown that it is convenient to use a compact double layer model for site-bound electrolyte ions and a diffuse layer to balance the net charge of the surface layer and the compact layer

$$\sigma_0 + \sigma_\beta + \sigma_d = 0 \tag{34}$$

In the development of the model, the ionization constants were estimated assuming that specific interactions were absent. However, it will be shown that estimates of the intrinsic ionization constants and intrinsic complexation constants can be derived from experimental titration data by assuming that specific interactions may be present and by applying the charge-potential equations of the Stern-Grahame model.

Surface Ionization Equilibria

Bolt [49] and Parks and de Bruyn [50] were among the first to apply potentiometric titration to study the acid-base surface equilibria and the edl at oxide-water interfaces. However, this technique had also been used much earlier by other workers, e.g., Baver [57], Mattson [52], Marshall [53-57] and Harward and Coleman [58] in the study of the ion exchange and surface acidity of functional groups on various clay minerals.

These studies established the importance of pH and the ionic strength of the electrolyte medium to the sign and magnitude of the surface charge. By analogy to the definition of the surface charge of classical colloids, the surface charge of oxides is determined from the proton condition, which allows estimation of the net uptake of H^+ and OH^-, i.e.,

$$\sigma_0 = \Gamma_{H^+} - \Gamma_{OH^-} = \mathscr{F}(C_{acid} - C_{base} + [OH] - [H^+])/A \quad (35)$$

where A = the surface area of the suspension,
C_{acid} and C_{base} = the analytical concentrations of strong acid and base after addition, and $\mathscr{F}$
= the Faraday constant.

The surface charge, σ_0 - pH curves at different constant concentrations of simple salts, e.g., NaCl, KNO_3, KCl, show a mutal intersection point that defines the point of zero charge, pH_{pzc}, where $\sigma_0 = \Gamma_{H^+} - \Gamma_{OH} = 0$ [18-23]. The magnitude and sign of the surface charge is measured with reference to the pH_{pzc}. The magnitude of σ_0 increases with electrolyte concentration, while the magnitude of the electrokinetic potential decreases [18,25].

Similar to the approach in the previous section, the surface ionization reactions may be represented by

$$SOH_2^+ \overset{K_{a_1}^{int}}{\rightleftharpoons} SOH + H_s^+ \quad (36)$$

and

$$SOH \overset{K_{a_2}^{int}}{\rightleftharpoons} SO^- + H_s^+ \quad (37)$$

SURFACE IONIZATION AND COMPLEXATION

If we assume that the concentration of protons at some location, i, in the electrical double layer is related to the bulk concentration by the Boltzmann distribution involving only electrostatic work, then

$$[H_i^+] = [H^+] \exp\left(-\frac{e\psi_i}{kT}\right) \tag{38}$$

and, hence,

$$K_{a_1}^{int} = \frac{[SOH][H^+]}{[SOH_2^+]} \exp\left(-\frac{e\psi_0}{kT}\right) \tag{39}$$

and

$$K_{a_2}^{int} = \frac{[SO^-][H^+]}{[SOH]} \exp\left(-\frac{e\psi_0}{kT}\right) \tag{40}$$

In an idealized planar surface model, ψ_0 is the mean potential in the plane of surface charge, σ_0, created by the amphoteric reactions of the surface sites with protons and hydroxyl ions.

In addition to these simple ionization reactions, Davis et al. [45] and others [59] have shown that counterion binding is important in determining the charge of polyelectrolytes [60], ion exchange resins [59], oxides [34,61] and polystyrene latexes [62]. These reactions of salt, say NaCl, with surface groups may be written as follows:

$$SOH_2^+Cl^- \xrightleftharpoons{^*K_{Cl}^{int}} SOH + H_s^+ + Cl_s^- \tag{41}$$

and

$$SOH + Na_s^+ \xrightleftharpoons{^*K_{Na}^{int}} SO^-Na_s^+ + H^+ \tag{42}$$

These species where counterions are associated with the surface sites have been termed ion pairs or surface complexes. The exact nature of the interaction is not always clear and deserves further experimental investigation. The formation of these complexes readjusts the equilibrium distribution of surface species in the acid-base reactions and, consequently, the net uptake of H^+ and OH^-. Since $\sigma_0 = \Gamma_{H^+} - \Gamma_{OH^-}$ represents the net of all reactions involving H^+ or OH^-, the surface charge may be written in terms of the surface sites in the following way:

$$\sigma_0 = B([SOH_2^+Cl^-] + [SOH_2^+] - [SO^-] - [SO^-Na^+]) \qquad (43)$$

The compact layer charge due to specific counterion interaction is then

$$\sigma_\beta = B([SO^-Na^+] - [SOH_2^+Cl^-]) \qquad (44)$$

where $B = 10^6 \mathscr{F}/A$ and σ_0 and σ_β have units of $\mu Coul\ cm^{-2}$. Electroneutrality of the whole interfacial region requires that

$$\sigma_0 + \sigma_\beta + \sigma_d = 0 \qquad (45)$$

where σ_d may be obtained from the Gouy-Chapman theory for planar surfaces. For symmetrical $z_+:z_-$ electrolytes,

$$\sigma_d = -11.74\sqrt{C}\ \sinh\left(\frac{ze\psi}{2kT}\right) \qquad (18)$$

at 25°C. For unsymmetrical electrolytes [25,26],

$$\sigma_d = -\frac{\psi_d}{|\psi_d|} 5.87\sqrt{C} \left[\frac{1}{z_+}\left\{\exp\left(-\frac{z_+ e\psi_d}{kT}\right) - 1\right\} - \frac{1}{z_-}\left\{\exp\left(-\frac{z_- e\psi_d}{kT}\right) - 1\right\}\right]^{1/2} \qquad (46)$$

where C is the concentration of electrolyte in equiv/dm³ and ψ_d is the mean potential at the start of the diffuse layer. Note that often the particle size is large compared to the "thickness" of double layer, so the flat plate used here is a reasonable approximation for the ψ_d, σ_d relationship.

Potential-determining ions, H^+ and OH^-, are confined to the mean surface plane and specifically adsorbed counterions to a second mean plane at a distance, β. The separation of charge layers allows the use of a molecular model for the charge and potential relations in the compact part of the edl, e.g.,

SURFACE IONIZATION AND COMPLEXATION

$$\psi_0 - \psi_\beta = \frac{\sigma_0}{C_1} = \frac{\beta}{\epsilon_1 \epsilon_0} \cdot \sigma_0 \quad (47)$$

and

$$\psi_\beta - \psi_d = -\frac{\sigma_d}{C_2} = -\frac{(d-\beta)}{\epsilon_2 \epsilon_0} \cdot \sigma_d \quad (48)$$

where C_1, C_2; ϵ_1, ϵ_2; and β, $d-\beta$ are the capacitance, dielectric function and thickness of the molecular condenser of the inner and outer compact layers, respectively

The equations that define the surface species in terms of solution composition and double-layer parameters are as follows:

$$[SOH_2^+] = [SOH][H^+] \exp\left(-\frac{e\psi_0}{kT}\right) \Big/ K_{a_1}^{int} \quad (49)$$

$$[SO^-] = K_{a_2}^{int}[SOH] \exp\left(\frac{e\psi_0}{kT}\right) \Big/ [H^+] \quad (50)$$

$$[SOH_2^+Cl^-] = [SOH][H^+][Cl^-] \exp\left(\frac{e\psi_\beta - e\psi_0}{kT}\right) \Big/ {}^*K_{Cl}^{int} \quad (51)$$

$$[SO^-Na^+] = {}^*K_{Na}^{int}[SOH][Na^+] \exp\left(\frac{e\psi_0 - e\psi_\beta}{kT}\right) \Big/ [H^+] \quad (52)$$

The mass balance on surface sites requires that the total number of sites per unit area, N_s, be given by

$$N_s = [SOH_2^+] + [SOH_2^+Cl^-] + [SOH] + [SO^-Na^+] + [SO^-] \quad (53)$$

N_s may be determined by a variety of experimental techniques, including isotopic exchange (e.g., ^3H), IR spectroscopic determination of surface hydroxyls and other methods.

This entire set of simultaneous equations can be solved at any pH (or analytical amount of added acid or base) with known values for the properties

N_s, $K^{int}_{a_1}$, $K^{int}_{a_2}$, $^*K^{int}_{Cl}$, $^*K^{int}_{Na}$, C_1, C_2 and electrolyte concentration. This is done by using $[\exp(-e\psi_0/kT)]$, $[\exp(-e\psi_\beta/kT)]$ and [SOH] as independent variables in the numerical procedure used for solution equilibria in the computer program MINEQL devised by Westall et al. [63]. The surface mass balance provides the convergence test equation for [SOH], and the convergence test equations for $\exp(-e\psi_0/kT)$ and $\exp(-e\psi_\beta/kT)$ are forms of the electroneutrality condition, Equations 4 and 5 and the charge-potential Equations 18, 47 and 48.

Thus, for stable colloid dispersions, mass law equations for both the surface and solution components of reactions are solved subject to the constraints of the charge balance and charge-potential relationships in the electrical double layer.

Determination of Intrinsic Ionization and Complexation Constants

The factors governing the ionization and potentiometric titration curves of oxide and polymer colloids are similar to those affecting the titration behavior and ionization of polyelectrolytes, which have been studied extensively in biochemical and medical research. The problems are simpler for inorganic and polymer colloids because it may be assumed that the geometry of the colloid does not alter significantly with the fractional ionization.

Schindler and others [2-5,43,44] have applied methods developed for polymer titration theory to study the ionization and titration of oxides. They assume, as done here, that the *ionization quotient* (not a constant) of a surface acid,

$$Q_a = \frac{[SO^-][H^+]}{[SOH_2^+]} \quad (54)$$

can be conveniently split into an *intrinsic acidity constant*, K^{int}_a, which is independent of the degree of ionization and complexation, and a factor that represents the electrostatic interaction of protons with the electric field, $\exp(+e\psi_0/kT)$, i.e., $Q_a = K^{int}_a \exp(+e\psi_0/kT)$.

In simple dilute electrolytes, in the absence of net specific adsorption, the diffuse double-layer theory requires that the surface potential becomes zero, $\psi_0 = 0$, at the point of zero charge, $\sigma_0 = 0$. By determining the acidity quotient as a function of the net charge or fractional ionization, $\alpha_\pm$, Schindler

SURFACE IONIZATION AND COMPLEXATION

and co-workers [43,44] have estimated K_a^{int} for SiO_2 and TiO_2 from the extrapolation of the $pQ_a - \sigma_0$ data to the point of zero charge and defining the intercept as pK_a^{int}.

Yates et al. [61] used the reported values of $\Delta pK_a^{int} = pK_{a_2}^{int} - pK_{a_1}^{int}$ for various oxides to obtain semiquantitative agreement of their specific site-binding model calculations with experimental observations of the surface charge, σ_0, and electrokinetic potentials, $\zeta \simeq \psi_d$. The bulk of their computations are for ΔpK_a^{int} equal to 3, which was considered appropriate for oxides such as TiO_2 [44], $\gamma\text{-}Al_2O_3$ and $\alpha FeOOH$ [2,34]. For SiO_2, the value of ΔpK_a^{int} was equal to 6.

However, the calculation procedure for the intrinsic acidity constants used by Schindler et al. [43,44] and Stumm et al. [2-5] was based on simple ionization of sites and diffuse double-layer theory, which is not consistent with the counterion site-binding model of Yates et al. [61]. The latter uses ionization and complexation of surface sites requiring the existence of a compact layer and diffuse layer of charge. In the estimation of the ionization constants, it was assumed that the [SO$^-$] and [SOH$_2^+$] sites could be directly related to the negative and positive surface charge density, e.g.,

$$\alpha_- = -\frac{\sigma_0}{N_s} = \frac{[SO^-]}{N_s} \quad \text{for pH} > pH_{pzc} \quad (55)$$

and

$$\alpha_+ = \frac{\sigma_0}{N_s} = \frac{[SOH_2^+]}{N_s} \quad \text{for pH} < pH_{pzc} \quad (56)$$

Thus,

$$pK_{a_1}^{int} = pH + \log\frac{\alpha_+}{1-\alpha_+} + \frac{e\psi_0}{2.303kT} = pQ_{a_1} + \frac{e\psi_0}{2.303kT} \quad (57)$$

and

$$pK_{a_2}^{int} = pH - \log\frac{\alpha_-}{1-\alpha_-} + \frac{e\psi_0}{2.303kT} = pQ_{a_2} + \frac{e\psi_0}{2.303kT} \quad (58)$$

Numerical values for $pK_{a_1}^{int}$ and $pK_{a_2}^{int}$ may be obtained by plotting the experimentally accessible quotients pQ_{a_1} and pQ_{a_2} as a function of α_+ and α_-, respectively. Intercepts at pQ_{a_1} (α_+=0) and pQ_{a_2} (α_-=0) give the intrinsic constants.

Davis et al. [45,46] tested the effect of electrolyte concentration on the intercepts pQ_{a_1} (α_+=0) and pQ_{a_2} (α_-=0) and found that there were large changes in the values of Q_{a_1}, α_+=0 and Q_{a_2}, α_-=0 with the increase of electrolyte concentration. This is demonstrated in Figure 4 for the data of Yates [34] on TiO_2 in aqueous solutions.

It was then shown that this variation could be due to the neglect of the surface complex sites, e.g., [SO$^-$Na$^+$] and [SOH$_2^+$Cl] in the approximation for estimating [SOH$_2^+$] and [SO$^-$] from the surface charge density σ_0 - pH titration curves. There are significant errors in this procedure if surface complexation is significant. An improved approximation is that

$$-\frac{\sigma_0}{B} \simeq [SO^-] + [SO^-Na^+] \qquad (59)$$

so that

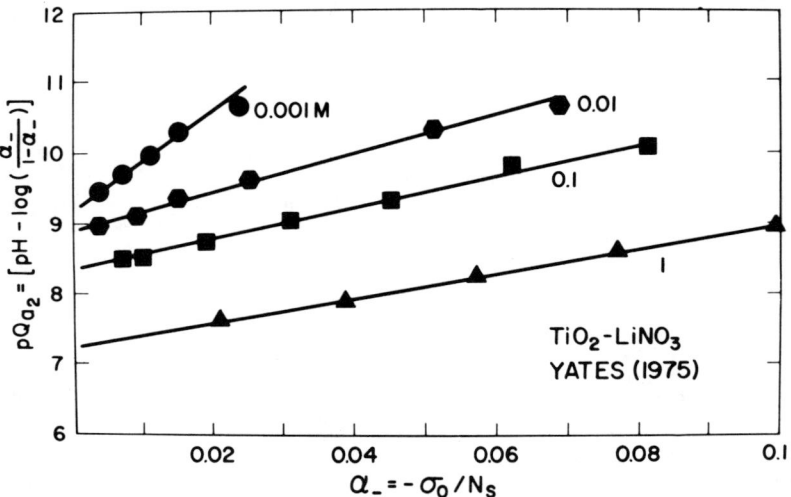

Figure 4. pQ_{a_2} as a function of fractional surface ionization for TiO_2 dispersed in aqueous $LiNO_3$ solutions at various ionic strengths. Data from Yates [34] reproduced by permission of the Journal of Colloid and Interface Science 63: 480 (1978).

SURFACE IONIZATION AND COMPLEXATION

$$[SOH] = \frac{N_s}{B} - |\sigma_0/B| = \frac{N_s}{B} - [SO^-] - [SO^-Na^+] \quad (60)$$

hence

$$-\frac{\sigma_0}{B} = [SO^-]\left(1 + [Na^+]\exp\left(-\frac{e\psi_\beta}{kT}\right)K_{Na}^{int}\right) \quad (61)$$

and

$$K_{a_2}^{int} = \frac{|\sigma_0|}{(N_s - |\sigma_0|)} \frac{[H^+]\exp\left(-\frac{e\psi_0}{kT}\right)}{\left[1 + [Na^+]K_{Na}^{int}\exp\left(-\frac{e\psi_\beta}{kT}\right)\right]} \quad (62)$$

In logarithmic form this gives

$$pK_{a_2}^{int} = pH - \log\frac{\alpha_-}{1-\alpha_-} + \frac{e\psi_0}{2.303kT} + \log\left[1 + [Na^+]K_{Na}^{int}\exp\left(-\frac{e\psi_\beta}{kT}\right)\right] \quad (63)$$

and

$$pK_{a_1}^{int} = pH + \log\frac{\alpha_+}{1-\alpha_+} + \frac{e\psi_0}{2.303kT} - \log\left[1 + [Cl^-]K_{Cl}^{int}\exp\left(\frac{e\psi_\beta}{kT}\right)\right] \quad (64)$$

where K_{Na}^{int} and K_{Cl}^{int} are the binding constants of Na^+ and Cl^- to SO^- and SOH_2^+ sites.

These equations account for the dependence of the intercepts, pQ_{a_2} ($\alpha=0$) and pQ_{a_1} ($\alpha_+=0$), on the electrolyte counterion concentration. Only in the limit of very low ionic strengths do the simpler Equations 56 and 57 give a good estimate of the intrinsic acidity constants $pK_{a_1}^{int}$ and $pK_{a_2}^{int}$.

To evaluate the importance of surface complex formation in the distribution of surface sites, it is necessary to evaluate the intrinsic complexation constants for the reactions

$$\text{SOH} + \text{Na}^+ \underset{}{\overset{*K^{int}_{Na}}{\rightleftharpoons}} \text{SO}^-\text{Na} + \text{H}^+ \qquad (65)$$

and

$$\text{SOH}_2^+\text{Cl}^- \underset{p^*K^{int}_{Cl}}{\rightleftharpoons} \text{SOH} + \text{H}^+ + \text{Cl}^- \qquad (66)$$

Considering just the data for negatively charged surfaces and making the assumption that at higher electrolyte concentrations complexed sites dominate ionized sites, $|\sigma_0| = \alpha_- \cdot N_s \doteq B[\text{SO}^-\text{Na}^+]$, then

$$p^*K^{int}_{Na} = pH - \log \frac{\alpha_-}{1-\alpha_-} + \log[\text{Na}^+] + \frac{e(\psi_0 - \psi_\beta)}{2.303kT} \qquad (67)$$

At the point of zero charge, Equation 47 gives $\psi_0 - \psi_\beta = 0$. Thus, the function $pQ_{a_2} = pH - \log \alpha_-/1 - \alpha_-$ is a function of both $\log_{10}[\text{Na}^+]$ and σ_0. However, $p^*K^{int}_{Na} = pQ_{a_2}$ in the limit that $\log[\text{Na}^+] = 0$, i.e., $[\text{Na}^+] = 1.0$ and $|\sigma_0| = \alpha_- \cdot N_s = 0$.

Another method to obtain $p^*K^{int}_{Na}$ is to plot the function $p^*Q_{Na} = pH - \log \alpha_-/1 - \alpha_- + \log[\text{Na}^+]$ as a function of α_- or σ_0 and extrapolate the data to the point of zero charge. When Davis et al. [45] used this type of plot for several oxides and electrolytes, most experimental points lay close to a single line for all ionic strengths studied. An example of this type of plot is shown in Figure 5 for TiO_2 in $LiNO_3$ electrolyte. The greatest deviations occur for the low-charge, low-ionic-strength points, where the approximation $\alpha_- N_s \doteq B[\text{SO}^-\text{Na}^+]$ is in error because the ionized $[\text{SO}^-]$ sites are also a significant fraction of the total sites.

Thus, $pK^{int}_{a_1}$ and $pK^{int}_{a_2}$ may only be obtained from the intercept $pQ_{a_1}(\alpha_+=0)$ and $pQ_{a_2}(\alpha_-=0)$ in the limit of very low ionic strength, i.e., as $I \to 0$. The surface complexation constants $p^*K^{int}_{Na}$ and $p^*K^{int}_{Cl}$ may be obtained from $p^*Q_{Na}(\alpha_-=0)$ or $p^*Q_{Cl}(\alpha_+=0)$ in the limit where $\log[\text{Na}^+] = 0$.

Using these intrinsic ionization and complexation constants in the edl model, Davis et al. [45] were able to compute the surface charge and zeta potential ($\zeta \equiv \psi_d$) to give very good agreement with the experimental data of Yates [34] for TiO_2 in $LiNO_3$ and KNO_3 electrolyte over a wide range of both pH and electrolyte concentration. These results are shown in Figures 6 and 7, where calculated curves are shown together with experimental points.

SURFACE IONIZATION AND COMPLEXATION

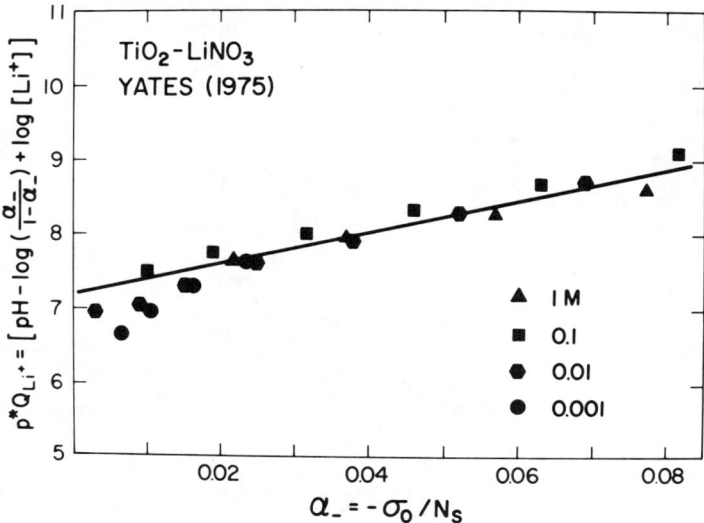

Figure 5. p^*Q_{Li} as a function of fractional surface ionization for TiO_2 dispersed in aqueous $LiNO_3$ solutions. From data of Yates [34] and reproduced with permission of the *Journal of Colloid and Interface Science* 63: 480 (1978).

The only variable parameters not directly accessible to experimental determination for oxides are the inner and outer layer capacitances, C_1 and C_2. The outer layer capacitance was assigned the value 20 $\mu F/cm^2$, which is in line with experimental values from $Hg\text{-}H_2O$ and $Ag\text{-}AgI\text{-}H_2O$ systems. The inner layer capacitance C_1 has been given values of 140 for $LiNO_3$ and 100 for KNO_3 electrolytes. These values are higher than measured for classical colloid surfaces and they imply a small separation of about 0.1 nm between the mean surface plane and the mean plane of surface complexed counterions. Table II shows some separation distances for various combinations of inner layer capacity and interfacial dielectric functions. Table III lists the surface ionization constants and surface complexation constants determined by Davis et al. [45] for a number of oxides in various electrolytes.

Surface Characterization of Polymer Colloids

In even more recent studies [62] of the ionization of surface functional groups on polystyrene lattices, an improved graphic method using a double extrapolation procedure has been developed and will be discussed in this section.

The methods used to characterize the surface functional groups of lattices are similar to those used for oxide colloids and clays, namely, potentiometric

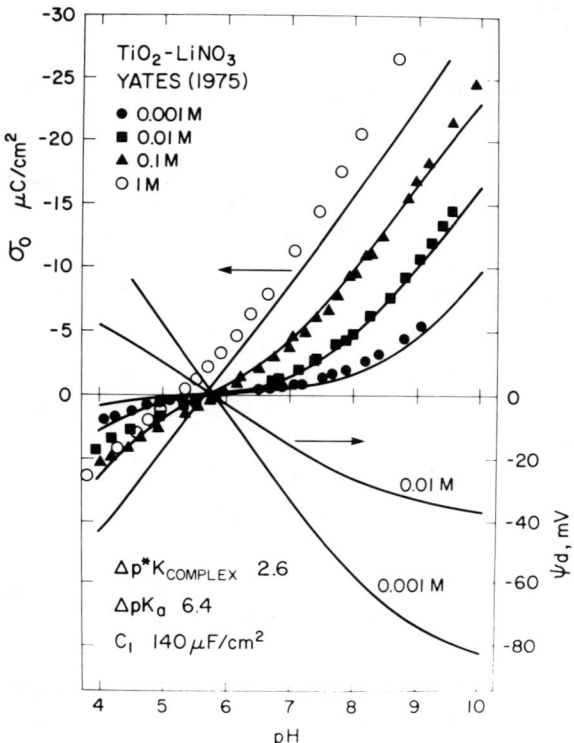

Figure 6. Surface charge density and diffuse layers potential of TiO_2 dispersed in aqueous $LiNO_3$ solutions as a function of pH. Solid lines are model calculations, points are data of Yates [34], reproduced by permission of the *Journal of Colloid and Interface Science* 63: 480 (1978).

acid-base titration, conductometric titration and electrokinetic potential measurements. These properties are used to assess the acidity and degree of ionization of the surfaces, so are useful in determining the importance of surface charge or diffuse layer potential in the colloid stability and rheological properties of the polymer colloids.

The types of functional groups may be strong acid, e.g., sulfonate $R\text{-}\bar{S}O_3 H^+$ and sulfate $R\text{-}\bar{S}O_4 H^+$; weak acid, e.g., carboxylic acids RCOOH; or very weak acids, e.g., ammonium groups $R_1 R_2 R_3 N^+H$. The surface may be monofunctional, e.g., one of RSO_3^-, RCOOH or $R_3 N^+H$, or multifunctional, e.g., the novel amphoteric or zwitterionic polystyrene lattices that mimic the edl properties of oxides [35,36]. These latices have different mole ratios of carboxylic acid and amine esters copolymerized into the polystyrene surface.

SURFACE IONIZATION AND COMPLEXATION

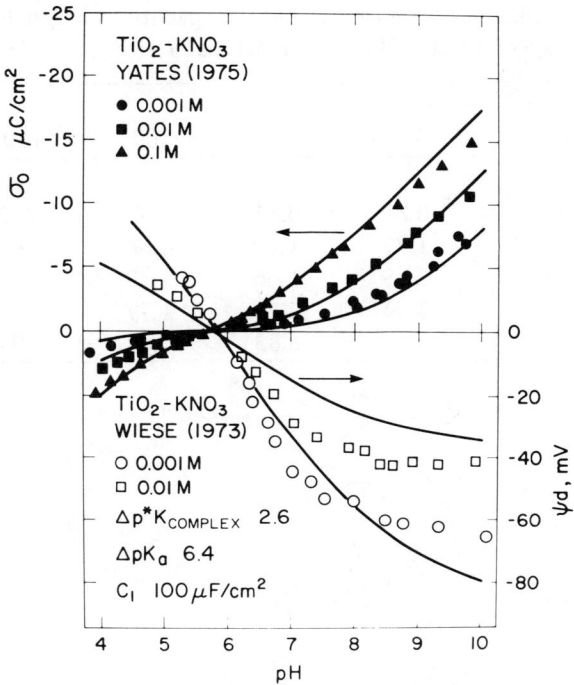

Figure 7. Surface charge density and diffuse layer potential of TiO_2 dispersed in aqueous KNO_3 solutions as a function of pH. Solid lines are calculated, points are experimental data of Yates [34] and Wiese and Healy [13-16], reproduced with permission of the *Journal of Colloid and Interface Science* 63: 480 (1978).

Since many of the properties of these colloid dispersions appeared similar to oxide dispersions, James et al. [62] applied the surface ionization and complexation model developed for oxides to polystyrene lattices and were able to describe quite accurately the extensive potentiometric and conductometric data of Stone-Masui and Watillon [37] and Yates [45].

The electrical double layer model is the same as used earlier. The potentials ψ_0, ψ_β and ψ_d and charge densities σ_0, σ_β and σ_d are related by the following equations:

$$\sigma_0 + \sigma_\beta + \sigma_d = 0 \tag{68}$$

$$\psi_0 = \psi_\beta + \frac{\psi_0}{C_1} \tag{69}$$

Table II. The Thickness of an Inner Layer Capacitor as a Function of Inner-layer Capacity (K_1) and Interfacial Dielectric (ϵ_{int})

ϵ_{int}	K_1 uF/cm²					
	30	50	80	100	140	200
5	1.47	0.88	0.55	0.44	0.32	0.22
10	2.94	1.77	1.10	0.88	0.63	0.44
15	4.41	2.65	1.66	1.33	0.95	0.66
20	5.88	3.54	2.21	1.77	1.26	0.88
80		14.1	8.83	7.07	5.04	3.54

$$K_1 = \frac{\epsilon_{int} \times \epsilon_0 \times 10^{6-4}}{\beta \times 10^{-10}} \; uf/cm^2$$

$$\epsilon_0 = 1/(36\pi 10^9)$$

$$\beta = \frac{\epsilon_{int}}{0.1131 \; K_1} \; \text{Å}$$

$$\psi_\beta = \psi_d - \frac{\sigma_d}{C_2} \tag{70}$$

$$\sigma_0 = (C_A - C_B + [OH^-] - [H^+])$$
$$= B([SOH_2^+] + [SOH_2^+ Cl^-] - [SO^-] - [SO^-Na]) \tag{71}$$

and if $N_s = B([SOH] + [SO^-] + [SO^-Na^+])$, i.e., the analytical total charge,

$$[SOH] = (N_s - |\sigma_0|)/B \tag{72}$$

and

$$[SO^-] = -\sigma_0 \bigg/ \left\{ B \left[1 + K_{Na}[Na^+] \exp\left(-\frac{e\psi_\beta}{kT}\right) \right] \right\} \tag{73}$$

giving

Table III. Parameters Used in Electrical Double Layer Modeling [45]

System	Ref.	Surface Area (m^2/g)	Surface Site Density (sites/nm^2)	$pK_{a_1}^{int}$	$pK_{a_2}^{int}$	$p*K_{cation}^{int}$	$p*K_{anion}^{int}$	$\log K_{cation}^{int}$	$\log K_{anion}^{int}$	C_1 ($\mu F/cm^2$)	ΔpK_a	$\Delta pK_{complex}$
γ-Al_2O_3/NaCl	[64]	117	8 [67]	5.7	11.5	9.2(?)	7.9(?)	2.3	2.2	100-120(?)	5.8	1.3(?)
α-FeOOH/KNO_3	[34]	48	16.8	4.2	10.8	8.9	6.1	1.9	1.9	100	6.6	2.8
α-FeOOH/NaCl	[65]	32	16.8 [34]	4.9	ND^a	ND	6.6	ND	1.7	ND	5.8(?)	ND
$Fe(OH)_3$/$NaNO_3$	[45]	600(?)	11 [34]	5.1	10.7	9.0	6.9	1.7	1.8	140	5.6	2.1
TiO_2/$LiNO_3$	[34]	20	12	2.6	9.0	7.1	4.5	1.9	1.9	140	6.4	2.6
TiO_2/KNO_3	[34]	20	12	2.6	9.0	7.1	4.5	1.9	1.9	100	6.4	2.6
TiO_2/$Mg(NO_3)_2$ S	[34]	20	12	2.6	9.0	12.5 ($MgOH^+$)	4.5	—	—	140	6.4	—
SiO_2/KCl	[66]	170	5 [68]	ND	6.5	6.8	ND	-0.3	ND	125	8.0(?)	-8.0(?)

^{a}ND = not determined

$\log K_{cation}^{int} = pK_{a_2}^{int} - p*K_{cation}^{int}$ $\Delta pK_a = pK_{a_2}^{int} - pK_{a_1}^{int}$

$\log K_{anion}^{int} = p*K_{anion}^{int} - pK_{a_1}^{int}$ $\Delta pK_{complex} = p*K_{cation}^{int} - p*K_{anion}^{int}$

$$K_a^{int} = \frac{-\sigma_0}{N_s - |\sigma_0|} \cdot \frac{[H^+] \exp\left(-\frac{e\psi_0}{kT}\right)}{\left[1 + [Na] \exp\left(-\frac{e\psi_\beta}{kT}\right) K_{Na}\right]} \quad (74)$$

and since $\alpha_- = -\sigma_0/N_s$,

$$pK_a^{int} = pH - \log\frac{\alpha_-}{1-\alpha_-} + \frac{e\psi_0}{2.3kT} + \log\left[1 + K_{Na}[Na] \exp\left(-\frac{e\psi_\beta}{kT}\right)\right] \quad (75)$$

Note that when the electrolyte concentration $[Na^+] = [NaCl]$ tends to infinite dilution, the last term on the right side of Equation 75 also tends to zero, so that

$$pK_a^{int}\Big|_{\lim c_{salt} \to 0} = pH - \log\frac{\alpha_-}{1-\alpha_-} + \frac{e\psi_0}{2.3kT} \quad (76)$$

In addition, $\psi_0 = 0$ at the pzc, i.e., $\alpha_- = 0$, thus

$$pK_a^{int}\Big|_{\lim\left(\begin{array}{c}c_{salt} \to 1.0\\ \alpha_- \to 0\end{array}\right)} = pH - \log\frac{\alpha_-}{1-\alpha_-} \quad (77)$$

The intrinsic complexation constant may be expressed in a similar manner if it is assumed that at high electrolyte concentration the surface complex sites dominate over the ionized sites, i.e.,

$$\alpha_- = \frac{-\sigma}{N_s} \simeq \frac{B[SONa^+]}{N_s} \quad (78)$$

This leads to the equation for $p^*K_{Na}^{int}$, i.e.,

$$p^*K_{Na}^{int} = pH - \log\frac{\alpha_-}{1-\alpha_-} + \log[Na^+] + \frac{e\psi_0 - e\psi_\beta}{2.3kT} \quad (79)$$

In the presence of surface complex charge, the term $\psi_0 - \psi_\beta$ becomes zero at the pzc, $\alpha_- = 0$, and the log[Na$^+$] term disappears for [Na] = 1.0, so that

$$p^*K_{Na}^{int} = pH - \log \frac{\alpha_-}{1-\alpha_-} \qquad (80)$$

$$\lim \begin{pmatrix} c_{salt} \to 1.0 \\ \alpha_- \to 0 \end{pmatrix}$$

From Equations 77 and 80 it can be seen that for both the intrinsic ionization constant, pK_a^{int}, and the surface complexation constant, $p^*K_{Na}^{int}$, the function $pQ_a = pH - \log \alpha_-/1-\alpha_-$ varies with both the degree of ionization, α_-, and the electrolyte concentration, [NaClO$_4$]. To enable unique determination of the intrinsic ionization constant pK_a^{int}, a double extrapolation plot of pQ_a against $\alpha_- + \sqrt{C_{NaClO_4}}$ is introduced. The data used are those of Stone-Masui and Watillion's reported pQ for carboxylated polystyrene as a function of α_- at various electrolyte additions.

These points are shown as circles in Figure 8. For each ionic strength, curves were drawn through the points to the zero charge condition, i.e., when $\alpha_- + \sqrt{C_{NaClO_4}} = \sqrt{C_{NaClO_4}}$. The extrapolated points thus obtained were then extrapolated to infinite dilution, so that $\alpha_- + $ NaClO$_4$ = 0. Only at this point does $pK_a^{int} = pQ$.

In addition, curves may be drawn through points at which the surface charge density is constant, i.e., $\sigma_0 = \alpha_- N_s = (\psi_0 - \psi_\beta) \cdot C_1$ = constant for various ionic strengths and then extrapolated to infinite dilution [NaClO$_4$] = 0, so that $\alpha_- + \sqrt{C_{NaClO_4}} = \alpha_-$. These extrapolations support the titration data for no added salt. Again, the extrapolated points may be extrapolated to the condition $\alpha_- + \sqrt{C_{NaClO_4}} = 0$. The double extrapolation gives a unique value for pK_a^{int} independent of the degree of ionization or the ionic strength of the electrolyte medium. The value thus obtained is $pK_a^{int} = 4.9 \pm 0.1$. This agrees quite closely to the observed pK_a values for several monomeric carboxylic acids.

To determine the surface complexation constant, the results of Stone-Masui and Watillon [37] were plotted in the form $pQ = pH - \log \alpha_-/1-\alpha_-$ as a function of $\alpha_- - \log$[NaClO$_4$] (Figure 9). Again, the double extrapolation technique can be employed. All the values of pQ for a given ionic strength can be extrapolated to the point of zero charge ($\alpha_- = \psi_0 - \psi_\beta = 0$), so that the coordinates of the points are $(-\log[NaClO_4], pQ_{\alpha=0})$ and these points are then extrapolated to the condition where log[NaClO$_4$] = 0, i.e., $\alpha_- - \log$[NaClO$_4$] = 0. The coordinates of this point are

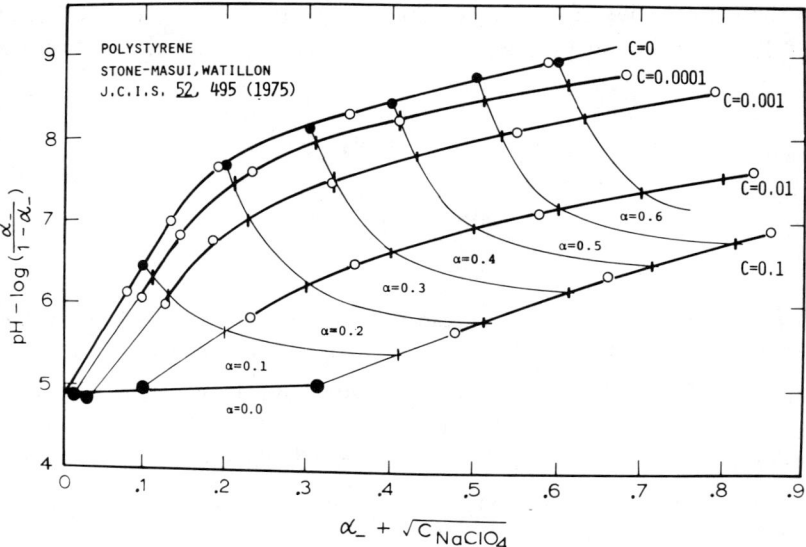

Figure 8. The variation of the surface ionization activity quotient as a function of surface charge and concentration of electrolyte solution. Lines are contours of constant electrolyte or constant surface charge density. The condition $\sigma_0/N_{s_e} = \alpha_- = (C)^{1/2} = 0$ gives pK_a^{int}. Reproduced with permission of the *Journal of Colloid and Interface Science* [62].

$$\begin{pmatrix} 0, pQ_{\alpha=0} \\ \log[Na]=0 \end{pmatrix} = p^*K_{Na}^{int}$$

In addition, points at constant fractional titration, i.e., $\sigma_0 = \alpha_- \cdot N_s =$ constant, can be extrapolated to the condition where $\log[NaClO_4] = 0$, so that $\alpha_- - \log[NaClO_4] = \alpha_-$. These points that have coordinates (α_{-i}, pQ_i) are extrapolated to the zero charge condition, $\alpha_- = 0$, so that again the coordinates of the final extrapolated point are unique, i.e., $(0, pQ = p^*K_{Na}^{int})$.

The results of the double extrapolation technique are shown in Figures 8 and 9 for the carboxylate polystyrene latex of Stone-Masui and Watillon [37]. The pK_a^{int} is 4.9±0.1 and $p^*K_{Na}^{int} = 4.4±0.1$. The difference between these values gives the logarithm of the intrinsic binding constant of Na^+ to SO^- sites for the following reaction:

$$Na_s^+ + SO^- \underset{}{\overset{K_{Na}^{int}}{\rightleftharpoons}} SO^-Na \qquad (81)$$

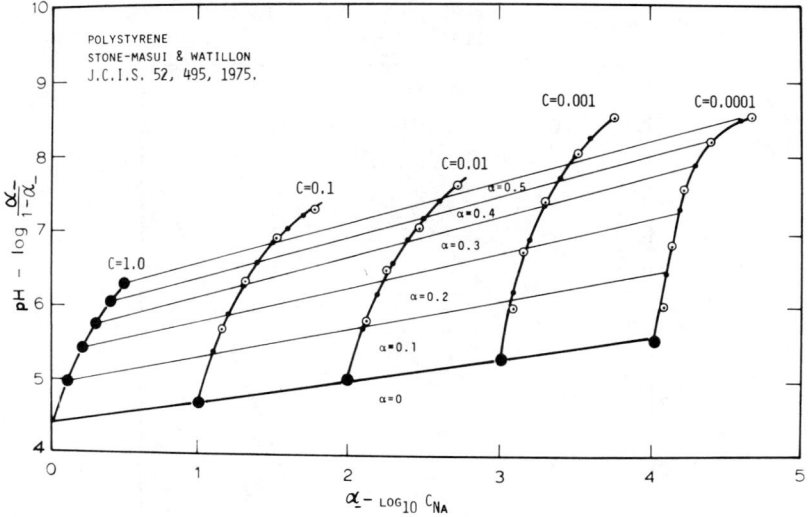

Figure 9. The variation of the surface ionization activity quotient as a function of surface charge and logarithm of electrolyte concentration. Lines are contours for constant electrolyte or constant charge density. The condition that $\sigma_0 = \log C = 0$ gives the surface complex acidity constant $p*K^{int}_{Na}$. Data of Stone-Masui and Watillon [37] and of James et al. [62], the latter reproduced with permission of the *Journal of Colloid and Interface Science*.

$$K^{int}_{Na} = 10^{pK^{int}_a - p*K^{int}_{Na}} = 10^{0.5 \pm 0.2} \quad (82)$$

$$\sim 3.2$$

The computation of the titration curves for latexes follows the same procedure as for the oxide calculations. The surface and solution equilibria are solved for each set of conditions specified by the equilibrium constants, surface area, capacitances C_1 and C_2, and for the analytical amounts of strong acid or base added during the simulated titration. Here, C_1 and C_2 are adjustable parameters. The value of C_2 is fixed at 20 $\mu F/cm^2$ in line with values for Ag-AgI-I and Hg-H_2O systems. The value of C_2 that gave good fit to the data of Stone-Masui and Watillon was about 80 $\mu F/cm^2$. Minimum estimates, but not exact values for C_1, may be obtained from the slope, $d\sigma_0/dpH$, of the surface charge curves, especially at the higher electrolyte concentrations.

Results of the computations simulating the titration of the latexes are shown in Figures 10 and 11. Figure 11 shows the dependence of pH on the added base for various added electrolyte concentrations. By exchange of Na^+

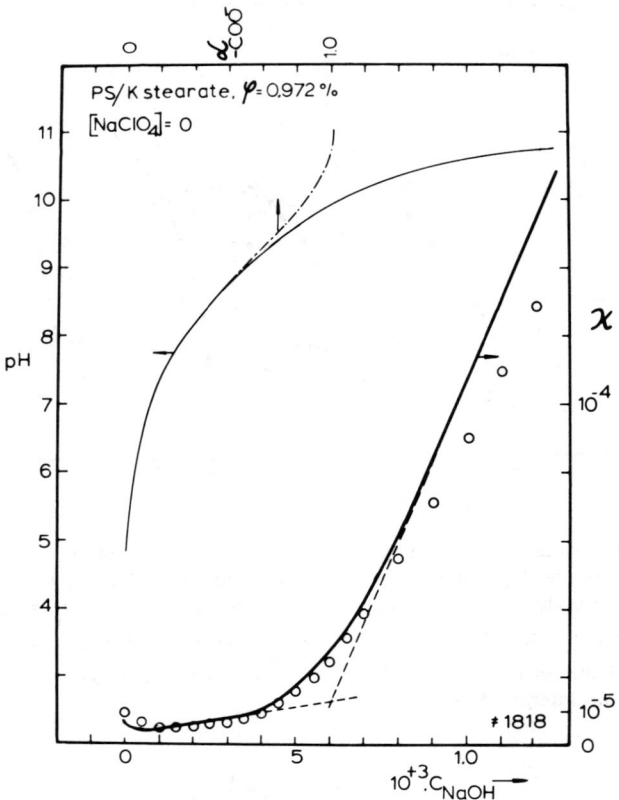

Figure 10. Simulated potentiometric and conductometric titration of a carboxylate polystyrene latex (Stone-Masui and Watillon [37]) with NaOH. From James et al. [62] with permission of the *Journal of Colloid and Interface Science*.

for H^+ at high salt concentrations, the surface groups show apparently enhanced acidity. This is due to the two types of ionization and surface complexation reactions discussed here. It should be noted that other surfaces or colloids that will give quite similar behavior include H-clays [51-57], humic acids [69,70] and ion exchange resins [60,69-77]. Figure 12 show the computed surface charge and diffuse layer potential ψ_d as a function of pH for various added electrolyte concentrations. The charge curves show good agreement with the results of Stone-Masui and Watillon. Although these workers did not report zeta potential, the calculated zeta potential is in quite reasonable agreement with some of their earlier data [78], for example for 10^{-4} electrolyte, equilibrated at pH 4, $\zeta = 62$; pH 6, $\zeta = 92$, and pH 10, $\zeta = 108$ mV, and data reported by Ottewill and Shaw [38] for their polystyrene latexes.

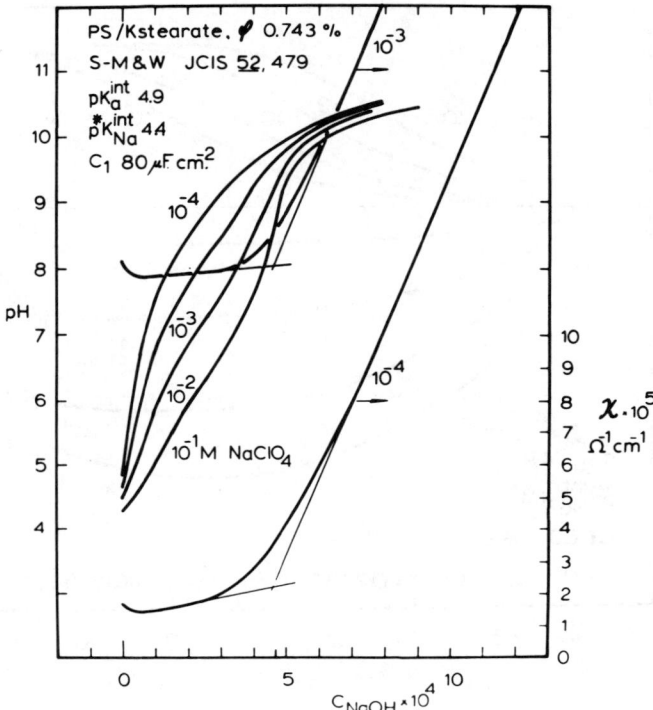

Figure 11. Simulated potentiometric and conductometric titration of a carboxylate latex [17] with NaOH in various NaClO$_4$ solutions. From James et al. [62] with permission of the *Journal of Colloid and Interface Science*.

James et al. [62] also reported calculations for model colloids with different functional groups that represented more acidic ionization and complexation covering the range from weak carboxylic acid groups to strong sulfonate or sulfate surface groups.

Specific Conductivity of Dilute Colloidal Dispersions

Once the ionic distribution between the surface and the solution has been determined, it is a simple matter to estimate the contribution of the solution species to the conductivity of these colloidal dispersions.

James et al. [62] summed the individual ionic equivalent conductances, i.e.,

$$\kappa = \sum_i^{\text{all solution species}} \frac{\lambda_i c_i z_i}{1000} \tag{83}$$

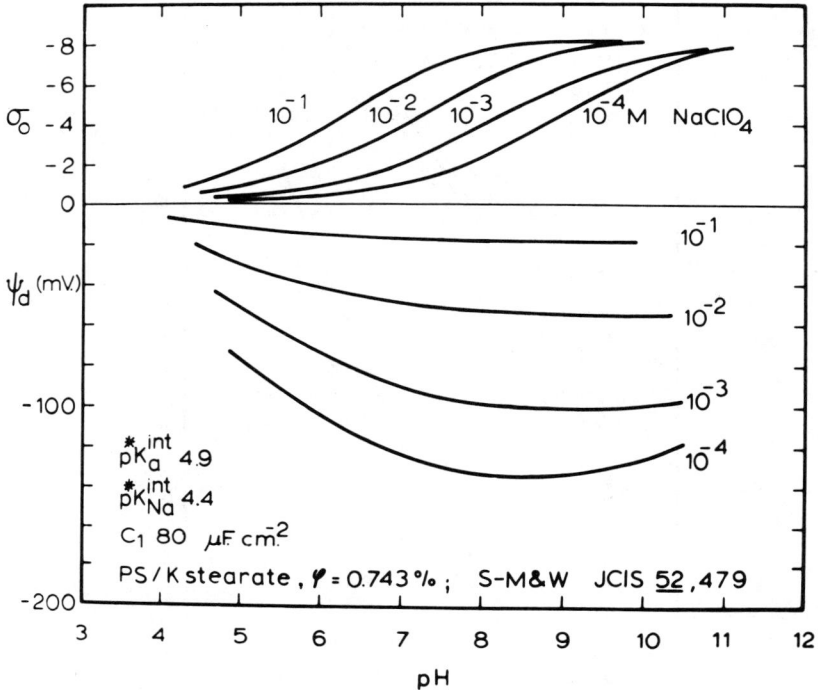

Figure 12. Simulated charge, σ_0, and diffuse layer potential, ψ_d, of a carboxylate latex as a function of pH and electrolyte concentration. From James et al. [62] with permission of the *Journal of Colloid and Interface Science*.

where κ = the specific conductivity (ohm^1/cm^1), and
C_i = the molar concentration of ion, i, charge, Z_i, and equivalent ionic conductance, λ_i.

This latter property was estimated from the limiting ionic equivalent conductance λ_i^0 and the ionic strength, I, by the equation, given by Robinson and Stokes [79]:

$$\lambda_i = \lambda_i^0 - \left[0.230 \, \lambda_i^0 + 30.32 |Z_i|\right] \frac{\sqrt{I}}{1 + 0.329 \times 10^8 \, a\sqrt{I}} \quad (84)$$

where $I = 1/2 \sum C_i Z_i^2$ and a is the ionic size parameter (cm).

To estimate the conductivity of the entire dispersion one would also have to account for 1. the contribution from the surface conductance and adsorbed ions, and 2. the nonadditivity of the individual ionic equivalent conductances in mixed electrolyte solutions [79-81]. However, in most surface characterization investigations by titration, the volume fraction of the colloid can be kept low, e.g., around 1%, so the contribution from surface conductance is relatively low.

The results of our calculations are shown on Figures 10 and 11 together with estimated potentiometric titration results. The computer simulated titrations agree very well with the experimental data.

For more complex surfaces, where there are several types of functional groups, the same approach can be extended [62]. For example, if in addition to N_s SOH sites there are also N_T TOH sites, then we need to define the ionization reactions and determine their equilibrium constants, as, for example, in the following:

$$TOH_2^+Cl \xrightleftharpoons{^*K_{Cl,T}^{int}} TOH + H_s^+ + Cl_s^- \tag{85}$$

and

$$TOH_2^+ \xrightleftharpoons{K_{a_T}^{int}} TOH + H_s^+ . \tag{86}$$

These additional variables, components and known properties can be included in the set of simultaneous equations solved by the computer program MINEQL [45,62,63].

CONCLUSIONS

The characterization of the surface behavior of oxide and polymer colloids has been discussed in terms of the dependence of measured properties, e.g., σ_o, ζ, Γ_i and κ as a function of pH, ionic strength and addition of acids or bases.

This general approach has used simple surface ionization and complexation reactions to form the electrical double layer at the colloid-water interface. The important characteristics of the colloids are their site number density, N_s, and the intrinsic equilibrium constants for the surface reactions. The estimation of the position of equilibrium requires selfconsistent solution of all the equations representing the solution and surface reactions and the presence of the electrical double layer.

Two methods have been employed: Firstly, in the case where surface charge is balanced only by a diffuse charge layer, the graphic methods developed by White et al. [28-30] provide a simple qualitative or semiquantitative demonstration of the dependence of the surface charge, σ_0, and surface potential, ψ_0, on the pH and electrolyte concentration. Using this model the surface potentials need not, or rarely, obey the Nernst equation for reversible electrode couples.

Secondly, using a more complicated but physically more realistic model, which involves ion binding of the electrolyte components to the charged surface sites, potentiometric and conductometric titrations and other properties have been simulated by the computer program MINEQL.

The agreement with experimental values of σ_0, ζ and κ is very good for those systems that have been studied in sufficient detail to report these properties simultaneously. Success in simulating a wide variety of properties creates confidence in the adequacy of the model and the chosen values of certain parameters. It is apparent from literature searches that such detailed studies are rare. If the models proposed are to be compared and contrasted, then much more experimental work on oxides, polymer colloids clays and other surfaces must be considered.

Relatively little work has been done to extend these models to the uptake of dilute ions, e.g., trace metals or organics, onto surfaces in the presence of excess inorganic salts. In this respect, studies by Davis [46] using the experimental data of Anderson [82] on the uptake of As on Al_2O_3, Γ_{As/Al_2O_3} and the zeta potential, ζ_{As/Al_2O_3}, are encouraging.

The overall approach taken here is to try to unify the apparently separate use of double layer theory and the use of ionization, ion exchange and surface-complexation theory to account for interfacial reactions. The models and approaches used here appear to have been quite successful; however, more study is required of the ionization of surfaces and the nature of the interaction between adsorbed ions and surface sites.

ACKNOWLEDGMENTS

Some of the studies discussed here have been made possible through support from the Australian-American Educational Foundation and the U.S. Geological Survey. This support is gratefully acknowledged.

Much of the work reviewed here was originally done in cooperation with Dr. James A. Davis of the Department of Civil Engineering at Stanford University. His cooperation was essential for the initiation of this approach to describing the electrical double layer properties of ionizable colloids.

REFERENCES

1. James, R. O., and T. W. Healy. *J. Colloid Interface Sci.* 40: 42, 53, 65 (1972).
2. Huang, C-P, and W Stumm. *J. Colloid Interface Sci.* 43: 409 (1973).
3. Stumm, W., C-P Huang, and S. R. Jenkins. *Croat. Chem. Acta* 42: 223 (1970).
4. Schindler, P. W., et al. *J. Colloid Interface Sci.* 55: 469 (1976).
5. Hohl, H., and W. Stumm. *J. Colloid Interface Sci.* 55: 281 (1976).
6. James, R. O., P. J. Stiglich, and T. W. Healy. *Disc. Faraday Soc.* 59: 142 (1975).
7. James, R. O., and T. W. Healy. *J. Colloid Interface Sci.* 40: 53 (1972).
8. Tewari, P. H., and W. Lee. *J. Colloid Interface Sci.* 52: 77 (1975).
9. Parfitt, G. D. *Croat. Chem. Acta* 45: 189 (1973).
10. Erikson, L., E. Matijevic, and S. Friberg. *J. Colloid Interface Sci.* 43: 591-598 (1973).
11. Matijevic, E. *J. Colloid Interface Sci.* 58: 374 (1977).
12. James, R. O., G. R. Wiese, and T. W. Healy. *J. Colloid Interface Sci.* 59: 381 (1977).
13. Wiese, G. R., and T. W. Healy. *J. Colloid Interface Sci.* 51: 427 (1975).
14. Wiese, G. R., and T. W. Healy. *J. Colloid Interface Sci.* 51: 434 (1975).
15. Wiese, G. R., and T. W. Healy. *J. Colloid Interface Sci.* 52: 458 (1975).
16. Wiese, G. R., and T. W. Healy. *J. Colloid Interface Sci.* 52: 452 (1975).
17. Grahame, D. C. *Chem. Rev.* 41: 441 (1947).
18. de Bruyn, P. L., and G. E. Agar. "Surface Chemistry of Flotation," in *Froth Flotation,* D. W. Fuerstenan, Ed. American Institute of Mining and Metallurgical Engineers, New York (1962).
19. Bérubé, Y. G., and P. L. de Bruyn. *J. Colloid Interface Sci.* 27: 305 (1968).
20. Bérubé, Y. G., and P. L. de Bruyn. *J. Colloid Interface Sci.* 28: 92 (1968).
21. Blok, L., and P. L. de Bruyn. *J. Colloid Interface Sci.* 32: 518, 527, 533 (1970).
22. Lyklema, J., and J. Overbeek. *J. Colloid Interface Sci.* 16: 595 (1961).
23. Wiese, G. R., R. O. James, D. E. Yates, and T. W. Healy. "The Electrochemistry of the Colloid-Water Interface," in *International Review of Science, Physical Chemistry series 2, Volume 6–Electro chemistry,* J. O'M. Bockris, Ed. (London: Butterworths, 1976).
24. Healy, T. W., and L. R. White. *Adv. Colloid Interface Sci.* 9: 303 (1978).
25. Hunter, R. J., and H. J. L. Wright. *J. Colloid Interface Sci.* 37: 564 (1971).
26. Wright, H. J. L., and R. J. Hunter. *Aust. J. Chem.* 26: 1183, 1191 (1973).
27. Levine, S., and A. L. Smith. *Disc. Faraday Soc.* 52: 290 (1971).
28. Healy, T. W., D. E. Yates, L. R. White, and D. Chan. *J. Electroanal Chem.* 80: 57 (1977).
29. Chan, D., J. W. Perram, L. R. White, and T. W. Healy. *J. Chem. Soc. Faraday Trans.* I, 71: 1046 (1975).
30. Chan, D., T. W. Healy, and L. R. White. *J. Chem. Soc. Faraday Trans.* I, 72: 2844 (1976).

31. Prieve, D. C., and E. Ruckenstein. *J. Colloid Interface Sci.* 60: 337 (1977).
32. Prieve, D. C., and E. Ruckenstein. *Colloid Interface Sci.*, Vol. IV *Hydrosols,* M. Kerker, Ed. (1976).
33. Prieve, D. C., and E. Ruckenstein. *J. Theoret. Biol.* 56: 205 (1976).
34. Yates, D. E. "On the Structure of the Oxide/Water Interface," Ph.D. Thesis, University of Melbourne, Australia (1975).
35. Homola, A., and R. O. James. *J. Colloid Interface Sci.* 59: 123 (1977).
36. James, R. O., A. Homola, and T. W. Healy. *J. Chem. Soc. Faraday Trans.* I, 73: 1436 (1977).
37. Stone-Masui, J., and A. Watillon. *J. Colloid Interface Sci.* 52: 479 (1975).
38. Ottewill, R. H., and J. N. Shaw. *Kolloid z.z. Polymere* 218: 34 (1967).
39. Yates, D. E., R. H. Ottewill, and J. W. Goodwin. *J. Colloid Interface Sci.* 62: 356 (1977).
40. Allen, L. H., and E. Matijević. *J. Colloid Interface Sci.* 33: 420 (1970).
41. Allen, L. H., E. Matijević, and L. Meites. *J. Inorg. Nucl. Chem.* 33: 1293 (1971).
42. Yates, D. E. "Potentiometric Titration of a Carboxyl Latex," in *Polymer Colloids,* NATO Advanced Study Institute, Trondheim, Norway (1975).
43. Schindler, P. W., and H. R. Kamber. *Helv. Chim. Acta* 51: 1781 (1972).
44. Schindler, P. W., and H. Gamsjäger. *Kolloid z.z. Polymere* 250: 759 (1972).
45. Davis, J. A., R. O. James, and J. O. Leckie. *J. Colloid Interface Sci.* 63: 480 (1978).
46. Davis, J. A., Ph.D. Thesis, Stanford University (1977).
47. Davies, J. T. *J. Colloid Interface Sci.* 11: 377 (1956).
48. Payens, Th. A. J. *Philips Res. Rep.* 10: 425 (1955).
49. Bolt, G. H. *J. Phys. Chem.* 61: 1166 (1967).
50. Parks, G. A., and P. L. de Bruyn. *J. Phys. Chem.* 66: 967 (1962).
51. Baver, L. D. *Soil Sci.* 29: 291 (1930).
52. Mattson, S., and L. Wiklander. *Soil Sci.* 49: 109 (1940).
53. Marshall, C. E., and W. E. Bergman. *J. Phys. Chem.* 46: 52 (1942).
54. Marshall, C. E., and W. E. Bergman. *J. Phys. Chem.* 46: 325 (1942).
55. Marshall, C. E., and C. A. Krinbill. *J. Phys. Chem.* 46: 1077 (1942).
56. Desphande, K. B., and C. E. Marshall. *J. Phys. Chem.* 63: 1659 (1959).
57. Desphande, K. B., and C. E. Marshall. *J. Phys. Chem.* 65: 33 (1961).
58. Harward, M. E., and N. T. Coleman. *Soil Sci.* 78: 181 (1954).
59. Helferrich, F. *Ion Exchange* (New York: McGraw-Hill Book Co., 1962).
60. Tanford, C. *Physical Chemistry of Macromolecules* (New York: John Wiley & Sons, Inc., 1961).
61. Yates, D. E., S. Levine, and T. W. Healy. *J. Chem. Soc. Faraday Trans.* I, 70: 1807 (1974).
62. James, R. O., J. A. Davis, and J. O. Leckie. *J. Colloid Interface Sci.* 65: 331 (1978).
63. Westall, J. C., J. L. Zachary, and F. M. M. Morel. Technical Note #18, Water Qual. Lab., Department of Civil Engineering, Massachusetts Institute of Technology (1976).
64. Huang, C-P. "The Chemistry of the Aluminum Oxide-Electrolyte Interface," Ph.D. Thesis, Harvard University, Cambridge, MA (1971).
65. Hingston, F. J., A. M. Posner, and J. P. Quirk. *Adv. Chem. Ser.* 79: 82 (1968).

66. Abendroth, R. P. *J. Colloid Interface Sci.* 34: 591 (1970).
67. Peri, J. B. *J. Phys. Chem.* 69: 211 (1965).
68. Armstead, C. G., and A. J. Tyler. *J. Phys. Chem.* 73: 3947 (1969).
69. Gilmour, J. T., and N. T. Coleman. *Soil Sci. Soc. Amer.* 35: 710 (1971).
70. Posner, A. M. *Trans. 8th Int. Cong. Soil Sci.* (1964), pp. 161-174.
71. Weiss, D. E., et al. *Aust. J. Chem.* 19: 561 (1966).
72. Weiss, D. E., et al. *Aust. J. Chem.* 19: 589 (1966).
73. Weiss, D. E., et al. *Aust. J. Chem.* 19: 765 (1966).
74. Weiss, D. E., et al. *Aust. J. Chem.* 19: 791 (1966).
75. Weiss, D. E., et al. *Aust. J. Chem.* 21: 2703 (1968).
76. Hamann, S. D., and C. H. J. Johnson. *Aust. J. Chem.* 21: 2695 (1968).
77. Hamann, S. D. *Aust. J. Chem.* 23: 1749 (1970); 24: 1979, 2439 (1968).
78. Watillon, A., and J. Stone-Masui. *J. Electroanal. Chem.* 37: 143 (1972).
79. Harned, H. S., and B. B. Owen. "The Physical Chemistry of Electrolytic Solutions," 3rd ed. (New York: Van Nostrand Reinhold Co., 1958).
80. Robinson, R. A., and R. H. Stokes. "Electrolyte Solutions," 2nd ed. (London: Butterworths, 1970).
81. Marion, G. M., and K. L. Babcock. *Soil Sci.* 122: 181 (1976).
82. Anderson, M. A., J. F. Ferguson, and J. Gavis. *J. Colloid Interface Sci.* 54: 391 (1976).

CHAPTER 7

ADSORPTION MODELS: A MATHEMATICAL ANALYSIS IN THE FRAMEWORK OF GENERAL EQUILIBRIUM CALCULATIONS

Francois M. M. Morel
Department of Civil Engineering
Massachusetts Institute of Technology
Cambridge, Massachusetts

Joseph G. Yeasted
Department of Civil Engineering
University of Pittsburgh
Pittsburgh, Pennsylvania

John C. Westall
Swiss Federal Institute of
 Technology
EAWAG
CH-8600 Duebendorf,
Switzerland

INTRODUCTION

At present, one of the major challenges for aquatic chemists—be they interested in oceans, lakes, groundwaters or sewage—is to account quantitatively for the chemical processes at the solid-liquid interface. In recent years, much experimental and theoretical work has focused on the question of adsorption in aqueous systems, resulting in the development of a whole array of adsorption models [1-7]. While models from various research groups were quite different from each other only a few years ago, a process of converging evolution has taken place. Effectively, most modern adsorption models share the same underlying paradigms; most are particularizations of the same "meta-model."

In this chapter, both the unity and the differences among adsorption models will be examined from the point of view of general equilibrium calculations for aquatic systems. We shall first discuss the principles common to such models and the possible variations on the basic themes, not as a systematic literature review, but as a simple exposé of the principal model elements.

The mathematical consequences of the differences among models will then be analyzed to understand the implications of particular model choices. The main focus of this analysis is the relationship between experimental data, model structure and parameter estimation. This will provide a method to understand to what degree model entities and model parameters have a fundamental and unique chemical and thermodynamic meaning, or to what degree they are arbitrary and model dependent. Finally, we shall examine how adsorption models can be fitted into the general scheme of equilibrium calculations to provide a more adequate thermodynamic description of aquatic systems.

ADSORPTION MODELS FOR NATURAL WATERS

Many equilibrium models designed to interpret or predict the interactions of various solutes with various suspended solids have been described. Such models are made to account for acid-base properties of solid suspensions, for adsorption isotherms, for electrical charge and electrical potential data. Since the most important adsorbing inorganic solids from the point of view of the chemistry of natural waters are hydroxides, oxides and clays, this discussion will focus on adsorption models for solids that contain oxygen atoms or hydroxyl groups at the solid-water interface. In addition, given our lack of knowledge about organic solutes in natural waters, we shall consider the major potential adsorbates to be simple inorganic cations and anions. To simplify the notation, consider a hydrolyzable divalent cation, Pb^{2+}, and a weak triprotic acid, H_3PO_4.

All adsorption models discussed here are effectively equilibrium thermodynamic models, i.e., they are defined by their stoichiometric and energetics expressions, not by the underlying mechanistic interpretation of the adsorption processes. For example, the mathematics of models in which specific reactive surface sites are considered are not necessarily different from those of models in which some surface area is "occupied" by an adsorbate. Also, from a thermodynamic point of view, it is a matter of semantics to argue whether adsorbates and surfaces react chemically. No matter what the origin of the surface-solute interaction, whether "physical" or "chemical," it is a process characterized by a particular free energy change and a particular stoichiometry, and we can always write it as a "chemical reaction." In a thermodynamic framework all adsorption models are effectively coordination models, so we shall use here the corresponding terminology, i.e., surface complexes, surface reactions, etc.

What sets surface adsorption apart thermodynamically from reactions among solutes is the variability of the electrostatic energy of interaction (the coulombic term) and the necessity to separate it from the total free energy change:

$$\Delta G_{adsorption} = \Delta G_{intrinsic} + \Delta G_{coulombic} \qquad (1)$$

The distinction between ΔG_{coul} and ΔG_{int} is necessary (as it is in the case of reactions with polyelectrolytes) because of the long-range nature of electrostatic interactions and the proximity of the various adsorption sites whose charges vary depending on their reactions with bulk solutes. ΔG_{coul} depends not only on the charges of the reacting species, but also on the charges of the neighboring species. Equation 1, which is the basis for the classical Stern theory [8], can also be considered to be the basis of all adsorption models discussed here. Such models then differ from each other in only three ways:
1. the set of surface species considered and the corresponding surface reactions;
2. the mathematical expression of free energy minimum, i.e., the "mass law" as a function of surface site "concentrations"; and
3. the formulation of the coulombic term.

Note that simple differences in the numerical values of ΔG_{int} are not considered here to make models different since, given sufficient experimental data and adequate choices for 1, 2 and 3 above, a single value of ΔG_{int} for each reaction ought to be obtainable. Also, note again that differences among models are ultimately based exclusively on algebraic relations and not on terminology, mechanisms or pictures of the surface structure.

SURFACE SPECIES

Although there are still many different views on the nature of surface species, i.e., the nature of the solute-solid interaction and the location of the adsorbates near the surface, there is increasing consensus on their formulation from a stoichiometric point of view. Three main types of surface species are considered: (1) protonated and hydrolyzed surface sites; (2) surface sites coordinated with major ions of the background electrolyte; and (3) surface-sorbate complexes.

1. Although many different symbols are used to describe acid-base surface species, most modern models of hydrous oxide surfaces consider three species: (1) a neutral one; (2) a negatively charged one by loss of a proton (hydrolysis); and (3) a positively charged one by gain of a proton:

$$\equiv SOH°; \quad \equiv SO^-; \quad \equiv SOH_2^+$$

The corresponding acid-base reactions can be written as follows:

$$\equiv SOH_2^+ = \equiv SOH^\circ + H^+; \quad K_{a1}$$

$$\equiv SOH^\circ = \equiv SO^- + H^+; \quad K_{a2}$$

With a suitable expression for the coulombic term (see below), these two acid-base reactions can be made to account for the acid-base properties of hydrous oxide surfaces as obtained from titration data.

2. As in all types of coordination studies, not only the ionic strength, but also the nature of the background electrolyte, exert an influence on the interaction of solutes with surfaces. The two basic approaches to this problem are similar to those used for studying coordination among solutes. In the first one, the quantitative description of the surface adsorption processes is made with the restriction of a particular background electrolyte composition (10^{-2} M NaCl, 3 M NaClO$_4$, seawater, etc.). The thermodynamic constants obtained in such an approach are conditional constants, valid only for the stated conditions. All possible effects of the background electrolyte on the adsorption process are effectively included in the constants. Note that this approach provides as valid a thermodynamic description of chemical process, including adsorption, as that based on extrapolation of equilibrium constants to zero ionic strength.

In the other approach, the effects of the background electrolyte are separated into those that can be accounted for on the basis of ionic strength effects (as in activity coefficient corrections and double layer compression) and those that are dependent on the nature of the electrolyte. The latter are described as coordination reactions between the surface and the major ions, much like "ion pairs" are sometimes used to account for the effective activities of major seawater ions. The type of surface species considered may be written as follows:

$$\equiv SONa^\circ \quad ; \quad \equiv SOH_2Cl^\circ$$

The advantage of this approach is to permit a more universal thermodynamic description of the adsorption process. Its major drawbacks are the added complexity (although this is hardly a problem with computer programs) and the extensive data requirements.

3. Adsorption models differ greatly in the species considered to account for the interaction of surfaces with specifically adsorbed solutes. However, these species are always a subset of the "reasonable" combinations of surface sites at various degrees of protonation or hydrolysis with solutes at various degrees of protonation or hydrolysis. For example, for a cation Pb^{2+}: $\equiv S-O-Pb^+$; $\equiv S-O-PbOH^\circ$; $\equiv S-O-Pb(OH)_2^-$;

$(\equiv S-O\genfrac{}{}{0pt}{}{H}{Pb})^{2+}$; $(\equiv\genfrac{}{}{0pt}{}{S-O}{S-O} Pb)°$; etc.

for an anion PO_4^{3-}: $\equiv S-O-PO_3H_2°$; $\equiv S-O-PO_3H^-$;

$(\equiv S-O\genfrac{}{}{0pt}{}{H}{PO_3H_2})^+$; $(\equiv\genfrac{}{}{0pt}{}{S-O}{S-O} PO_2H)°$; $(\equiv\genfrac{}{}{0pt}{}{S-O}{S-O} PO_2)^-$; etc.

Although the species written above appear to imply particular reaction mechanisms (coordination with the surface oxygen atom for Pb^{2+} and sharing of surface oxygen atoms for PO_4^{3-}) they need not be considered in this way. For example, the same species in a general model with surface component S^- and solutes M^{2+} and L^{3-} would be simply written as follows:

SM^+; $SMOH°$; $SM(OH)_2^-$; SMH^{2+}; $S_2M°$; etc. and

$SOH_4°$; SLH_3^-; SLH_5^+; $S_2LH_5°$; $S_2LH_4^-$; etc.

In the end, the actual choice of surface species considered in a particular adsorption model is always a compromise between simplicity and the necessity to fit experimental data. The more extensive the data, the more likely it is that a model will have to include several surface species. Note that the major effect of the choice of species such as those considered above is to change the total concentration of surface complexes (i.e., the adsorption) as a function of pH. The actual choice of, and the thermodynamic constants for, surface complexes in a given model is thus highly dependent on the prior characterization of the acid-base properties of the surface, the consideration of acid-base reactions in solution, and the precise formulation of the coulombic interaction term, as will be discussed later.

Although the examples of specifically sorbed species presented so far have focused exclusively on hydrolyzed metals and protonated ligands, there is no reason why any metal ligand complex cannot also be considered to adsorb if experimental data warrant it. Note also that there is no reason, a priori, to rule out the existence of several types of surface sites, each with its own acid-base and coordination characteristics. Such generalizations are easy to implement in practice when needed, but are not appropriate for this discussion.

MASS LAW EQUATIONS

Following the original treatment of Stern [8], all modern adsorption models express the thermodynamic relations for reactions between solutes and surfaces as typical mass law expressions. For example,

$$K_{a1} = \frac{\equiv SOH^\circ \; H^+}{\equiv SOH_2^+} \qquad (2)$$

$$K_{Pb} = \frac{\equiv SOPb^+}{\equiv SO^- \; Pb^{2+}} \qquad (3)$$

The activities of the solutes H^+ and Pb^{2+} are taken as their bulk activities (the coulombic term is included in K_{a1} and K_{Pb}) and are often replaced by their concentrations, effectively including the activity coefficients in the adsorption constants. The activities of the surface species can be expressed in a great variety of units. Some are based on the number of sites occupied by the species, e.g., site concentration (moles of sites per liter of solution), site density (moles of sites per surface area), or mole fractions (moles of sites per total number of surface sites or moles of sites per total number of moles in the system). And some are based on the surface area occupied by the species, e.g., surface concentration (area of sites per liter of solution) or surface density (area of sites per total surface area). The multiplicity of possible representations creates confusion in comparing different models. If the specific surface area occupied by the various types of surface species is considered identical, all representations are equivalent (e.g., K_{a1} and K_{ads} are independent of the units chosen) and the conservation equation can be written as a mole balance or as an expression for the total surface area. For example,

$$A = A_{\equiv SO^-} + A_{\equiv SOH^\circ} + A_{\equiv SOH_2^+} + A_{\equiv SOPb^+} \cdots \qquad (4)$$

$$A = a_{\equiv SO^-}(\equiv SO^-) + a_{\equiv SOH^\circ}(\equiv SOH^\circ) + \ldots \qquad (5)$$

where A = the total area, m^2/l,
$(\equiv SO^-)$, etc. = the site concentrations, mol/l, and
a_{SO^-}, etc. = the specific site areas (m^2/mol).

This yields

$$S_T = (\equiv SO^-) + (\equiv SOH^\circ) + \ldots \qquad (6)$$

if

$$a_{\equiv SO^-} = a_{\equiv SOH^\circ} = \ldots = a = A/S_T.$$

However, if the specific surface area occupied by each type of surface species is considered different, the equilibrium constants have to be modified when changing from units based on area to units based on number of sites:

$$K_{Pb} = \frac{(\equiv SOPb^+)}{(\equiv SO^-)(Pb^{2+})} \qquad (7)$$

$$K_{Pb}^A = \frac{A_{\equiv SOPb^+}}{A_{\equiv SO^-} \cdot (Pb^{2+})} \qquad (8)$$

and, therefore,

$$K_{Pb} = \frac{a_{\equiv SO^-}}{a_{\equiv SOPb^+}} \cdot K_{Pb}^A \qquad (9)$$

and the conservation equation has to be expressed as a total surface area expression, including the specific site areas as "stoichiometric" coefficients (Equation 5).

Because it is by far the simplest and the most convenient representation to include in general equilibrium calculations, we shall consider here that surface activities are given by site concentrations (mol/l) and that the conservation equation can be written as a mole balance (Equation 6). Experimental data to support the use of different specific site areas for various surface species is simply lacking.

A fundamental difference among models with respect to the thermodynamic expressions of surface reactions is encountered when considering polydentate species. For example, with a bidentate species such as, $\begin{matrix}\equiv S-O\\ \equiv S-O\end{matrix}Pb$, some write the mass law with a square dependence on the surface site concentration, as follows:

$$K = \frac{((\equiv SO)_2 Pb)}{(\equiv SO^-)^2 \cdot (Pb^{2+})} \qquad (10)$$

Some write it with a linear dependence on it, as if dealing with a bidentate ligand:

$$K' = \frac{((\equiv SO)_2 Pb)}{((\equiv SO)_2{}^{2-})(Pb^{2+})} \qquad (11)$$

The second expression has a rather intuitive appeal. Doubling the solid concentration will undoubtedly no more than double the amount of adsorbate. If one considers the probability of finding two adjacent sites free to react with a potential absorbate compared to the probability of finding one free site, it is obviously unity at low adsorbate density and must decrease sensibly only at very high adsorbate densities (simple model calculations indicate near 90% saturation). Although a precisely correct formulation must be more complicated, the simple mass law with an exponent of unity (Equation 11) appears adequate, given experimental uncertainties and model flexibility. In a model of arsenate adsorption on aluminum oxide in which the decrease in maximum adsorption as a function of pH is explained by invoking bidentate and tridentate surface species [9], it was found necessary to use an exponent of unity in the various mass law expressions to fit the experimental data.

FORMULATION OF COULOMBIC INTERACTIONS

Given the value of the electrical potential, Ψ, at the locus of adsorption, compared to a bulk potential of zero, the coulombic energy of interaction can always be written as follows:

$$\ln K_{coul} = \frac{-1}{RT} \Delta G_{coul} = -\Delta Z \frac{-F}{RT} \Psi \qquad (12)$$

where ΔZ is the net change in charge number of the surface species due to the adsorption reaction. For example:

$$\equiv SOH_2{}^+ = \equiv SOH^\circ + H^+ \;;$$

$$\Delta Z = -1; \; K_{a1} = K_{a1}^{int} K_{coul} = K_{a1}^{int} \exp(\frac{F}{RT} \Psi) \qquad (13)$$

$$\equiv SO^- + Pb^{2+} = \equiv SOPb^+ \;;$$

$$\Delta Z = +2; \; K_{Pb} = K_{Pb}^{int} K_{coul} = K_{Pb}^{int} \exp(\frac{-2F}{RT} \Psi) \qquad (14)$$

The coulombic energy, ΔG_{coul}, is simply the energy required to bring an ion from the bulk solution to the surface at a potential, Ψ. The difficulty, of course, is in determining the correct potential, Ψ, since, unlike the situation in classical electrode systems, the surface potential of suspended solids cannot be determined directly.

According to the description of the acid-base chemistry of the surface presented above, it is clear that an ideal Nernstian coulombic term with H^+ as the potential-determining ion is not applicable here:

$$\Psi = 2.3 \frac{RT}{F} (pH_{pzc} - pH) \qquad (15)$$

Such an ideal expression corresponds effectively to a constant activity of the potential-determining ion at the surface, as can sometimes be imposed by solubility relations. Here, the hydrogen ion activity at the surface, as determined from acid-base surface reactions, is far from constant. Further, all specifically sorbed ions may be considered to some degree to be potential-determining ions, and the surface potential is the result of all concomitant reactions at the interface.

Because of this complexity, modern models of adsorption on oxide surfaces all approach the question of the coulombic term on an empirical, rather than theoretical, basis. The surface potential(s), Ψ, is taken to be a unique function of the surface charge(s), σ, and the expression $\Psi(\sigma)$ is determined by fitting the given adsorption model to experimental data (see the next section). Remarkably, virtually all models end up with a simple proportionality between Ψ and σ, as follows:

$$\Psi = \frac{1}{C} \sigma \qquad (16)$$

where C is the surface capacitance. Since C is found to be approximately constant (but model dependent), at least for surface charges of a given polarity (e.g., C^- for $\sigma < 0$; C^+ for $\sigma > 0$), this empirical expression shall be used in the mathematical formulations of the models.

The major difference among models, in terms of the expression for coulombic interactions, stems from their geometric description of the adsorption process. If all ions (H^+, Pb^{2+}, Na^+, Cl^-) are considered to react with the surface in the same surface plane, the same potential, Ψ, is considered to apply to all of them, and the surface charge is determined simply by the sum of the charges of the surface species:

$$A/F \cdot \sigma = (\equiv SOH_2^+) - (\equiv SO^-) + (\equiv SOPb^+) \ldots \text{etc.}$$

where σ = the specific surface charge density in the adsorption plane,
A = the total surface area, and
F = the Faraday constant.

On the contrary, if different ions are considered to react with the surface in different surface planes, based on their chemical nature or on their size, different potentials, Ψ_1, Ψ_2, etc., must be defined for each of the adsorption plane. A surface charge must be calculated for each plane and a capacitance determined for each adsorption layer.

Regardless which approach is taken, it is essential that the coulombic term be defined and estimated coherently with the other specifics of the adsorption model, i.e., the choice of surface species. For example, in an adsorption model that does not consider surface coordination with major electrolytes, the surface charge has to be entirely accounted for by acid-base surface species and by specifically adsorbed ions. In such a case, the coulombic term can be obtained directly as a function of pH, $\Psi(pH)$, if the specific adsorption is far from saturating the surface, and has effectively no effect on the surface charge. With this limitation, such expressions $\Psi(pH)$ are perfectly coherent with the general approach discussed here.

MATHEMATICAL FORMULATIONS

As noted earlier, differences among models have to be examined ultimately on the basis of differences in their mathematical formulations. Presented here is the complete set of equations for three models that span the range of variability discussed in the preceding sections. For contrast, a fourth model will be outlined that does not fit into the general framework presented in this chapter, that is, one which is not part of the "metamodel."

First Model

In the first and simplest model, let us not consider the coordination of major electrolytes with the surface, and take all ions (H^+ and Pb^{2+} for the following expressions) to be adsorbed in the same surface plane as indicated schematically in Figure 1. Equation 16 and the following define the model:

$$A/F \cdot \sigma = (\equiv SOH_2^+) - (\equiv SO^-) + (\equiv SOPb^+) \qquad (17)$$

$$S_T = (\equiv SOH_2^+) + (\equiv SOH^\circ) + (\equiv SO^-) + (\equiv SOPb^+) + 2((\equiv SO)_2 Pb^\circ) \qquad (18)$$

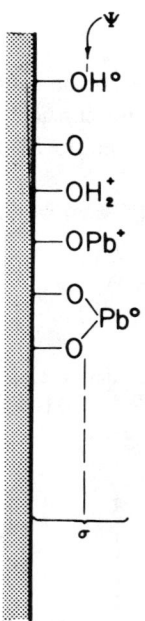

Figure 1. First model: fixed ionic strength; no surface coordination of major electrolytes; all surface species at the same potential.

$$K_{a1}^{int} \cdot \exp(\frac{F}{RT}\Psi) = \frac{(\equiv SOH^{\circ})(H^+)}{(\equiv SOH_2^+)} \quad (19)$$

$$K_{a2}^{int} \cdot \exp(\frac{F}{RT}\Psi) = \frac{(\equiv SOH^-)(H^+)}{(\equiv SOH^{\circ})} \quad (20)$$

$$K_{Pb1}^{int} / \exp(\frac{2F}{RT}\Psi) = \frac{(\equiv SOPb^+)}{(\equiv SO^-)(Pb^{2+})} \quad (21)$$

$$K_{Pb2}^{int} / \exp(\frac{2F}{RT}\Psi) = \frac{((\equiv SO^-)_2 Pb^{\circ})}{((\equiv SO)_2^{2-})(Pb^{2+})} \quad (22)$$

The choice of surface species $\equiv SOPb^+$ and $(\equiv SO)_2 Pb^{\circ}$ is arbitrary here, and, in general, this choice is determined by the necessity to fit experimental data. If A and S_T can be obtained independently, this model contains five constants (C, K_{a1}^{int}, K_{a2}^{int}, K_{Pb1}^{int}, K_{Pb2}^{int}) to be determined from acid-base titrations and adsorption isotherm data.

Second Model

In the second model, let us keep the simplicity of the unique adsorption plane, Figure 2, but introduce the consideration of surface coordination of major electrolytes. Equations 17 and 19-22 are applicable and, in addition,

$$S_T = (\equiv SOH_2^+) + (\equiv SOH^\circ) + (\equiv SO^-) + (\equiv SONa^\circ) + (\equiv SOH_2Cl^\circ)$$

$$+ (\equiv SOPb^+) + 2((\equiv SO)_2Pb^\circ) \qquad (23)$$

$$K_{Na}^{int}/\exp(\frac{F}{RT}\Psi) = \frac{(\equiv SONa^\circ)}{(\equiv SO^-)(Na^+)} \qquad (24)$$

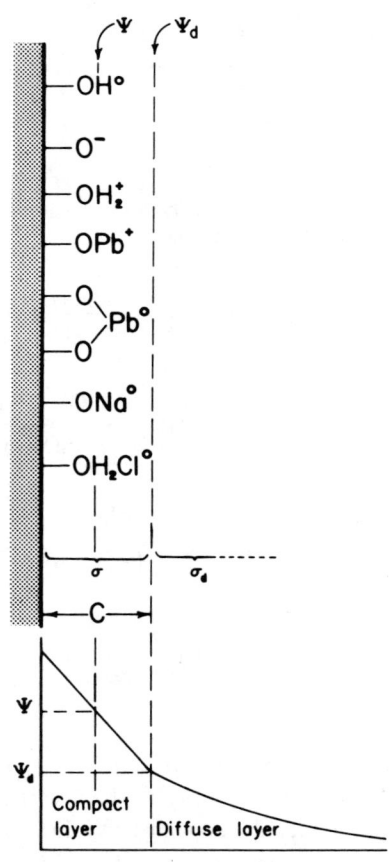

Figure 2. Second model: variable ionic strength; surface coordination of major electrolytes; all surface species at the same potential.

ADSORPTIVE MODELS

$$K_{Cl}^{int} \cdot \exp(\frac{F}{RT}\Psi) = \frac{(\equiv SOH_2 Cl^\circ)}{(\equiv SOH_2^+)(Cl^-)} \quad (25)$$

Although in this model the same simple expression directly relating surface potential and surface charge, Equation 16, is generally applicable, the possibility of changing the background electrolyte and, hence, the ionic strength, has to be considered. The resulting change in the diffuse layer thickness must then be accounted for in the potential-charge relationships:

$$\Psi = \Psi_d + \frac{1}{2}\frac{\sigma}{C} \quad (26)$$

and

$$\sigma_d + \sigma = 0 \quad (27)$$

Using the expression relating charge, σ_d, and potential, Ψ_d, at the boundary of the compact and the diffuse layer as a function of the double layer thickness, $1/\kappa$, for small potentials,

$$\sigma_d \simeq \frac{\epsilon\kappa}{4\pi} \cdot \Psi_d \quad (28)$$

yields

$$\Psi = (\frac{1}{2C} - \frac{4\pi}{\epsilon\kappa})\sigma \quad (29)$$

The more complex expression for large potentials is also readily obtained. Equations 17, 19-25 and 29 completely define an adsorption model in which all adsorption reactions are equivalent (same coulombic term), including adsorption of the background electrolyte ions.

Third Model

The third and most complicated model to be presented here is basically that of Davis [5], who integrated the modeling approach of Bowden et al. [3] with the site-binding model of Yates et al. [2]. In this model, H$^+$ and OH$^-$ are considered to react in an inner adsorption layer (effectively *in* the solid), while specifically sorbed ions, including major electrolytes, are in a layer adjacent to the surface (Figure 3). In effect, the compact adsorption layer, the Stern layer, is split into two parts.

Two different capacitances are assumed for the inner and outer regions of the compact layer, and the expressions of the charge in the inner layer must now include the surface complexes of the major electrolyte ions:

$$A/F \cdot \sigma_1 = (\equiv SOH_2^+) + (\equiv SOH_2 Cl^\circ) - (\equiv SO^-) - (\equiv SONa^\circ)$$

$$- 2((\equiv SO)_2 Pb^\circ) - (\equiv SOPb^+) \qquad (30)$$

$$A/F \cdot \sigma_2 = (\equiv SONa^\circ) - (\equiv SOH_2 Cl^\circ) + 2(\equiv SOPb^+) + 2((\equiv SO)_2 Pb^\circ) \qquad (31)$$

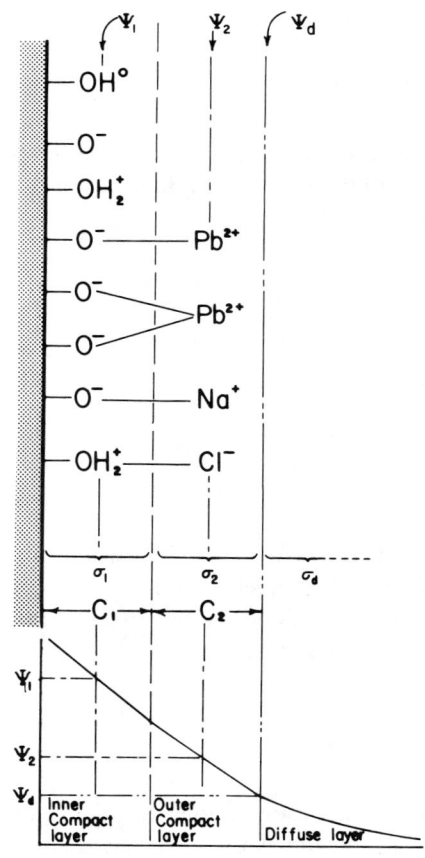

Figure 3. Third model: variable ionic strength; surface coordination of major electrolytes; separation of surface species into two inner layers with two different potentials.

$$S_T = (\equiv SOH_2^+) + (\equiv SOH^\circ) + (\equiv SO^-) + (\equiv SOH_2 Cl^\circ) + (\equiv SONa^\circ)$$
$$+ (\equiv SOPb^+) + 2((\equiv SO)_2 Pb^\circ) \tag{32}$$

$$K_{a1}^{int} \exp\left(\frac{F}{RT}\Psi_1\right) = \frac{(\equiv SOH^\circ)(H^+)}{(\equiv SOH_2^+)} \tag{33}$$

$$K_{a2}^{int} \exp\left(\frac{F}{RT}\Psi_1\right) = \frac{(\equiv SO^-)(H^+)}{(\equiv SOH^\circ)} \tag{34}$$

$$K_{Na}^{int}/\exp\left(\frac{F}{RT}\Psi_2\right) = \frac{(\equiv SONa^\circ)}{(\equiv SO^-)(Na^+)} \tag{35}$$

$$K_{Cl}^{int} \exp\left(\frac{F}{RT}\Psi_2\right) = \frac{(\equiv SOH_2 Cl^\circ)}{(\equiv SOH_2^+)(Cl^-)} \tag{36}$$

$$K_{Pb1}^{int}/\exp\left(\frac{2F}{RT}\Psi_2\right) = \frac{(\equiv SOPb^+)}{(\equiv SO^-)(Pb^{2+})} \tag{37}$$

$$K_{Pb2}^{int}/\exp\left(\frac{2F}{RT}\Psi_2\right) = \frac{((\equiv SO)_2 Pb^\circ)}{((\equiv SO)_2^{2-})(Pb^{2+})} \tag{38}$$

$$\Psi_2 = \Psi_d + \frac{\sigma_2}{2C_2} \tag{39}$$

$$\Psi_1 = \Psi_d + \frac{\sigma_2}{C_2} + \frac{\sigma_1}{2C_1} \tag{40}$$

and, for small potentials,

$$\sigma_d = -(\sigma_1 + \sigma_2) = \frac{\epsilon\kappa}{4\pi}\Psi_d \tag{41}$$

Fourth Model

Although they have different conceptual bases, all expressions of the electrostatic interaction terms presented so far consider constant capacitance(s) for adsorption layer(s), and this is ultimately justified and evaluated empirically on the basis of experimental data. More drastic empirical approaches to the problem are possible. For example, Anderson [4,10] has proposed a model for arsenate adsorption based on a single Langmuir isotherm (thus presumably on only one type of surface reaction):

$$S + X = SX \quad ; \quad K_{int} \cdot K_{coul} \quad (42)$$

Therefore,

$$\Gamma = \Gamma_{max} \frac{(X) K_{int} \cdot K_{coul}}{1 + (X) K_{int} \cdot K_{coul}} \quad (43)$$

The "coulombic constant" K_{coul} is then obtained on the basis of extensive isotherm and electrophoretic data in the form of two empirical plots: K_{coul} vs $(pH_{iep} - pH)$ and pH_{iep} vs Γ_{iep}. In this approach, the "coulombic" term accounts for all pH effects, including electrostatic interactions at the surface, acid-base chemistry of the surface, and even acid-base chemistry of the anion in the bulk solution (X is the total anion concentration in solution). Although such amalgamation of terms may limit the range of applicability of the resulting model, it illustrates how the separation between acid-base and electrical properties of the surface is model dependent, and thus somewhat arbitrary.

CHOICE OF SURFACE SPECIES:

Fitting Acid-Base Titration Data

Given a particular adsorption model and experimental adsorption data, the problem of obtaining suitable model parameters is, in principle, a simple matter of optimization. The model parameters should be determined by minimizing the differences between model predictions and experimental measurements. We shall discuss later how experimental data can be fitted by a given model with various combinations of parameters and how sensitive a model is to the values of its parameters. This section illustrates how different models can, on the basis of identical data, result in widely different choices of surface species, in widely different evaluations of surface parameters, which are ostensibly identical.

Although a simultaneous fit of all available data for a given surface and a given adsorbate is, in principle, preferable, it is instructive to break down this procedure into two consecutive steps: (1) the fitting of acid-base titration data by acid-base surface species, coulombic expression and, if considered, surface complexes of major electrolyte ions; and (2) fitting adsorption isotherm data, both as a function of sorbate concentration and of pH, by specific surface-sorbate complexes. If electrical charge or electrical potential data are available in conjunction with either set of data, they can be used simultaneously for determining model parameters.

Let us consider the data of Hohl and Stumm [11] on acid-base titrations of γ-Al_2O_3 in 0.1 M $NaClO_4$. In Figures 4 and 5 are plotted the calculated apparent acidity quotients, Q^+ and Q^-, as a function of excess or deficiency of surface protons, q. These parameters, whose meaning will become apparent in the following discussion, are obtained directly from the titration data, given a total number of sites S_T:

$$q = (Acid) - (Base) - (H^+) + (OH^-) \tag{44}$$

$$q > 0 \quad ; \quad q^+ = q \quad ; \quad Q^+ = \frac{(S_T - q^+)(H^+)}{q^+} \tag{45}$$

$$q < 0 \quad ; \quad q^- = -q \quad ; \quad Q^- = \frac{q^-(H^+)}{(S_T - q^-)} \tag{46}$$

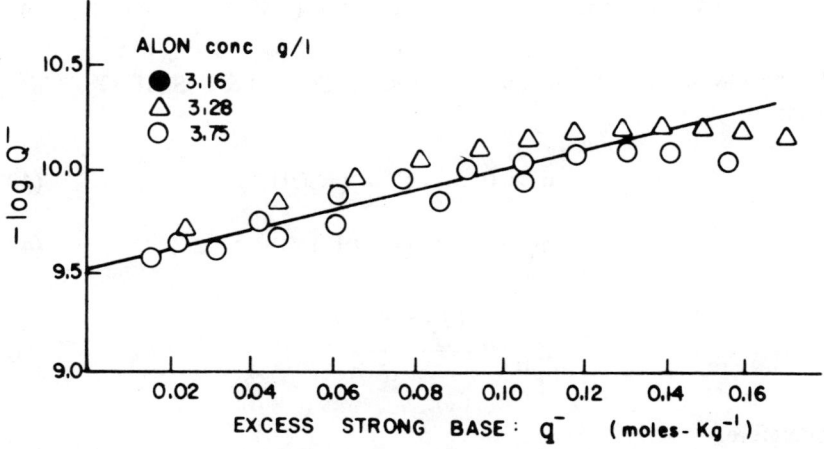

Figure 4. Base titration of γ-Al_2O_3. See text for definition of parameters [11].

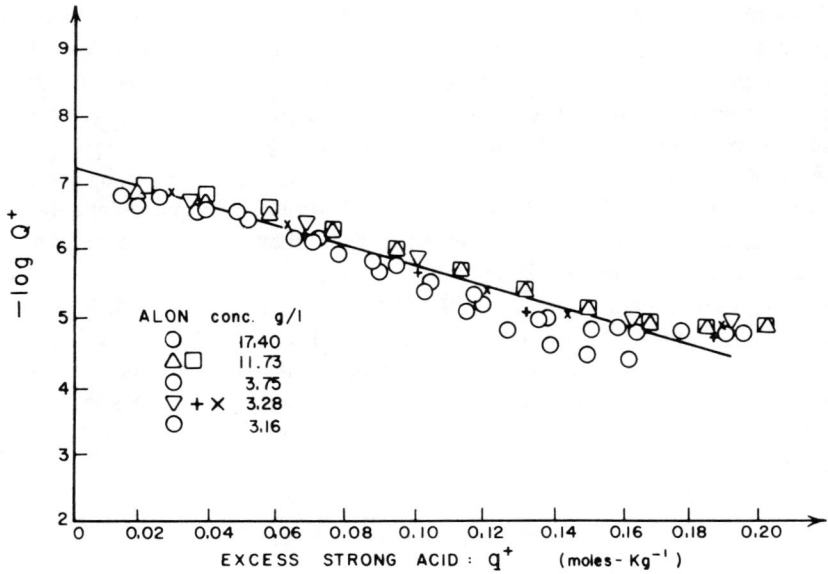

Figure 5. Acid titration of γ-Al_2O_3. See text for definition of parameters [11].

The interpretation of q^+, q^-, Q^+ and Q^-, the obtainment of thermodynamic data from Figures 4 and 5, and the definition of the coulombic correction term from these figures depend on the chosen model. Consider the first model of the previous section, which includes no surface complex of the background electrolyte. The proton balance of the system is given by

$$(Acid) - (Base) = (H^+) - (OH^-) + (\equiv SOH_2^+) - (\equiv SO^-) \quad (47)$$

In the low pH region, the approximation $(\equiv SOH_2^+)$ & $(\equiv SOH^\circ) \gg (\equiv SO^-)$ leads to

$$q > 0 \ : \ q^+ \simeq (\equiv SOH_2^+) \quad (48)$$

$$S_T - q^+ \simeq (\equiv SOH^\circ) \quad (49)$$

$$K_{a1} \simeq \frac{(S_T - q^+)(H^+)}{q^+} \quad (50)$$

Therefore,

$$K_{a1} \simeq Q^+ \qquad (51)$$

Similarly, in the high pH region,

$$(\equiv SO^-) \ \& \ (\equiv SOH^\circ) \gg (\equiv SOH_2^+)$$

$$q < 0 \ : \ q^- \simeq (\equiv SO^-) \qquad (52)$$

$$K_{a2} \simeq \frac{q^-(H^+)}{(S_T - q^-)} \qquad (53)$$

$$K_{a2} \simeq Q^- \qquad (54)$$

According to this model, the plots of Q vs q are plots of apparent acidity constants vs surface charge. The extrapolated values of Q^+ and Q^- for $q = 0$ yield the intrinsic constants K_{a1}^{int} and K_{a2}^{int}. Given a value for the specific surface area A, the function $\Psi(\sigma)$ is obtained directly from the plots, and the slopes of the lines give the surface capacitance. For example,

$$K_{a1} = K_{a1}^{int} \exp\left(\frac{F}{RT} \Psi\right) \qquad (55)$$

Therefore,

$$\Psi = \frac{RT}{F} \ln Q^+/K_{a1}^{int}$$

$$\text{and} \quad \sigma = q^+ \cdot F/A \qquad (56)$$

Consider now the second and third models in the previous section which include surface complexes of the background electrolyte. Assuming no complex formation for ClO_4^-, Na^+ is the only complex-forming ion. The proton balance becomes

$$(Acid) - (Base) = (H^+) - (OH^-) + (\equiv SOH_2^+) - (\equiv SO^-) - (\equiv SONa^\circ) \qquad (57)$$

In the low pH range, for $q > 0$, the approximations are the same as those above. At high pH, the appropriate approximations are not as straightforward and depend on the importance of the sodium surface species. Let us assume $(\equiv SONa^\circ) \gg (\equiv SO^-) \& (\equiv SOH_2^+)$. Therefore,

$$q < 0 \quad : \quad q^- \simeq (\equiv SONa^\circ) \tag{58}$$

$$^*K_{Na^+} = \frac{(\equiv SONa^\circ)(H^+)}{(\equiv SOH^\circ)(Na^+)} = \frac{q^-(H^+)}{(S_T - q^-)(Na^+)} \tag{59}$$

$$^*K_{Na^+} = Q^-/(Na^+) \tag{60}$$

While the plot of Q^+ vs q^+ yields the first acidity constant, unless ClO_4^- surface complexes are considered, the plot of Q^- vs q^- yields the intrinsic surface complexation constant for sodium by extrapolation to $q^- = 0$ (and division by (Na^+)). In these models, the determination of the second acidity constant K_{a2} necessitates additional titration data at low Na^+ concentrations.

Although both the second and third models lead to the same interpretation of the extrapolated value of Q^-, the expression of the coulombic term is different.

1. In the second model, the constant $^*K_{Na^+}$ is independent of coulombic effects since it corresponds to a reaction with no net charge change (exchange of H^+ for Na^+ at the surface). In order to calculate the surface potential as a function of q^-, it is necessary to obtain the value of the second acidity constant at each pH; for example from data at low Na^+ concentrations:

$$\Psi = \frac{RT}{F} \ln K_{a2}/K_{a2}^{int} \tag{61}$$

The calculation of the surface charge which is given by the concentration of negative surface groups also necessitates an estimation of K_{a2}. This is because the surface proton deficiency, which is dominated at high pH by the sodium surface complex, is unrelated to the surface charge:

$$A/F \cdot \sigma = (\equiv SOH_2^+) - (\equiv SO^-) \tag{62}$$

2. In the third model, conversely, the surface charge σ at high pH is still directly given by the proton deficiency:

$$A/F \cdot \sigma_1 = (\equiv SOH_2^+) - (\equiv SO^-) - (\equiv SONa^\circ) \tag{63}$$

However, there are now two potentials to be determined: Ψ_1, which applies to H^+ exchange, and Ψ_2, which applies to Na^+ binding; and another surface charge: σ_2. The curve Q^- vs q^- yields the difference $(\Psi_1 - \Psi_2)$:

$$*K_{Na} = *K_{Na}^{int} \exp \frac{F}{RT} (\Psi_1 - \Psi_2) \tag{64}$$

and, therefore, a relationship between the surface capacitances and the surface charges:

$$\Psi_2 - \Psi_1 = \frac{\sigma_1}{2C_1} - \frac{\sigma_2}{2C_2} \tag{65}$$

As in the second model, additional acid-base data provide the means to calculate σ_2 as a function of pH:

$$A/F \cdot \sigma_2 = (\equiv SONa^\circ) = -A/F \cdot \sigma_1 + (\equiv SOH_2^+) - (\equiv SO^-) \tag{66}$$

and C_1 is usually calculated given a fixed (arbitrary) value for C_2.

It is now clear that the determinations of the surface acidity constants, the coulombic expression and the complexation constants for the major ions with the surface are completely interdependent and are very much a function of the model considered. Acidity constants determined by one author on the basis of one model cannot be used by another author for another model. For example, Hohl and Stumm [11] determined the second intrinsic acidity constant for $\gamma\text{-}Al_2O_3$ to be $10^{-9.5}$. With similar data, Davis [5] determined the same constant to be $10^{-11.5}$. Although those constants are widely different, both yield a good fit of experimental data if they are used with the appropriate corresponding models. The models themselves are equivalently valid thermodynamic representation of adsorption processes, and neither constant can be said to be correct or incorrect on the basis of available data. It remains that a uniform convention for the choice of an adsorption model would result in a universal set of constants, thus avoiding much of the existing confusion.

Fitting Adsorption Isotherms

Adsorption isotherms for specific sorbates on oxide surfaces can be obtained experimentally as a function of sorbate concentration and pH. From the standpoint of modeling, the more adsorption data there are to be fitted, the more likely it is that several surface species will have to be invoked in the model.

Consider the situation where only one set of adsorption data at fixed pH and varying sorbate concentration is available. Such data usually have the form of a Langmuir isotherm. In this case, one surface species only need be considered in the adsorption model. From the point of view of the general model presented here, this is expected. At a fixed pH, the acid-base chemistry of the surface is fixed (say $\equiv SOH_2^+$ and $\equiv SOPb^+$ as the dominant species), as is that of the bulk solutes (say Pb^{2+} as the major sorbing species). The adsorption process can then be described as a unique reaction ($\equiv SOH_2^+ + Pb^{2+} = \equiv SOPb^+ + 2H^+$), and the Langmuir isotherm is merely a reflection of the corresponding mass law equation. Of course the predictive ability of such a "model" is limited to the pH of the original data.

Such simple results suggest an approach to adsorption models in which a single adsorption reaction is considered, but the adsorption constant is defined as a function of pH. This is precisely the approach taken in the fourth model discussed earlier. In cases where specific adsorption data are obtained near saturation of the surface, the effect of the sorbate on the surface charge becomes large and, even with fixed pH, the adsorption constant needs to be modified as a function of the degree of surface saturation itself. This is taken into account in the fourth model by the effect of adsorption on pH_{iep}.

The particular type of surface species that needs to be included in a model to fit specific adsorption data as a function of pH is highly dependent on the model itself, i.e., on the chosen mathematical description of the acid-base chemistry of the surface by a combination of intrinsic acid-base properties, coulombic interactions and coordination of background electrolytes. Consider for example, the data of Hohl and Stumm [11] on adsorption of Pb^{2+} by $\gamma\text{-}Al_2O_3$ in 0.1 M $NaClO_4$. Figure 6 shows how three different models give a satisfactory fit of the data. The first model is that of Hohl and Stumm, which includes a description of the acid-base property of the surface according to the first model. However, in their model, the surface coordination reactions of Pb^{2+} are considered to be independent of surface charge and no coulombic term is included in the corresponding mass law expressions:

$$\equiv SOH^\circ + Pb^{2+} = \equiv SOPb^+ + H^+; \quad {^*K_{Pb}} = {^*K_{Pb}^{int}}$$

According to this model, the important surface species is written as $\equiv SOPb^+$. $(\equiv SO)_2 Pb^\circ$ is also included but its role is minor in the range of values considered.

The second model retains the description of the acid-base properties of the surface of the Hohl and Stumm model but includes the coulombic term in the expression of Pb^{2+} surface complexation as explained in the first model. In this case, the major surface species that need to be considered to fit the data

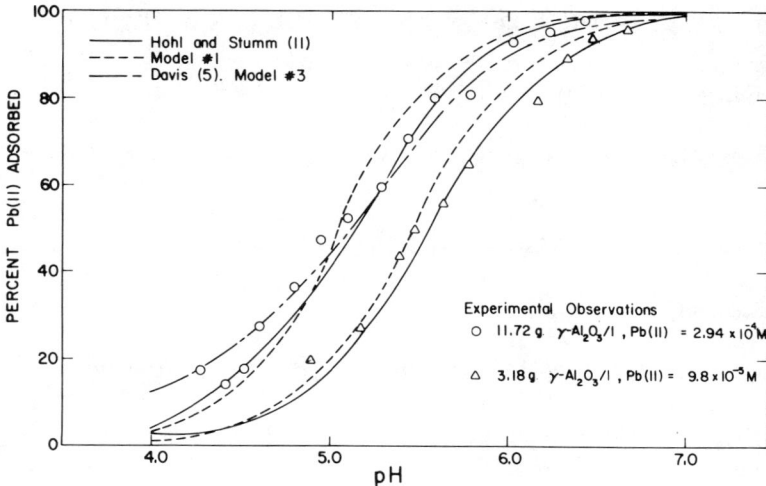

Figure 6. Adsorption of lead onto γ-Al_2O_3 in 0.1 M $NaClO_4$. Fitting of experimental data by various models [11].

is $\equiv SOHPb^{2+}$. Addition of other surface species in this model would permit improving the fit on Figure 6, but this is not important for the purpose of this discussion. Finally, the third model is that of Davis, which is essentially the third model in the previous section. To obtain a good fit of the data, Davis has to consider two surface species, $\equiv SOPb^+$ and $\equiv SOPbOH°$, the second one being dominant above pH 5.

What is the major cause for these differences in the major surface species among models: $\equiv SOPb^+$, $\equiv SOHPb^{2+}$, $\equiv SOPbOH°$? Consider the effective increase in adsorption with pH in each model around the mid-point of adsorption (ca. pH 5.5 where $\equiv SOH_2^+$ is the major acid-base species in all models). Given that the surface is far from saturation, the effect of lead adsorption on the surface charge is negligible and for the first two models the coulombic expression can be given directly as a function of pH (on the basis of the raw data of Figure 4) as being approximately

$$pK_{coul} \simeq 2/3 \, (pH - pH_{iep}) \qquad (67)$$

The reaction of formation of the major surface species in the various models yields, then, the dependency of adsorption on pH. First,

$$\equiv SOH_2^+ = \equiv SOH + H^+; \quad K = K_{a1}^{int} \cdot K_{coul}$$

$$\equiv SOH + Pb^{2+} = \equiv SOPb^+ + H^+; \quad K = {^*K}^{int}_{Pb}$$

Therefore,

$$\equiv SOH_2^+ + Pb^{2+} = \equiv SOPb^+ + 2H^+; \quad K = {^*K}^{int}_{Pb} \cdot K^{int}_{a1} \cdot K_{coul}$$

$$(\equiv SOPb^+) \propto H^{-2} \cdot H^{+2/3} \propto H^{-4/3} \tag{68}$$

And second,

$$\equiv SOH_2^+ + Pb^{2+} = \equiv SOHPb^{2+} + H^+; \quad K = {^*K}^{int}_{Pb}/K_{coul}$$

Therefore,

$$(\equiv SOHPb^{2+}) \propto H^{-1} \cdot H^{-2/3} \propto H^{-5/3} \tag{69}$$

Although the dependency of the coulombic terms on pH in the third model is not straightforward because it involves different potentials for H^+ and Pb^{2+}, we can take as a rough approximation that the term for H^+ is similar to that of the other models ($K_{coul_1} \propto H^{2/3}$) and that the one for Pb^{2+} is negligible ($K_{coul_2} \propto H^0$):

$$\equiv SOH_2^+ + Pb^{2+} + H_2O = \equiv SOPbOH^\circ + 3H^+; \quad K = {^*K}^{int}_{Pb} \cdot K^2_{coul_1}/K_{coul_2}$$

so that

$$(\equiv SOPbOH^\circ) \propto H^{-3} \cdot H^{+4/3} \propto H^{-5/3} \tag{70}$$

In all models, the formulation of the principal surface species is thus determined by the need to obtain the appropriate dependency of adsorption on pH (the correct slope on a log concentration vs pH graph), given a particular formulation for the coulombic term. All three models achieve roughly the same results by including very different surface species. This is because the respective expressions of the coulombic term are so different. It illustrates how the inclusion of such surface species in adsorption models should be viewed as an expedient data-fitting device, not as a description of chemical entities.

DETERMINATION OF MODEL PARAMETERS

We have seen in the two preceding sections how the choice of surface species and the values of the adjustable parameters depended critically on the choice of a particular adsorption model. This section addresses the question of parameter determination when a particular model has been chosen a priori.

Let us first consider the fitting of the acid-base titration data of Hohl and Stumm using the first model, which involves only one adsorption plane and a unique fixed surface capacitance. The purpose is to demonstrate the sensitivity of model predictions to the choice of the adjustable parameters and the interdependence of the various parameters. The optimal set of parameters, as determined by the computer method of Westall and Hohl [12] yields the curve shown in Figure 7. The fit of the model to the data points is truly excellent.

The effect on the model-generated titration curve of changing the value of one of the adjustable parameters, while the others are maintained at their optimum values, is seen in Figure 8. In each graph the solid lines are the curves generated from the optimal parameter values (and thus correspond to the experimental data points), and the broken lines are the curves generated with one of the parameters at a nonoptimal value. In Figure 8A, an increase in the exchange capacity by 50% causes an increase in buffer capacity and a flattening of the titration curve; in Figure 8B, a decrease in capacitance causes increased surface potential for a given charge, hence an increased resistance of the surface to protonation/deprotonation and a steeper titration curve. In Figure 8C, moving the surface acidity constants closer together increases the buffer intensity near the pzc and makes the titration curve flatter in this region. These effects can be made to compensate for one another, as shown in Figure 8D. The deviations from the optimal (experimental) curve induced in Figures 8A, 8B and 8C can be corrected by readjusting the other parameters, and parameter values far from the optimum can be used to very nearly reproduce the optimal curve (and the experimental data).

In this example, the exchange capacity was assumed to be an adjustable parameter, which adds another degree of freedom to the adjustment procedure. But even when the exchange capacity is taken as a known quantity, there is still a considerable interdependence between the values found for the capacitance and the values of the formation constants.

Consider now the fitting of adsorption data of Pb^{2+} on γ-Al_2O_3 [11]. To illustrate simply the sensitivity of the model predictions to the values of the parameters, let us use the model of Hohl and Stumm, which is most free from the influence of the coulombic expression. The effect of changing the formation constants for the major surface species by one order of magnitude,

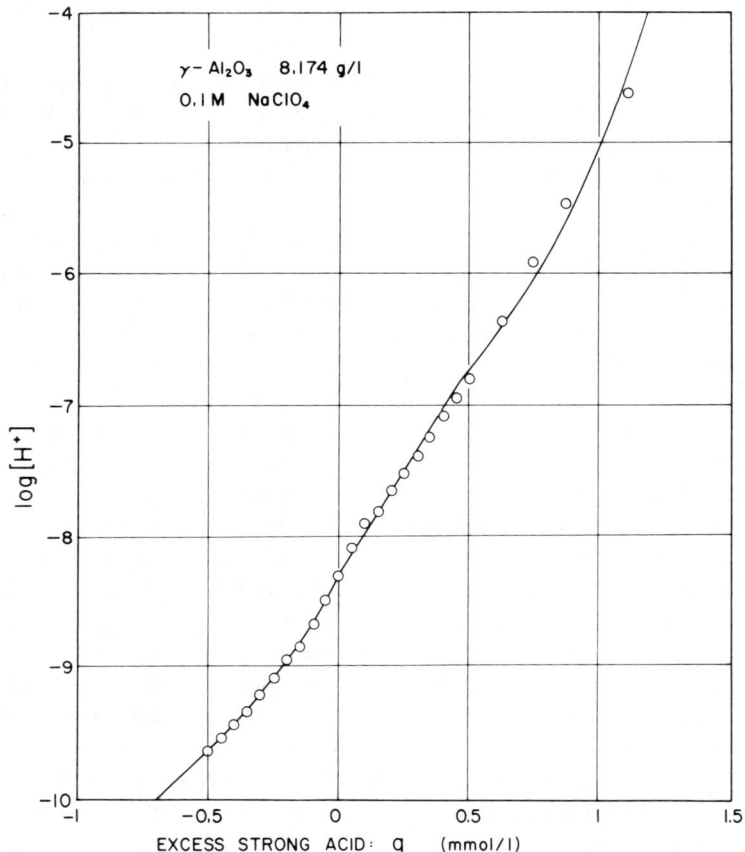

Figure 7. Optimal fitting of γ-Al_2O_3 acid base titration data by the first model. Experimental data are those of Hohl and Stumm [11].
Model parameters: S_T = 1.12 mmol/l; C = 1.06 F/m^2; $\log K_{a1}$ = -7.40 $\log K_{a2}$ = -9.25.

which is dramatic, and for the minor surface species, which is negligible, is shown in Figure 9. Again, such effects of changes in model parameters can be compensated for by modifying other parameters. For example, a decrease in the formation constant of the major surface species by a factor of ten can be compensated for by an increase of a factor of ten in the total site concentration, as shown in Figure 10. This is possible here because the surface is far from being saturated. This figure also illustrates the compensating effects (for the fitting of adsorption data) of decreasing the surface capacitance by a factor of two and decreasing simultaneously the surface acidity constants. In

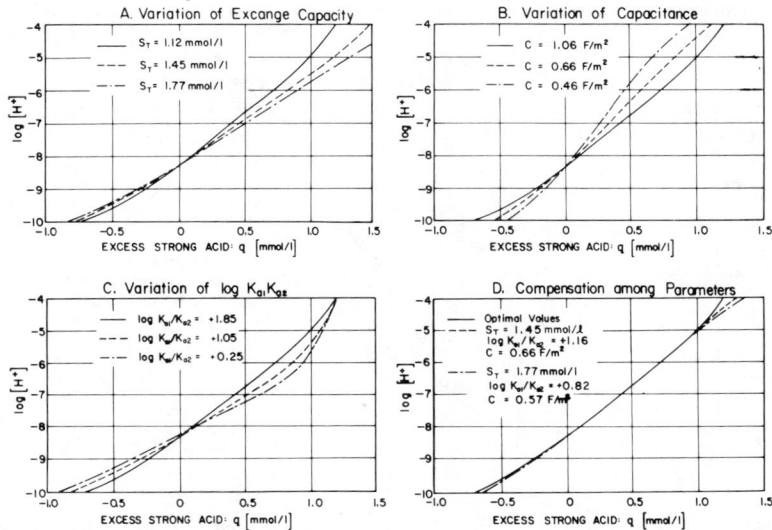

Figure 8. Effect of varying the parameters of the first model on the theoretical acid-base titration of γ-Al$_2$O$_3$. In all graphs the parameters are set at their optimal values, unless otherwise noted. In each figure, the solid line corresponds to the optimal fit of Figure 7.

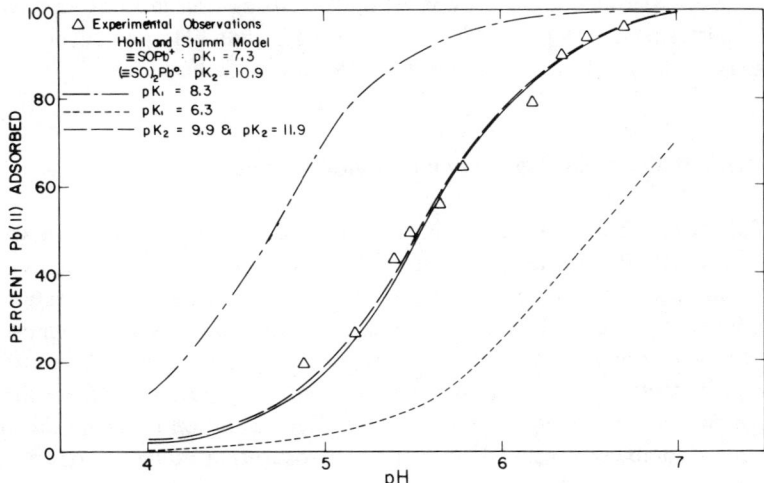

Figure 9. Effect of varying adsorption constants (Hohl and Stumm model [11]) on the theoretical adsorption isotherm of Pb^{2+} on γ-Al$_2$O$_3$.

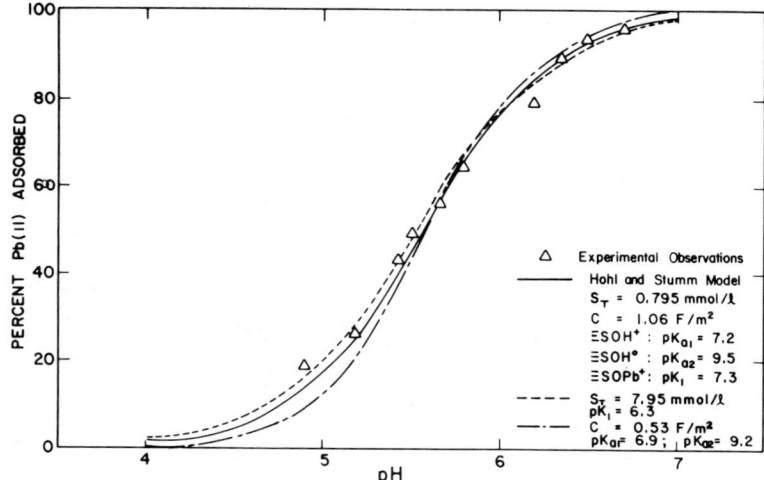

Figure 10. Compensation among model parameters (Hohl and Stumm model [11]) for fitting the adsorption isotherm of Pb^{2+} on γ-Al_2O_3.

general, once the model has been chosen and the total site concentration, the acidity constants and the coulombic terms are fixed, the choice of the principal surface species and the value of the corresponding constants are well determined by experimental adsorption data. The fitting procedure is effectively well constrained. Extensive data sets covering a wide pH and concentration range typically result in the need to include more surface species.

Inclusion of Models in General Equilibrium Programs

Most adsorption data are obtained with simple chemical systems, in which solution equilibria are easily calculated. However, extension of these data to model the chemistry of natural waters requires simultaneous consideration of complex solution equilibria and a surface adsorption process. It is then necessary to include adsorption models as part of general equilibrium calculations.

In the thermodynamic framework presented so far, such inclusion is straightforward in principle if the method of computation for chemical equilibrium is completely general. In effect, all adsorption models that we have considered are defined by one surface component (say, $\equiv SO^-$) in addition to the solution components, and any number of surface species (e.g., $\equiv SO^-$, $\equiv SOH^\circ$, $\equiv SOH_2Cl^\circ$, $\equiv SOPb^+$) in addition to the solution species. For each surface species a mass law can be written as a function of the components. For example,

$$(\equiv SOH°) = K_1 (\equiv SO^-)(H^+) \tag{71}$$

$$(\equiv SOPb^+) = K_2 (\equiv SO^-)(Pb^{2+}) \tag{72}$$

$$(\equiv SOH_2Cl°) = K_3 (\equiv SO^-)(H^+)^2 (Cl^-) \tag{73}$$

and a mole balance constraint for the surface component can be written in the usual form

$$S_T = \Sigma \text{ all surface species} = (\equiv SO^-) + (\equiv SOH°) + \ldots$$
$$+ (\equiv SOH_2Cl°) + \ldots + (\equiv SOPb^+) \ldots + 2((\equiv SO)_2Pb°) + \ldots \tag{74}$$

In effect, these expressions are strictly similar to those used for defining the typical equilibrium problem. In general equilibrium programs such as MINEQL [13], any adsorption model can thus be included simply by adding the necessary component and species. This is so because all adsorption models considered are effectively coordination models.

There is, however, one major difficulty remaining. Although the formal inclusion of the adsorption model as part of the equilibrium problem is straightforward, the solution scheme becomes somewhat more complicated since the adsorption constants (K_1, K_2, K_3) typically include a coulombic term, which is a function of adsorption and the final equilibrium composition. Depending on the complexity of the problem, this difficulty can be resolved in various ways:

1. Fixed Adsorption Constants. (Fixed pH – Low Adsorption Density).

In many instances, equilibrium calculations, including adsorption, are made for a fixed given pH. If the adsorption density is low, the surface charge, hence the coulombic term, is then fixed a priori and the computation can be made in the usual way with fixed given equilibrium constants. For example, one may include the function $\Psi(pH)$ in the input data and multiply the intrinsic constants appropriately for each input pH to compute metal adsorption (far from surface saturation).

2. Iteration on the Coulombic Term.

In cases where the pH is not fixed and/or the adsorption density is high, an obvious extension of the previous approach is to iterate on the value of the coulombic term. An initial guess of the surface charge is made to obtain a first estimation of Ψ; at the end of the equilibrium computation the surface charge and, thus, the coulombic term are recalculated from the system composition, and the equilibrium computation is started anew. The process can be repeated iteratively until it converges. Convergence is easily obtained when the conditions that control the surface charge (pH and specific adsorption) are well known or estimated a priori. The simplicity of this approach is particularly appealing and is the one used for the similar problem of ionic strength corrections of equilibrium constants. However, unlike in the case of ionic strength corrections, the equilibrium constants and, hence, the results of the equilibrium computations, can be quite sensitive to the value of the surface charge. Unless the initial value is very good, convergence is then very slow or impossible.

3. The Coulombic Term as a Dummy Component.

An elegant solution to the problem of iteration on the coulombic term is obtained by using simultaneously for Ψ the solution scheme that is used for the equilibrium computation itself [14]. If one considers mass law expressions for adsorption, it appears that the exponential of the dimensionless potential, $X = \exp(-F/RT \cdot \Psi)$, may be considered as a component with a stoichiometric coefficient of Z_i for each species, Z_i being the difference in charge between the surface species considered and the surface component. For example,

$$\equiv SOPb^+ = K_{Pb}^{int} \cdot (X)^{+2} (\equiv SO^-)(Pb^{2+}) \qquad (75)$$

The equivalent of a mole balance expression for X is obtained from the relation between surface charge and potential:

$$\Sigma Z_i(i) = A/F \cdot \sigma = A/F \cdot C \cdot \Psi \qquad (76)$$

where C is the capacitance of the adsorption layer. We then define

$$\lambda = 2.3 AC \frac{RT}{F} \qquad (77)$$

$$\Sigma Z_i(i) + \lambda \cdot \log(X) = 0 \qquad (78)$$

Since λ is fixed and given in all adsorption models considered, this expression, except for the last term, has the usual linear form of the mass balance equations. The usual iteration scheme used for all components can then be used also for X with minor modifications of the formulae. In MINEQL this scheme involves calculation of the various partial derivatives and use of Newton's method. This method can easily be generalized to several surface adsorption layers, each with its own surface potential and, hence, its corresponding dummy component.

CONCLUSIONS

In this chapter we have attempted to contribute to the converging evolution of adsorption models by presenting the basic concepts common to all models and examining in some detail the mathematical consequences of particular differences among them. At this point, a prime goal for aquatic chemists must be to obtain a general and unified thermodynamic description of adsorption processes similar to our catalogs of soluble and solid species with free energies and equilibrium constants. We hope to have made it clear that this will be achieved only if it is realized how arbitrary some of the choices are and if there is agreement on a particular adsorption model that has a common set of conventions and is applicable to a wide range of solid-solute interactions.

The difficulty in including a quantitative description of adsorption processes in general equilibrium models for aquatic systems does not lie in the fundamental principles, which are effectively agreed on, nor in the implementation of a numerical solution scheme. Adsorption models of any complexity can now be included in general chemical equilibrium computer programs. The difficulty lies in the uniform acceptance of a particular adsorption model with which to extract a coherent set of stoichiometric and thermodynamic information from experimental data and with which to systematically describe solute-solid interactions.

REFERENCES

1. James, R. O., and T. W. Healy. "Adsorption of Hydrolyzable Metal Ions at the Oxide/Water Interface. I. Co(II) Adsorption on SiO_2 and TiO_2 as Model Systems," "II. Charge Reversal of SiO_2 and TiO_2 Colloids by Adsorbed Co(II), La(III), and Th(IV) as Model Systems," "III. A Thermodynamic Model of Adsorption," *J. Colloid Interface Sci.* 40: 42, 52, 65 (1972).
2. Yates, D. E., S. Levine and T. W. Healy. "Site-binding Model of the Electrical Double Layer at the Oxide/Water Interface," *Chem. Soc. Faraday Trans. I* 70: 1807 (1974).
3. Bowden, J. W., A. M. Posner and J. P. Quirk. "Ionic Adsorption on Variable Charge Mineral Surfaces. Theoretical Charge Development and Titration Curves," *Aust. J. Soil Res.* 15: 121 (1977).
4. Malotky, D. T., and M. A. Anderson. "The Adsorption of the Potential Determining Arsenate Anion on Oxide Surfaces," *J. Colloid Interface Sci.* 55: 281 (1976).
5. Davis, J. A., III. "Adsorption of Trace Metals and Complexing Ligands at the Oxide/Water Interface," Ph.D. Thesis, Stanford University (1977).
6. Schindler, P. W., and B. Fürst, et al. "Ligand Properties of Surface Silanol Groups," *J. Colloid Interface Sci.* 55: 469 (1976).

7. Stumm, W., H. Hohl and F. Dalang. "Interaction of Metal Ions with Hydrous Oxide Surfaces," *Croat. Chem. Acta* 48: 491 (1977).
8. Stern, O. *Physik. Chem.* 30: 508 (1924).
9. Yeasted, J. G. "The Modeling of Lake Response to Phosphorus Loadings: Empirical, Chemical and Hydrodynamic Aspects," Ph.D. Thesis, Department of Civil Engineering, Massachusetts Institute of Technology, Cambridge, MA (1978).
10. Anderson, M. A., J. F. Ferguson and J. Gavis. "Arsenate Adsorption on Amorphous Aluminum Hydroxide," *J. Colloid Interface Sci.* 54: 391 (1976).
11. Hohl, H., and W. Stumm. "Interaction of Pb^{2+} with Hydrous γ-Al_2O_3," *J. Colloid Interface Sci.* 55: 281 (1976).
12. Westall, J., and H. Hohl. "A General Method for the Computation of Equilibria and Determination of Equilibrium Constants for Adsorption at Hydrous Oxide Surfaces," paper presented at Symposium "Chemical Modeling—Speciation, Sorption, Solubility, and Kinetics" ACS Annual Meeting, Miami, September 10-15, 1978.
13. Westall, J., J. L. Zachary and F. Morel. "MINEQL, A Computer Program for the Calculation of Chemical Equilibrium Composition of Aqueous Systems," Technical Note No. 18, Ralph M. Parsons Laboratory, Massachusetts Institute of Technology, Cambridge, MA, 1976.

CHAPTER 8

ADSORPTION OF FREE AND COMPLEXED METALS FROM SOLUTION BY ACTIVATED CARBON

Alan J. Rubin and Danny L. Mercer
   Department of Civil Engineering
   The Ohio State University
   Columbus, Ohio 43210

INTRODUCTION

The presence of heavy metals in water supplies is of increasing concern for a number of environmental and health-related reasons [1-4]. It is not surprising then that activated carbon, which is so effective for removing organics at trace levels, would also be examined for metals removal. Activated carbon is a readily available, inexpensive adsorbent with a very large specific surface area. Unfortunately, it has little net surface charge and is thus ineffective for adsorbing free hydrated metal ions. It seems possible, however, that effective adsorption of metals by activated carbon might be obtained by complexing the metal with an organic molecule prior to contacting the carbon. The complexing agent of choice must be highly adsorbable by activated carbon and able to complex strongly with the metal. It would be expected that the resultant metal-organic complex would display adsorption properties more closely related to those of the organic molecule than of the free metal ion. This would allow the removal of low concentrations of metals by operation of carbon columns similar to those now employed for the removal of trace organics. In fact, it seems feasible that the simultaneous removal of organics and chelated metals could be carried out. Gardiner's [5] data suggest that a significant percentage of cadmium present in sewage and river waters is complexed by organic ligands. This complexed cadmium is more or less resistant to removal by chemical precipitation. As a result, cadmium solubility calculated from simple solubility product considerations will be greatly exceeded if complexing species are present. The consequence of metal complexation by organic ligands in natural waters and wastewaters is that additional or alternative treatment to chemical precipitation is necessary to achieve the low metals concentrations often required. Adsorption represents such an alternative process.

There are a number of such studies described in the literature. In general, it has been found that the removal of free hydrated metal cations by carbon is slight, but reports on the effects of complexing the metals are frequently contradictory. Moreover, there are problems in interpreting adsorption results reported in the engineering literature. Both the Freundlich and Langmuir equations are used, although almost never in the same report. It is also usual that the applicability of the equation to the system being examined is not evaluated. An additional problem is that in determining the parameters of the Langmuir equation, the linear estimation technique most often used gives the most unreliable results.

These facets of adsorption for metals removal with activated carbon will be examined in this chapter. Three metals—cadmium, zinc and lead—and several different activated carbons were used, although most of the studies were performed with Cd(II) and one particular carbon. These metals were selected in part because they have environmental significance, but mainly because they have been used in other studies and can be compared, as well as because they differ widely in their ability to hydrolyze.

Adsorption Models

The Langmuir and Freundlich isotherms are the equations most frequently used to represent data on adsorption from solution. The Freundlich equation has the form:

$$X = KC^{1/n} \qquad (1)$$

where C is the concentration of adsorbate at equilibrium and K and $1/n$ are constants. Estimation of the constants is possible by the simple transformation of the equation:

$$\log X = \log K + 1/n \log C \qquad (2)$$

This equation is popular because log-log equations are easy to use and are relatively insensitive, lending themselves to imprecise data. However, the

Freundlich equation does not express X as a linear function of C at low concentrations, a common experimental observation, and it does not provide for a maximum value of X [6]. As a result, the Freundlich equation predicts infinite surface coverage at infinite concentration, a condition that does not occur. Therefore, its application is limited to ranges of intermediate surface coverages.

Langmuir [7] developed a quantitative model that has been widely applied to describe experimental adsorption data. He predicted that under equilibrium conditions and constant temperature, simple adsorption should obey a function of the form

$$X = \frac{X_m bP}{1 + bP} \quad (3)$$

where X_m = a maximum surface coverage representing the formation of a monomolecular layer on the surface of the adsorbent,
  P = the equilibrium gas pressure, and
  b = a constant related to the energy of adsorption.

Substituting $1/K$ for b and multiplying both the numerator and denominator by K gives an equation that is identical in form to the Michaelis-Menton equation of enzyme kinetics:

$$X = \frac{X_m P}{K + P} \quad (4)$$

A plot of X as a function of P passes through the origin and is nearly linear at low pressures. As the pressure increases, adsorption approaches the limiting value X_m. Thus, both weaknesses of the Freundlich equation are corrected. A small value of K (i.e., 1/b) means that the absorbent is effective at low gas pressure; a high value of X_m indicates a large absorbent surface area.

Langmuir pointed out that this equation is strictly applicable only to the adsorption of gas molecules by simple crystalline materials with homogeneous surfaces such as mica and platinum. According to the theory, these simple adsorbents have only one elementary type of adsorption site with a single adsorption energy potential. Langmuir conceded that most adsorbent surfaces are substantially less homogeneous. He stated that adsorption of gas molecules onto solids that contain more than one elementary type of adsorption site should follow a related, but more complex function. The implication of adsorbent heterogeneity is that if the equilibrium gas pressure is varied over several orders of magnitude, X can show a continuous increase. The exact shape of the adsorption isotherm for a heterogeneous adsorbent will depend on the distribution of the K values or, more specifically, on the frequency distribution of the adsorption energy sites on the adsorbent. Adsorption onto activated carbon, which is a highly heterogeneous solid, should follow this type of function.

Despite these limitations, the simple Langmuir equation is often used to describe adsorption from solution by heterogeneous solids. Replacement of the equilibrium gas pressure with the equilibrium adsorbate concentration gives

$$X = \frac{X_m C}{K + C} \quad (5)$$

Mathematical procedures for fitting nonlinear functions to experimental data are laborious and time consuming, and they are frequently carried out by digital computers. However, transformations allow evaluation of K and X_m using the relatively simple graphic or linear least square technique. A linear form of Equation 5 is obtained by taking the reciprocal of both sides. This results in an equation of the form y = ax + b:

$$\frac{1}{X} = \frac{K}{X_m}\left(\frac{1}{C}\right) + \frac{1}{X_m} \quad (6)$$

On multiplication by C and rearrangement, a second linear form results:

$$\frac{C}{X} = \frac{1}{X_m}(C) + \frac{K}{X_m} \quad (7)$$

Multiplication of Equation 6 by $X \cdot X_m$, followed by rearrangement, yields a third linear form:

$$X = -K\left(\frac{X}{C}\right) + X_m \quad (8)$$

Three additional linear forms of the Langmuir equation can be generated by interchanging the variables:

$$\frac{1}{C} = \frac{X_m}{K}\left(\frac{1}{X}\right) - \frac{1}{K} \quad (9)$$

$$C = X_m \frac{C}{X} - K \tag{10}$$

$$\frac{X}{C} = -\frac{1}{K} X + \frac{X_m}{K} \tag{11}$$

Any of these may be used to evaluate K and X_m from experimental data using graphic or linear least squares analysis. As pointed out by Dowd and Riggs [8], the ability of these equations to accurately predict the true least squares values of K and X_m are not the same. As it contains two reriprocal quantities subject to experimental error, Equation 6 is less reliable than either Equation 7 or 8, which should have similar predictive values. Equation 8 possesses the added advantage that K and X_m are obtained directly from the slope and intercept of the least squares line, eliminating the need for further calculations. Equations 9-11, although analogous to Equations 6-8, should not be expected to predict the same values of K and X_m from experimental data. In fact, for experimental isotherm data, all six regression equations calculate different values.

Literature Review

To make quantitative comparisons between adsorption systems using the simple Langmuir equation, it is necessary to have an idea of the ranges of values of K and X_m that constitute effective adsorption. The following review is by no means exhaustive, but is only intended to point out some of the more important factors that influence adsorption from aqueous solution. Also, the problems associated with using the simple Langmuir equation to describe adsorption of heterogeneous solids are illustrated briefly.

As stated previously, activated carbon is highly effective for the adsorption of large organic compounds from water. For the adsorption of alkyl benzenesulfonate detergents by activated carbon, Morris and Weber [9] reported X_m values between 160 and 400 µmol/g. The reported K values were quite small, ranging from 0.5 to 4 µM. Such small values reflect the ability of carbon to adsorb low levels of these materials. For the adsorption of phenol onto activated carbon, they calculated an X_m equal to 1090 µmol/g and a K equal to 9.3 µM at equilibrium phenol concentrations between 5 and 210 µM. For equilibrium phenol concentrations ranging from 1000 to 140,000 µM, they calculated X_m equal to 4500 µmol/g and K equal to 5000 µM. These values illustrate two important points. First, when using the simple Langmuir equation to model adsorption by a heterogeneous solid, the calculated X_m value may increase significantly as the equilibrium adsorbate concentration is increased over several orders of magnitude. Second, since K is nothing more than the equilibrium adsorbate concentration at which X is equal to one half X_m, the calculated value of K also depends on the range of concentrations employed in the study. Both these observations are the result of the inadequacy of the simple Langmuir equation to model adsorption by heterogeneous adsorbents.

O'Connor and Renn [10] reported on the adsorption of zinc(II) by river silt. They noted that the adsorption process is pH dependent, adsorption increasing as the pH was elevated. They fit their adsorption data to the Freundlich equation obtaining X equal to $1.59C^{0.614}$. Fitting their data to the linearized Langmuir model, one calculates X_m equal to 109 µmol/g and K equal to 42 µM.

Posselt et al. [11] demonstrated the high adsorptive capacity of colloidal hydrous manganese dioxide for cations. They noted that the pH of the isoelectric point (iep) of MnO_2 is quite low, ranging from pH 2.8 to pH 4.5. At higher pH, colloidal manganese dioxide has a net negative charge. These workers demonstrated that the mechanism of adsorption by MnO_2 is primarily electrostatic by comparing the adsorption of cationic, anionic and uncharged organic compounds. Only the organic cation showed any significant adsorption, and it was adsorbed to about the same extent as the metal cations investigated. For metals including calcium, magnesium and silver, X_m values ranging from 1000 to 3000 µmol/g and K values ranging from 35 to 170 µM were reported.

In a later study, Posselt and Weber [12] investigated the removal of trace cadmium by adsorption onto hydrous oxides of manganese, iron and aluminum. The iep of iron and aluminum hydroxides is much higher, ranging from 5 to 8.5. Therefore, MnO_2 is a much more effective adsorbent for cations. They reported X_m values ranging from 1370 µmol/g at pH 5 to 2200 µmol/g at pH 8.3 for the adsorption of Cd^{2+} onto MnO_2. Calculated values of K were on the order of 0.04 µM. Gadde and Laitinen [13] studied the adsorption of several metals on hydrous manganese oxide and hydrous ferric oxides. In general, adsorption followed the order $Pb^{2+} > Zn^{2+} > Cd^{2+}$ and was pH dependent.

Smith et al. [14] reported the use of activated carbon to remove mercury from caustic soda plant effluent. They showed that methyl mercury chloride, an organic mercury compound, is highly adsorbed by activated carbon. Nelson et al. [15] showed that enhanced adsorption of Fe(III) by activated carbon occurs when complexed by chloride ions.

Huang and Wu [16] reported the removal of chromate anions from solution by calcinated coke. Their work showed that low pH favors the adsorption of chromate anions by this adsorbent. Fitting their adsorption data collected at pH 2 to Equation 8, we calculated X_m equal to 42.8 µmol/g and K equal to 345 µM for the 10-20 mesh calcinated coke. Using the 100-200 mesh adsorbent, X_m and K are 99 μmol/g and 158 μM, respectively. These results indicate that chromate ions are only sightly adsorbed by calcinated coke even at high equilibrium concentrations. Table I is a summary of the Langmuir adsorption parameters just discussed.

O'Connor et al. [17,18] demonstrated that the adsorption of Cd(II) and Hg(II) can be enhanced by chelating agents. The increased adsorption of cadmium by activated carbon as a result of complexation with EDTA was observed under certain conditions. In these experiments, the initial cadmium concentration was maintained at 0.45 μM (0.05 mg/l), the calcium bicarbonate concentration at $10^{-3} M$ and the pH at 7. Several disodium ethylenediamineletraacetate (EDTA) concentrations were investigated, and carbon doses of 10-100 mg/l were employed. In the absence of EDTA, only about 50% of the initial Cd^{2+} concentration was removed by 80 mg/l of activated carbon. At an EDTA:Cd^{2+} molar ratio of 0.1, almost 90% of the cadmium was removed by carbon doses of 40 mg/l or larger. Using an EDTA:Cd ratio of 1.0, the removal of cadmium was only about 70%; at a ratio of 10, only 20% of the cadmium was removed, which is less than half the removal that resulted when no EDTA was added. In all these experiments, the calculated X values are small, on the order of 5-10 μmol Cd/g, of carbon. It is interesting that the greatest removal of cadmium occurred at an EDTA:Cd^{2+} ratio of 0.1, instead of 1.0, which is the stoichiometric ratio of the stable Cd-EDTA complex.

METHODS AND MATERIALS

Four activated carbons were examined. Three were manufactured by Westvaco Chemical Division in Covington, Virginia, including Nuchar WV-L, Aqua Nuchar and Nuchar S-A. The fourth was Darco HDC manufactured by ICI United States, Inc., of Wilmington, Delaware. The Darco carbon was a lignite-based material, while the WV-L and Aqua Nuchar carbons had bituminous coal matrices. The remaining carbon was wood based. All of these carbons were produced by a high-temperature activation process (1800-2000°F) under reducing conditions. The carbons were obtained in powdered form with the exception of Nuchar WV-L, which was an 8 x 30-mesh granular material. A 50- to 200-mesh powdered activated carbon was produced from the granular WV-L by grinding in an electric blender followed by careful sizing through U.S. Standard sieves. This particular carbon was used for most of the studies. Suspensions of the powdered carbons were prepared each day of use by mixing a weighed amount with distilled water in a volumetric flask.

Stock solutions of 0.01 M were prepared from the reagent grade hydrated nitrate salts of cadmium, zinc and lead; 0.001 M 1,10-phenanthroline solu-

Table I. Representative Langmuir Adsorption Parameters

K (µM)	X_m (µmol/g)	pH	Adsorbate Solute	Equilibrium concentration range, µM	Adsorbent Solid	Concentration mg/l	Reference
0.5-4.0	160-400		ABS detergents	0.5-50	Activated carbon	50	9
9.3	1090		Phenol	5-210	Activated carbon	50	9
5000	4500		Phenol	1000-140,000	Activated carbon	50	9
42	109	7.3	Zn^{2+}	2.9-159	River silt	171	10
35-170	1000-3000		Ca^{2+}, Mg^{2+}, Ag^+	35-170	MnO_2	86	11
0.04	1370	5	Cd^{2+}	10	MnO_2	0.91-12.7	12
0.04	2200	8.3	Cd^{2+}	0.8	MnO_2	0.0091-0.0637	12
345	42.8	2	CrO_4^{2-}	100-6000	10-20 mesh calcinated coke	5000	16
158	99	2	CrO_4^{2-}	100-6000	100-200 mesh calcinated coke	5000	16

tions were prepared from the reagent solid (G. Frederick Smith Chemical Co.). EDTA solutions were prepared by drying the dihydrate at 80° for four days and cooling in a desiccator, after which 37.21 g/l were dissolved. Three replicates of EDTA solution standardized against standard calcium solution had an average concentration of 0.0994 M. Just prior to the start of an experiment, metal iron and/or chelating agent working solutions were prepared by volumetric dilution of the appropriate stock solutions. Sodium chloride or lithium perchlorate solutions were used for ionic strength adjustment. Acetate and phosphate buffer solutions were each prepared at two concentrations. A buffer solution of 1.0 M was used in adsorption experiments in which the carbon dose was 5000 mg/l. With doses of 500 mg/l or less, 0.1 M buffer was used. Experiments were run at 24 ($\pm$ 3)°C.

EDTA and 1,10-phenathroline concentrations were determined by complexiometric titration with standard Cd^{2+} solution. The course of the reaction was monitored potentiometrically using an Orion (model 94-48A) cadmium ion electrode coupled with an Orion model 90-01 single junction reference electrode and Orion model 801A "Ionanalyzer." The Gran's plotting procedure, which facilitates potentiometric endpoint determination using only four or five values from the titration curve, was employed. All concentrations were calculated to the nearest 0.1 μmol/l. Lead was determined by a similar procedure using a Lazaar (model IS-146) ion-selective electrode. Zinc was determined with a procedure involving an Orion (model 94-29) cupric ion electrode. Since the ion sensed by the electrode (cupric ion) was absent from the sample, an indicator was prepared by titrating a 0.01 M copper solution with tetraethylenepentamine (TEPA) exactly to the endpoint. Of this 0.01 M CuTEPA solution, 1 ml was added to a 100-ml zinc sample and subsequently titrated with TEPA solution.

A Fortran program was used to calculate the adsorption variables from raw data that are needed to plot Langmuir and Freundlich isotherms. Computer analysis using the "Statistical Analysis System" developed at North Carolina State University [19] was carried out to investigate the ability of the various linearized forms of the Langmuir equation to predict the least squares values of the Langmuir parameters K and X_m for fitting isotherms to experimental data points. The Biomedical Computer Program X85, "Nonlinear Least Squares" [20], was used to obtain the best "unbiased" estimates of the Langmuir parameters.

The isoelectric points for each of the activated carbons were determined by titration with acid or base solutions adjusted to different ionic strengths with sodium chloride. A sample of activated carbon was washed prior to titration with double distilled water and dried at 105°C. The washing was repeated five times to ensure removal of impurities. A 0.5-g sample of carbon in 100 ml of solution was then titrated while passing nitrogen gas through the solution to purge the CO_2. The titrant was 0.01 M HCl with three titrations performed, each at a different ionic strength. Equilibrium was attained before recording the pH. This procedure was repeated using 0.01 M NaOH as the titrant. From the data generated, the hydrogen ions or hydroxide ions adsorbed is determined by the difference between total added base or acid and the equilibrium OH^- and H^+ concentrations in solution.

EXPERIMENTAL RESULTS

Effect of pH, EDTA and Adsorbent Dose

A few experiments were run to compare the rate of cadmium adsorption by 8- to 10-mesh granular Nuchar WV-L and the same carbon ground to 50-200 mesh. The rapid adsorption rate of powdered carbon, its similar adsorptive capacity to granular forms and the ease of quantitative carbon dosing using a well-mixed slurry make powdered carbon ideally suited for batch adsorption studies. As a result, all further experiments were conducted using 50-200 mesh powdered carbon. Samples were shaken for 24 hours to provide adequate reaction time for the attainment of equilibrium in all samples.

Figure 1A shows the percentage removal of free Cd^{2+} ions by powdered Nuchar WV-L activated carbon as a function of pH. Three carbon doses are

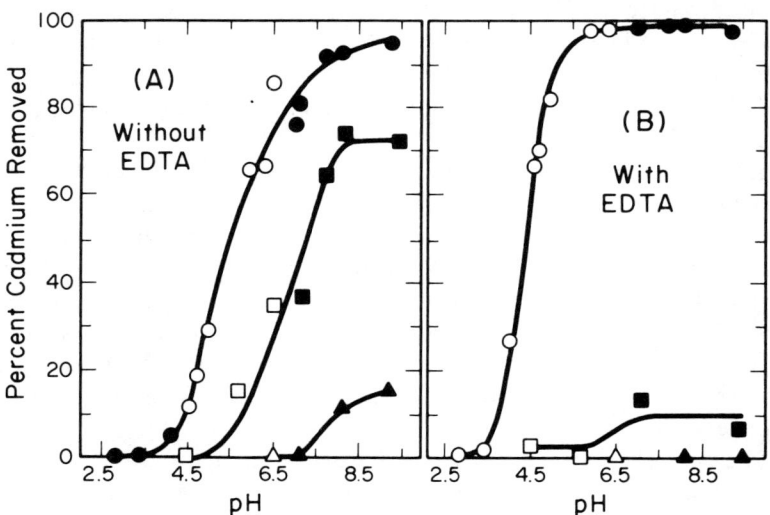

Figure 1. Removal of cadmium at different carbon doses as a function of pH in the absence and presence of EDTA. Circles represent 5000; squares represent 500; and triangles represent 50 mg/l of carbon. Initial Cd^{2+} and EDTA concentrations were each 9.8 μM. Open symbols are acetate- and blackened symbols are phosphate-buffered systems.

compared using an initial cadmium concentration of 9.8 μM. Open symbols represent acetate-buffer samples and blackened symbols are for phosphate-buffered samples. The data show, as expected, that increasing the adsorbent dose and pH results in a greater percentage removal of cadmium. No significant effect due to the type of buffer was observed. Figure 1B gives the percentage cadmium removal from a solution that was initially 9.8 μM in both cadmium and EDTA. For a carbon dose of 5000 mg/l (circles), the presence of EDTA enhanced the removal of cadmium over most of the pH range. At 500 (squares) and 50 (triangles) mg/l carbon, however, EDTA suppressed the removal of cadmium.

Comparison of Isotherms and Models

Isotherms for the adsorption of cadmium ion by Nuchar WV-L are presented in Figures 2-6. Carbon doses in these experiments were 500 mg/l or less. In the figures, open symbols are for carbon doses of 500 mg/l; blackened symbols represent lower doses. Unless otherwise specified, the Langmuir parameters X_m and K were determined from the data using the nonlinear least squares program discussed earlier. Freunlich parameters k and 1/n were calculated from a linear least squares analysis of the data transformed according to Equation 2. These adsorption isotherms represented in the figures by the solid and broken lines, respectively. The calculated parameters are summarized in Table II.

Table II. Summary of Experimental Adsorption Isotherm Parameters for Nuchar WV-L Powdered Activated Carbon[a]

Adsorbate	pH	Langmuir		Freundlich		Ligand:Cd Ratio
		X_m	K	k	1/n	
Cadmium	5.7	6	5	1.6	0.34	0
Cadmium	7.1	–	–	3.7	0.54	0
Cadmium	8.1	247	37	8.4	0.78	0
EDTA	7.1	14	5	3.3	0.41	–
Cd-EDTA	7.1	–	–	3.2	0.50	0.1
Cd-EDTA	7.1	14	7	3.7	0.33	0.5
Cd-EDTA	7.1	5	4	1.8	0.24	1.0
1,10-phenanthroline	7.1	1131	12	57.1	0.83	–
Cd-phenanthroline	7.1	544	12	29.8	0.85	1.0
Cd-phenanthroline	8.1	684	16	18.7	1.03	1.0

[a] Range of initial concentrations: 50-500 mg/l activated carbon; 5-50 μM cadmium; 1-50 μM EDTA; and 30-90 μM phenanthroline. X_m is in μmol/g and K is in μM.

Figure 2 compares the adsorption of uncomplexed (free) Cd^{2+} at pH 5.7, 7.1 and 8.1 as a function of the equilibrium metal concentration. The importance of solution pH on cadmium adsorption is shown clearly. The calculated Langmuir and Freundlich isotherms for the pH 5.7 data are virtually identical, and the solid line in the figure represents both isotherms. The nonlinear least squares program was unable to successfully fit the pH 7.1 data to the Langmuir equation. This can be understood by close examination of the plotted data. Instead of bending toward the abscissa at higher concentrations, the data appear to be slightly concave upward. This also explains why the calculated pH 7.1 Freundlich isotherm shows such deviation from the experimental points at higher concentrations. For the pH 8.1 data, the calculated Langmuir parameters are 247 μmol/g and 37 μM for X_m and K, respectively, which fall beyond the range of the experimental points plotted in the figure. This indicates that the experimental adsorption data are relatively linear, being well below the plateau region of the isotherm. The portion of the Langmuir isotherm corresponding to submonolayer surface coverage is precisely the range where the Freundlich equation gives a good fit, as shown in the figure.

The seemingly contradictory observation indicated by Figure 1, that EDTA enhances cadmium adsorption by activated carbon at high carbon doses and suppresses cadmium adsorption at lower carbon doses, should be explainable by a comparison of the adsorption isotherms for free Cd^{2+} ion and the cadmium-EDTA complex. The pH 7.1 Freundlich adsorption iso-

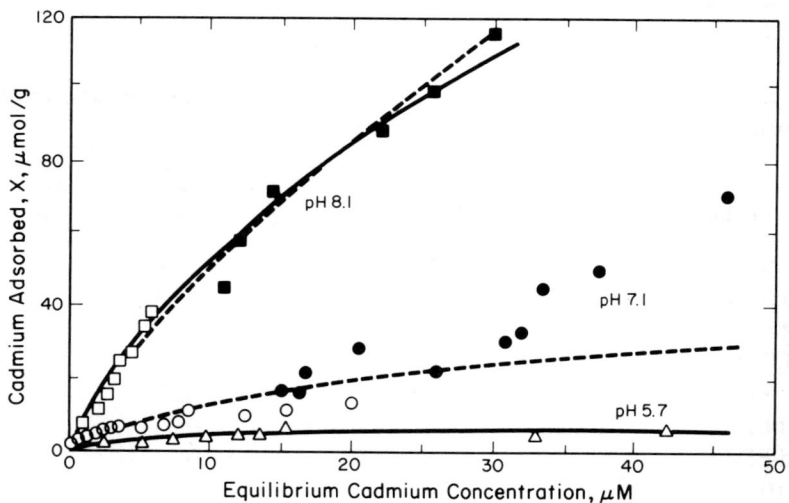

Figure 2. Adsorption of cadmium on Nuchar WV-L at pH 5.7, 7.1 and 8.1. Solid line is calculated Langmuir isotherm; broken lines are calculated Freundlich isotherms. Carbon dose: 500 mg/l (open symbols) and <500 mg/l (blackened symbols).

therms for free EDTA and the Cd-EDTA complex (measured as cadmium) are shown in Figure 3. The free Cd^{2+} ion adsorption isotherm at pH 7.1 from Figure 2 is included for comparison. The extensive scatter of the EDTA adsorption data (circles) results from the poor precision of the analytical technique used to measure the residual EDTA. The ordinate of Figure 3 is expanded four times relative to Figure 2. It is apparent from the figure that EDTA is less extensively adsorbed at pH 7.1 by Nuchar WV-L than is Cd^{2+} over most of the range of equilibrium adsorbate concentrations examined. The Cd-EDTA complex is adsorbed to an even lesser extent, being only about as adsorbable as free Cd^{2+} ion at pH 5.7. The isotherms in Figure 3 corroborate the data of Figure 1 (and Table II) showing, at carbon doses of 50 or 500 mg/l, that free Cd^{2+} ion is more adsorbable than the Cd-EDTA complex over the equilibrium adsorbate concentration range between 5 and 50 μM.

To further study the suppressive effect of EDTA on the adsorption of cadmium by activated carbon, the ratio of EDTA to cadmium in the test solution was varied. The results are shown in Figure 4. The free Cd^{2+} and the complexed cadmium (EDTA to metal ratio of 1.0) isotherms presented in Figure 3 are included for comparison. The circles and squares represent an EDTA:Cd ratio of 0.1 and 0.5, respectively. Figure 4 indicates that the principal effect of EDTA is to suppress the adsorption of cadmium by Nuchar WV-L, and that the extent of suppression is proportional to the EDTA concentration.

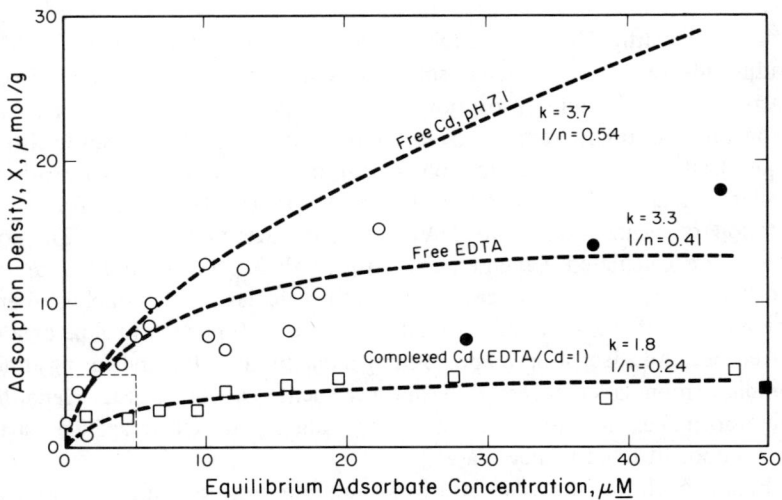

Figure 3. Adsorption of free EDTA and complexed cadmium by Nuchar WV-L at pH 7.1. Circles are EDTA; squares are complexed cadmium. Carbon dose: 500 mg/l (open symbols) and < 500 mg/l (blackened symbols).

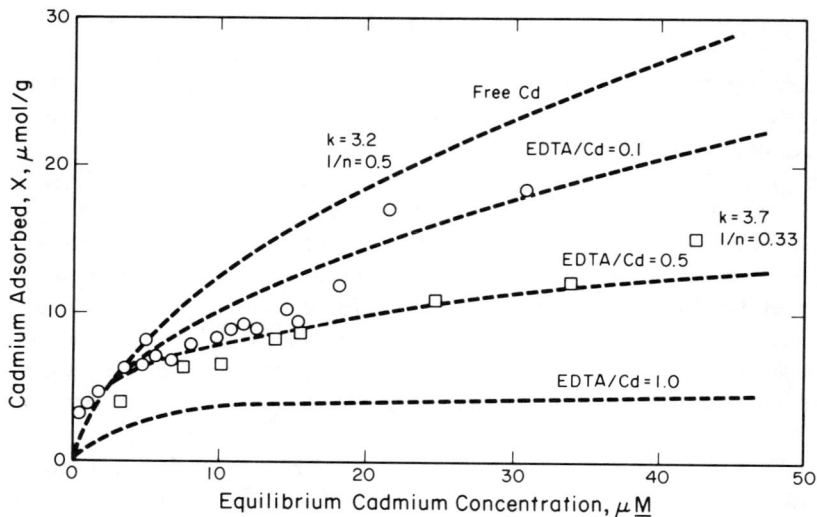

Figure 4. Effect of the molar ratio of EDTA:cadmium on adsorption by Nuchar WV-L at pH 7.1. Circles are EDTA:Cd = 0.1; squares are EDTA:Cd = 0.5. Carbon doses are 500 mg/l.

Effect of 1,10-Phenanthroline

1,10-phenanthroline is relatively insoluble in water and, hence, should be readily adsorbed from aqueous solution by activated carbon. Figure 5 summarizes the equilibrium adsorption data for experiments at pH 7.1. The scatter in the data points resulted from the poor precision of the analytical technique used to measure residual phenanthroline. The extent of adsorption is similar in magnitude to that reported by Morris and Weber [9] for the adsorption of benzenesulfonate detergents onto activated carbon. The solid line is the calculated Langmuir isotherm, with X_m equal to 1131 μmol/g and K equal to 12 μM. The dashed line is the calculated Freundlich isotherm. The poor fit of the Freundlich equation to the 1,10-phenanthroline data occurred because adsorption is approaching a maximum value, indicating almost complete monolayer coverage of the adsorbent. Of course, the Freundlich equation makes no allowance for a maximum surface converage, so usually gives a poor fit to curvilinear data.

Figure 6 shows the adsorption of cadmium from a solution containing equimolar concentrations of the metal and 1,10-phenanthroline at pH 7.1 (circles) and pH 8.1 (squares). Carbon doses of 50 mg/l were employed in both cases. The triangles represent data at pH 7.1 and carbon doses of 500

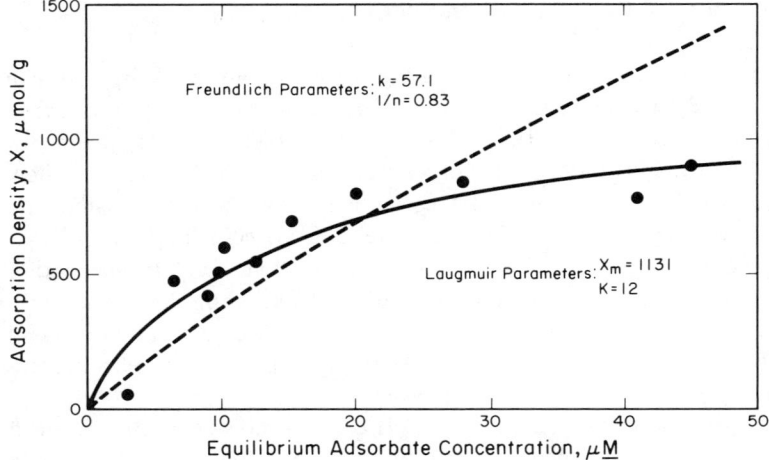

Figure 5. Adsorption of 1,10-phenanthroline by 50 mg/l Nuchar WV-L. Dashed line is Freundlich isotherm; solid line is Langmuir isotherm.

mg/l. The upper and lower dashed lines are the Freundlich isotherms for the adsorption of Cd^{2+} presented earlier.

Table II summarizes the adsorption isotherm parameters, calculated from the experimental data. It is apparent that uncomplexed 1,10-phenanthroline is adsorbed to a much greater extent than free cadmium ion or EDTA.

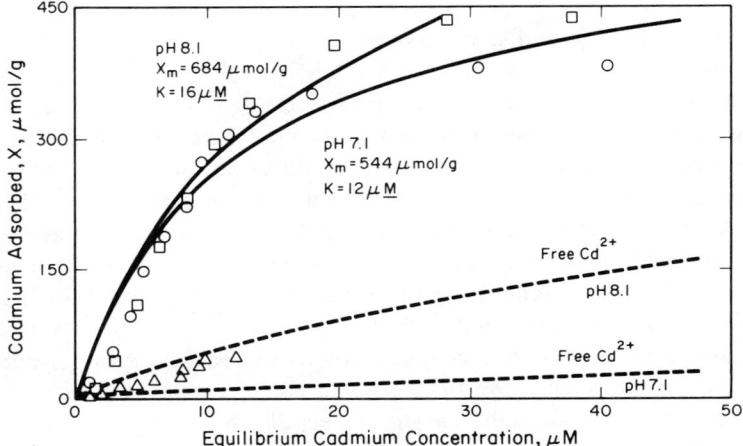

Figure 6. Adsorption from a 1,10-phenanthroline-cadmium equimolar mixture by Nuchar WV-L. Triangles are pH 7.1 and 500 mg/l carbon. Circles are pH 7.1 and 50 mg/l carbon. Squares are pH 8.1 and 50 mg/l carbon.

Comparison of Other Carbons and Metals

A few experiments were run using three carbons in addition to Nuchar WV-L and with zinc and lead for comparison with cadmium. Using the procedure described by Parks and DeBruyn [21], the carbons were titrated with acid and base solutions at different ionic strengths to determine their isoelectric points. The results of these studies, along with the manufacturer's values for their specific surface areas, are summarized in Table III.

The adsorptive capacity of each of the carbons for 1,10-phenanthroline at pH 7.1 was also determined. The initial phenanthroline concentrations ranged from 5 to 1000 μmol/l, and 50 mg/l of powdered carbon was used for each experiment. The X_m values correlated directly with the surface area of the respective carbons (data not shown).

X_m values for Cd, Zn and Pb at pH 6.5 and 8.0 were also determined for the four carbons and included in Table III. There was no significant adsorption of the metals on Nuchar S-A, the carbon with the highest isoelectric point. Adsorption was greatest on the carbon with the lowest isoelectric point, but with the smallest specific surface area. The relative capacities of the carbons for the three metals followed the sequence reported by Gadde and Laitinen [13]. The pK_1 for Cd, Zn and Pb are 10.3, 9.17 and 7.86, respectively [22].

DISCUSSION AND CONCLUSIONS

Estimation of Langmuir Parameters

For this research, the Langmuir parameters X_m and K were calculated using a nonlinear regression method with computer iteration. The procedure most commonly applied, however, involves a linear regression on one of the various transformations of the Langmuir equation. In particular, the double-reciprocal form (Equation 6) is widely used. There are some pitfalls associated with regressions on these linear equations [8]. In addition, the limiting assumptions of the simple Langmuir model are not always valid for a given solute-solid system. Frequently this is obvious, as, for example, with data generated when using heterogeneous adsorbents or when there are significant lateral interactions between adsorbate molecules at the solid surface [23,24].

Both aspects of using the Langmuir equation to calculate X_m and K, its linearization and its applicability, were examined using either the phenanthroline or the Cd-phenanthroline adsorption data shown in Figures 5 and 6. These systems were chosen since both the organic chelating agent and its metal complex were extensively adsorbed, minimizing the relative error of the calculated

Table III. Comparison of Metals and Carbons

Activated Carbon	Specific Surface Area (m^2/g)	Isoelectric Point[a]	X_m at pH 6.5			X_m at pH 8.0		
			Cd	Zn	Pb	Cd	Zn	Pb
Darco HDC	650	3.8	3.2	5.5	9.2	178	340	870
Nuchar WV-L	1000	4.3	3.2	5.5	9.2	160	310	821
Aqua-Nuchar	1000	6.2	2.0	2.7	5.6	125	220	620
Nuchar S-A	1500	8.3	<0.3	<0.3	<0.3	<10	<10	<10

[a] Isoelectric pH of carbons by the titration procedure described by Parks and de Bruyn [21]. X_m is in $\mu mol/g$.

adsorption densities. Both exhibited typical Langmuir behavior insofar as the adsorption densities reached a plateau of limiting values; the Cd-phenanthroline data, however, were "untypical" in that the isotherm formed an "S" curve [23].

Statistical Considerations

As presented by Snedecor and Cocharn [25], the simple linear regression equation has the mathematical form R = α + β + E. The assumptions involved are as follows:

1. For each independent variable, I, there is a normally distributed population of dependent or response variables, R, from which the sample value is drawn;
2. The population of R for each I has a mean or average value, μ, that lies on the straight line $\mu = \alpha + \beta(I-\bar{I})$ such that $(I-\bar{I})$ is equal to i;
3. The standard deviation (σ_R) of all R populations is equal; and
4. The independent variable is known with infinite precision. Usually, however, the independent variable can be measured precisely enough so that the standard deviation is negligibly small.

A plot of adsorption density against equilibrium adsorbate concentration which conforms to the Langmuir model closely approximates all of the criteria for application of the linear regression equation except, of course, for criterion 2 above, since the relationship between C and X is not linear. Transformation linearizes the relationship but also tends to alter it so that one or more of requirements of the linear regression model are no longer satisfied. This can be illustrated by Table IV in which are listed several values of C and X calculated from the Langmuir equation using X_m and K set equal to 6.0 and 2.0, respectively. Therefore, the tabulated X values represent "exact" Langmuir adsorption quantities. By assigning a small but constant standard deviation to the C values of ± 0.05 and a somewhat larger but constant standard deviation to the X values of ± 0.5, the tabulated values can be considered "experimental" quantities. The various transformed variables listed in Table IV were then calculated for three values of X and C. Close examination of the transformed values reveals that the standard deviations are not constant from observation to observation. For example, for X equal to 1.20 ± 0.5, the range of 1/X is from 0.59 to 1.43, but for X equal to 4.80, the range of 1/X is only 0.19 to 0.23. The linear regression equation is not strictly applicable to the analysis of such transformed data. Consequently, linear regression analysis of "experimental" data using the six linear Langmuir forms (Equations 6-11) will result in a unique solution of X_m and K for each form. These solutions will only be estimates of the least squares best fit values of X_m and K obtained from a nonlinear least squares analysis. However, except

Table IV. Transformed Langmuir Adsorption Values

C	X	1/C			1/X			C/X			X/C		
$\sigma_c = \pm 0.05$	$\sigma_X = \pm 0.5$	Min.	Avg.	Max.	Min.	Avg.	Max.	Min.	Avg.	Max.	Min.	Avg.	Max.
0.50	1.20	1.82	2.00	2.22	0.59	0.83	1.43	0.26	0.42	0.79	1.27	2.40	3.78
2.00	3.00												
4.00	4.00												
6.00	4.50												
8.00	4.80	0.12	0.13	0.13	0.19	0.21	0.23	1.50	1.67	1.87	0.53	0.60	0.67
12.00	5.14												
14.00	5.25	0.07	0.07	0.07	0.17	0.19	0.21	2.43	2.67	2.96	0.34	0.38	0.41

Notes:
$1/C_{min} = 1/(C + \sigma_C)$ $1/C_{avg} = 1/C$ $1/C_{max} = 1/(C - \sigma_C)$
$1/X_{min} = 1/(X + \sigma_X)$ $1/X_{avg} = 1/X$ $1/X_{max} = 1/(X - \sigma_X)$
$(C/X)_{min} = (C - \sigma_C)/(X + \sigma_X)$ $(C/X)_{avg} = C/X$ $(C/X)_{max} = (C + \sigma_C)/(X - \sigma_X)$
$(X/C)_{min} = (X - \sigma_X)/(C + \sigma_C)$ $(X/C)_{avg} = X/C$ $(X/C)_{max} = (X + \sigma_X)/(C - \sigma_C)$

for rounding errors, all seven forms will calculate the same values of X_m and K from "exact" Langmuir adsorption values. For a given set of experimental data, no a priori judgment as to which of the six linear forms will give the best estimates of X_m and K seems possible because it depends on the interaction of the following factors:

1. the magnitude of all of the X and C values used in the analysis;
2. the distribution of the experimental X values about the "true" isothermal line; and
3. the distribution of data points along the concentration axis.

Figure 7 shows plots of X against C or their transforms for the Table IV values and the associated isotherm. The triangle, square and blackened circle represent the values of X for which the transformed variables are listed in the table. The error bars of the dependent variable are equal for all values of X, and the error of the independent variable C is almost negligible. Thus, except for the linear relationship, all the requirements of the linear regression model are satisfied. Figure 7B shows the C/X against C plot for the three values of C/X given in the table. Note that the error bars for C/X are neither constant between observations nor uniform about their average values. Thus, two requirements of the linear regression equation are violated. Figure 7C is similar

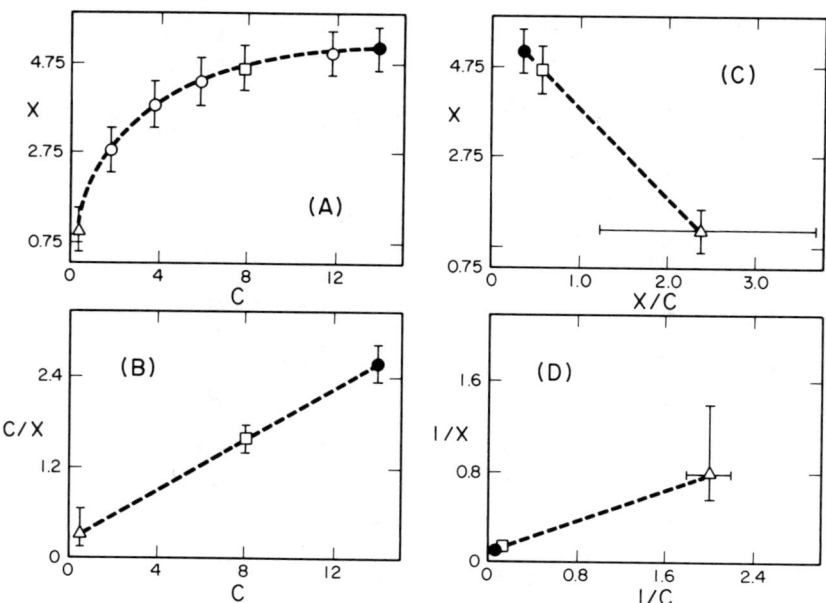

Figure 7. The effect on data of three linearized forms of the Langmuir equation. The values are from Table IV.

in that both contain one variable with an experimental quantity in the denominator. Therefore, the uncertainties associated with plots of X vs X/C and C/X vs C are analogous. The problems with the double reciprocal form, Figure 7D, are obvious. Not only do low values of X result in high values of 1/X with large errors, but the inverse of small values of C result in significant uncertainty of the plotted independent variable, 1/C. Further, the point corresponding to the largest values of X and C, the ones with the smallest relative error, are compressed near the origin. The result is that small values of X are the most influential in the determining the regression line.

Since the numerical quantities used in the regression analyses vary from form to form, comparison of correlation coefficients is not a valid criterion for determining which regression best fits the data and, therefore, calculates the most accurate estimate of X_m and K. The plots in Figure 8 which employ the 1,10-phenanthroline adsorption data presented in Figure 5, illustrates these points. The open circle is for the value of X of 45 μmol/g and for C equal to 2.8 μM. The effect of this data point on the calculated regression line is shown in the figure for the various transformations. The solid lines are the regression lines with the open point included; the dashed lines are the regressions with the open point deleted from the analysis. For the plot of C/X vs C (Figure 8B), the deletion of the open point significantly increased the

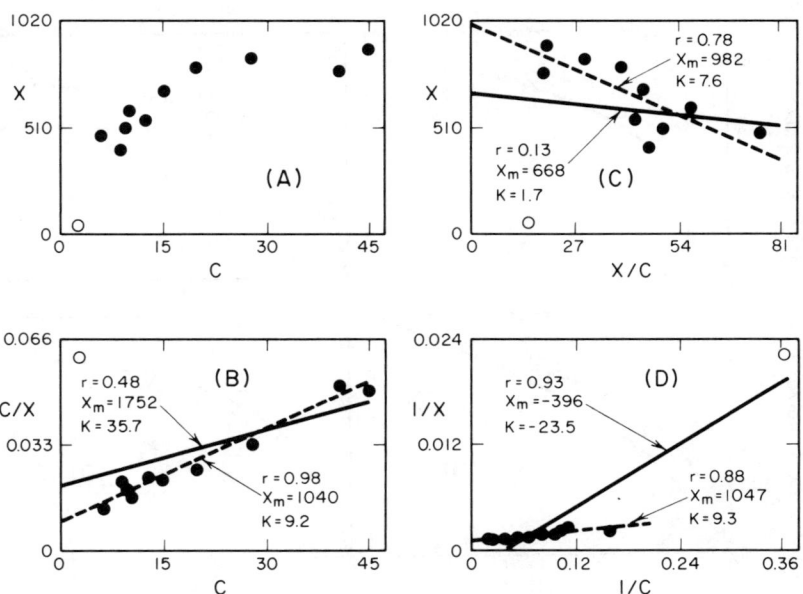

Figure 8. Langmuir parameters for 1,10-phenanthroline data calculated by three linear forms of the equation. Same data as Figure 5. Solid line is regression including all data; dashed line with open circle deleted.

correlation coefficient, r, and resulted in much smaller estimates of X_m and K being calculated. The results with X vs X/C in Figure 8C were similar, the correlation coefficient increasing on deletion of the low point. The double reciprocal form, however, showed a higher correlation coefficient with the low point included even though negative values of X_m and K resulted from the calculations. Interestingly, when the open point was deleted, plots of C/X vs C and 1/X vs 1/C resulted in almost identical estimates of X_m and K.

Figure 9 shows the Langmuir isotherms generated from the X_m and K values calculated in Figure 8. The isotherm in Figure 9A is for the nonlinear least squares values of X_m and K calculated using all the data points. Figure 9B shows the fitted isotherms from Figure 8B, and so on. The solid and dashed lines have the same meaning as in the previous figure. Note that when the low point was deleted, all the linear forms gave a reasonably good fit to the data. The values of X_m ranged from 982 to 1131 μmol/g and K varied between 7.6 and 11.9 μM for the four "best-fit" isotherms plotted in Figure 9.

Table V includes estimates of the Langmuir parameters calculated from all six linear regression models (as well as the nonlinear model) for the phenanthroline and Cd-phenanthroline (pH 7.1 and pH 8.1 at 50 mg/l carbon) adsorption data. The calculated X_m and K values are listed for the regression analyses using all data points (ALL) and, in some cases, for the regression analyses in which the low point was deleted (DELETE).

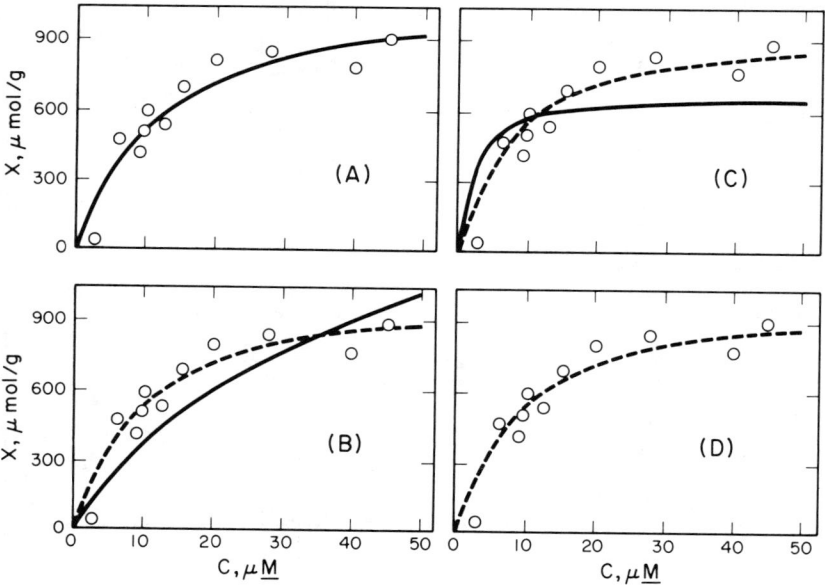

Figure 9. Langmuir isotherms for 1,10-phenanthroline data calculated from various estimated values of X_m and K. Same notation as Figure 8.

Table V. Summary of Calculated Langmuir Parameters Using Various Regression Forms of the Langmuir Equation

Absorbate[a]	X vs C		C/X vs C		1/X vs 1/C		X vs X/C		C vs C/X		1/C vs 1/X		X/C vs X	
	X_m	K	X_m	K	X_m	K	X_m	K	X_m	K	X_m	K	X_m	K
1,10-Phen (ALL)	1131	11.6	1752	35.7	−396	−23.5	668	1.7	425	−5.1	−282	−19.6	4881	105.6
1,10-Phen (DELETE)			1040	9.2	1047	9.3	982	7.6	1002	8.0	1330	15.4	1198	12.6
Phen-Cd pH 7.1 (ALL)	544	11.6	639	19.1	−1134	−57	374	6.3	481	11.3	−1036	−52	1672	63.8
Phen-Cd pH 7.1 (DELETE)			577	15.0	−3401	−155	427	7.8	482	10.2	−1335	−65.4	1137	38.9
Phen-Cd pH 8.1 (ALL)	684	16.2	1790	835	−162	−16.2	245	−0.2	144	−5.4	−134	−14.3	−39.80	−1880

[a] See text; refers to possible deletion of a data point. The units of X_m and K are $\mu mol/g$ and μM, respectively.

Figures 8 and 9 and Table V illustrate the instability of the linear regression forms of the Langmuir equation to the inclusion or exclusion of data and demonstrate the poor ability of the linear forms to accurately predict the true least squares values of X_m and K. As shown by the resultant isotherms plotted in Figures 9B and 9C, some linear transforms of the Langmuir equation give an uneven weight to individual data points. When the same data (with the low point included) were analyzed using the double reciprocal plot, which gives the most weight to the lowest values of X and C, negative X_m and K were calculated. Consequently, it is difficult to predict under these circumstances which linear form will best handle a given set of experimental data. Dowd and Riggs [8] presented evidence suggesting that on the average, the double reciprocal plot is the one most subject to error. Whenever possible, however, it is apparent that a nonlinear regression of the data should be used.

Model for Heterogeneous Adsorbents

Careful examination of the calculated isotherms in Figure 6 reveals that the simple Langmuir model fits the experimental points rather poorly over most of the range of the plotted data. It predicts too large an X at low and high values of C and too small an X at intermediate values of C.

For adsorption onto a heterogeneous solid, the energy released during adsorption, Q, is usually a nonlinear decreasing function of the surface coverage, Θ [6]. This suggests that there is a distribution of adsorption energy sites on the adsorbent so that at very low adsorbate concentration only the most energetic sites are able to adsorb. At higher concentrations, the driving force is greater and the less energetic sites become available for adsorption. The fact that Q is not a linear function of Θ indicates that the number of adsorption sites, f(Q), is probably not distributed equally over all values of Q. The distribution of energy sites on a heterogeneous adsorbent is thus one of the possible factors in determining the overall shape of the associated iostherm. There are other explanations for the S-curve, including lateral interaction of the adsorbate at the adsorbent surface. It is quite conceivable that the flat, almost two-dimensional shape of the phenanthroline molecule allows their orientation and close packing on adsorption.

Several attempts have been made to incorporate an adsorption energy distribution function into a quantitative model for the adsorption of gases by heterogeneous solids. Most of these start with the adsorption equation given by

$$\Theta(P,T) = \int_0^\infty \Theta(Q,P,T) f(Q) dQ \qquad (12)$$

where $\Theta(P,T)$ = the fractional surface coverage, X/X_m, as a function of pressure and temperature,
$\underline{\Theta}(Q,P,T)$ = the fractional surface coverage as a function of adsorption energy, pressure and temperature, and
$f(Q)dQ$ = the adsorption energy distribution function.

Adamson [6] has shown that substitution of the distribution function, $f(Q) = ke^{-\alpha Q}$, into the adsorption equation (letting $\underline{\Theta}(Q,P,T)$ be the Langmuir model) and integration between zero and infinity yields the Freundlich equation. Therefore, the failure of the Freundlich equation to adequately fit adsorption data above minimal surface coverage can be viewed from the standpoint that an unrealistic $f(Q)$ function was assumed. Integration of the adsorption equation using a normal distribution, $f(Q) = Ce^{-\beta Q^2}$ ($\underline{\Theta}(Q,P,T)$ is the Langmuir model) and using concentrations instead of pressure, results in

$$X = \frac{X_m C^n}{K' + C^n} \qquad (13)$$

To preserve the original definition of the constant K and to keep its units the same as C gives

$$X = \frac{X_m C^n}{K^n + C^n} \qquad (14)$$

Some of the data suggest such a normal distribution. Therefore, it seems reasonable to attempt a fit to Equation 14. The Cd-phenanthroline adsorption data given in Figure 6 were used at several values of n. An excellent fit was obtained using n equal to 2, as shown in Figure 10. The solid line is the isotherm calculated from the parameters. Compare with Figure 6 (n=1). The Langmuir parameters for both curves in Figures 6 and 10 are also compared in Table VI. The values of X_m generated for the data by the modified equation are lower and more realistic. The K's are also acceptable in that they reflect the values of C on the curves at $\frac{1}{2}X_m$. The difficulty with Equation 14 is finding the exact magnitude of n. In the present case, n=2 was probably fortuitous.

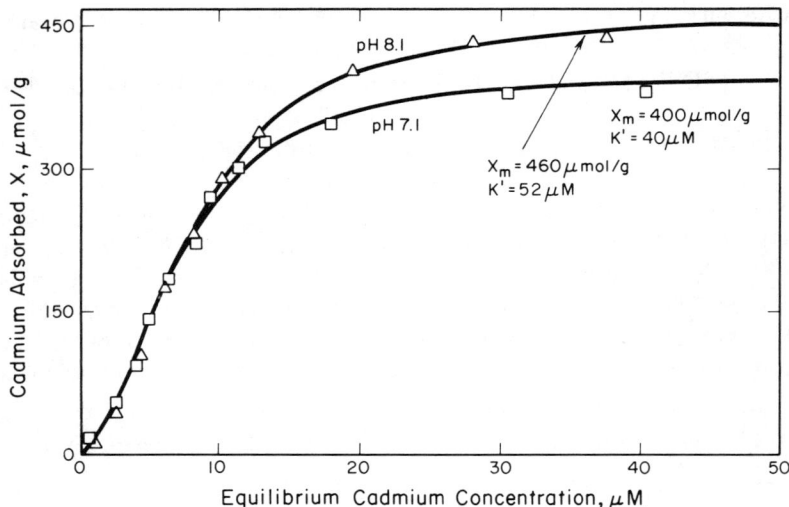

Figure 10. Fit of the modified Langmuir equation to the 1,10-phenanthroline-cadmium adsorption data. The value of n is 2.0. 50 mg/l Nuchar WV-L activated carbon. Compare with Figure 6.

Table VI. Comparison of Parameters for Simple and Modified Langmuir Isotherms

pH	Parameters	n=1	n=2
8.1	X_m	648	460
	K	16	7.2
7.1	X_m	544	400
	K	12	6.3

Effect of Adsorbent

In some preliminary studies the rate of cadmium adsorption by an 8- to 10-mesh granular carbon was compared to the rate of uptake by the same carbon ground to 50-200 mesh. The powdered carbon reached equilibrium much faster than the granular form (approximately 6 hours as opposed to 95 hours). The equilibrium capacities of the two were nearly identical, however. Morris and Weber [9], in discussing the effects of adsorbent particle size, concluded that intraparticle diffusion is often the rate-limiting step in adsorption by activated carbon. Consequently, small particles will adsorb faster than larger ones because the mean diffusion path decreases with adsorbent size. Also, for adsorbents that have a large internal surface area relative to external surface area, such as activated carbon, there is a negligible increase in total adsorptive capacity as a result of grinding. Powdered carbon was used in all subsequent studies, and attempts were made to assure that equilibrium adsorption had been attained.

Figures 1A and 2 show that the equilibrium adsorption of free Cd^{2+} ion by Nuchar WV-L-activated carbon is strongly affected by the pH of the solution, as is the adsorption of zinc and lead. Metal cation adsorption is enhanced by increasing the pH. Even at fairly high pH, however, adsorption of free Cd^{2+} ions by activated carbon was slight compared to the reported adsorption of Cd^{2+} by materials such as manganese dioxide. The Langmuir parameters X_m and K for the adsorption of free Cd^{2+} ion by Nuchar WV-L activated carbon at pH 8.1 were 247 μmol/g and 37 μM, respectively; the corresponding values reported by Posselt and Weber [12] for Cd^{2+} adsorption by colloidal MnO_2 at pH 5 were 1370 μmol/g and 0.04 μM. At pH 8.3 Posselt and Weber reported an X_m value of 2200 μmol/g (see Table I). These values indicate that in the pH range of 8.1 to 8.3 colloidal MnO_2 has an adsorptive capacity for free Cd^{2+} ions nearly 10 times higher than does the activated carbon. As demonstrated by numerous workers, including Posselt et al. [11], the adsorption of cations follows an electrostatic mechanism. Consequently, the difference in Cd^{2+} adsorption by these two materials can be largely accounted for by the surface charge characteristics of the adsorbents. It seems reasonable, therefore, that if an activated carbon could be produced with a large negative surface charge, it would be a highly effective adsorbent for cationic metals.

The effect of surface charge was also examined by comparing activated carbons with different isoelectric pH values (see Table III). Carbons with the lowest isoelectric points will be relatively more negative at any given pH above the iep. Accordingly, it was found that X_m for a specific metal increased with decreasing iep. Nuclear WV-L had a relatively low iep, so was a good choice for comparison with other adsorbents. Hence, the conclusion is that activated carbon is not an effective adsorbent for uncomplexed metals in solution.

From the data presented in Table III it can be concluded that the sequence of adsorption of metals on activated carbon is the same as with other charged adsorbents. Generally, the more acid the metal (the greater its tendency to hydrolyze), the greater will be its adsorption.

Effect of Chelating Agent

Figure 1B shows that EDTA increases the adsorption of cadmium when a relatively high carbon dose is employed (5000 mg/l) and reduces cadmium adsorption at lower carbon doses (500 mg/l or less). Figure 3 compares the adsorption isotherms for free Cd^{2+}, free EDTA and the Cd-EDTA complex.

Over the equilibrium adsorbate concentration range between 5 and 50 μM, EDTA was less extensively adsorbed than free Cd^{2+} at pH 7.1, presumably because of the high aqueous solubility of EDTA. The Cd-EDTA complex was adsorbed to an even lesser extent, being only about as adsorbed as free Cd^{2+} ion at pH 5.7 (Figure 2). This is consistent with the data presented in Figure 1 indicating that EDTA suppressed cadmium adsorption at 500 and 50 mg/l carbon. Figure 4 illustrates that the suppressive effect of EDTA on cadmium adsorption is proportional to the EDTA:Cd^{2+} molar ratio over the equilibrium adsorbate concentration range from 5 to 50 μM. Figure 11 shows the portion of Figure 3 enclosed by the dotted lines after expanding the scale five

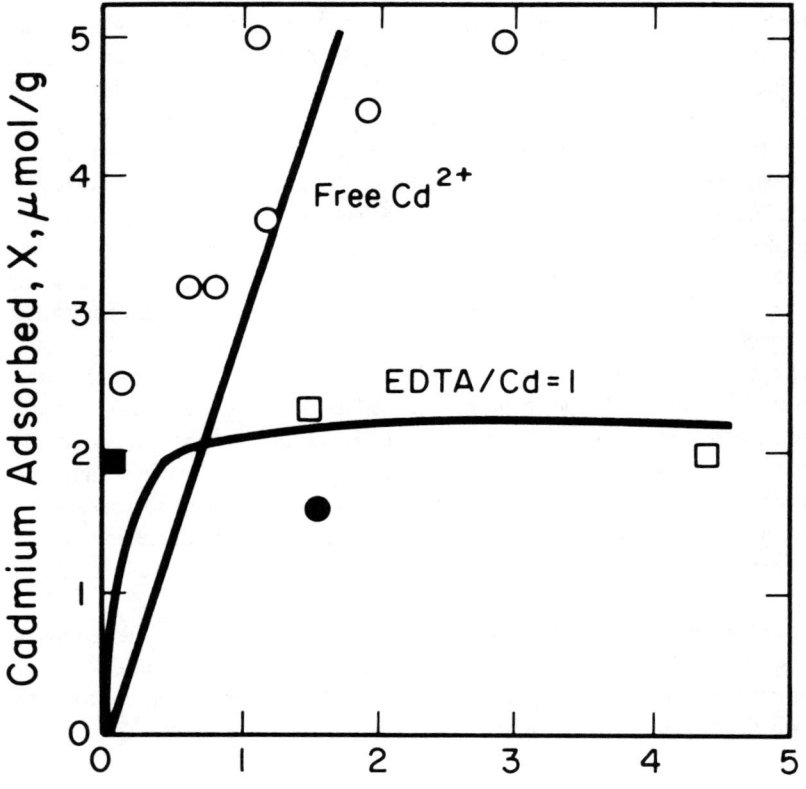

Figure 11. Shape of isotherms for low-equilibrium adsorbate concentrations. Open circles are Cd^{2+} adsorption data from Figure 2; blackened square is from Figure 1A (5000 mg/l); open squares are EDTA-Cd adsorption data from Figure 3; blackened square is from Figure 1B (5000 mg/l). All are at pH 7.1 on Nuchar WV-L.

times. The blackened symbols are data from Figure 1, and the isotherm lines represent a visual fit to the data. The crossing of the two isotherms at low equilibrium adsorbate concentration, as depicted in Figure 11, accounts for the observation that under certain conditions EDTA can enhance the adsorption of cadmium by activated carbon. Also, this interpretation is consistent with the data of O'Connor et al. [17], who reported the increased adsorption of the Cd-EDTA complex over that of free Cd^{2+} ion at equilibrium concentrations less than, or equal to, 0.45 μM. If the isothermal relationships given in the figure are essentially correct, the usefulness of EDTA to enhance cammium adsorption to activated carbon is quite limited at best.

1,10-Phenanthroline was highly adsorbed by all four activated carbons, as could have been predicted from its low aqueous solubility. The calculated X_m value for the 1,10-phenanthroline data presented in Figure 6 is 1131 $\mu mol/g$ and is of the same order of magnitude as X_m values reported by Morris and Weber [9] for the adsorption of ABS detergents by activated carbon. At equilibrium adsorbate concentrations between 30 and 40 μM, 1,10-phenanthroline was more than 50 times more adsorbable on Nuchar WV-L than EDTA and about 30 times more adsorbable than free Cd^{2+} ion at pH 7.1, based on a comparison of the isotherms presented in Figures 2 and 5.

Figure 6 shows that at pH 7.1 and 8.1 the adsorption of cadmium from a 1:1 M mixture with 1,10-phenanthroline was about 4-10 times greater than the corresponding adsorption of free Cd^{2+} ions. Also, the adsorption of the Cd-phenanthroline complex seemed less insensitive to pH over most of the equilibrium adsorbate concentration range investigated. The result of complexation is that adsorption becomes less dependent on electrostatic considerations and more dependent on the solubility of the resultant complex. Because each Cd^{2+} ion can complex with up to three phenanthroline molecules in the presence of excess phenanthroline, ratios of phenanthroline to Cd^{2+} ion greater than one result in more extensive complexation of the metal. These higher order complexes have an even lower charge density, so should be more highly adsorbed by activated carbon. This effect was not examined during the study, however.

In summary, at very low surface coverage, as when an excess of adsorbent is used, EDTA appears to enhance the sorption of cadmium. Under more realistic conditions, that is, at high surface coverages, the effect of this soluble chelating agent is to suppress adsorption of the metal. Therefore, it must be concluded that the use of EDTA is detrimental to metals removal. It can be further concluded that complexation with relatively insoluble chelating agents such 1,10-phenanthroline can significantly promote metals removal by activated carbon.

As is evident from the present study, there are several limitations to the Langmuir model for representing equilibrium adsorption data. It was also shown, however, that the Freundlich equation is even less successful in this regard. The chief utility of the Langmuir isotherm is in the use of its parameters X_m and K for the comparison of adsorbents and adsorbates.

ACKNOWLEDGMENTS

This project was supported in part by the Civil Engineering Department and the College of Engineering of The Ohio State University, as well as by the Office of Water Research and Technology (Project A-044-Ohio). Some of the experiments described in this report were performed by Mary Rozich, whose contribution is also gratefully acknowledged.

REFERENCES

1. Gardner, G. R., and P. P. Yenick. *J. Fish. Res. Bd. Can.* 27: 2185 (1970).
2. Warnick, S. L., and H. L. Bell. *J. Water Poll. Control Fed.* 41: 280 (1969).
3. Flick, D. F., H. F. Kraybill and J. M. Domitrof. *Environ. Res.* 4: 71 (1971).
4. Corrill, L. S., and J. E. Huff. *Environ. Health Pers.* 18: 181 (1976).
5. Gardiner, J. *Water Res.* 8: 22 (1974).
6. Adamson, A. W. *Physical Chemistry of Surfaces*, 2nd ed. (New York: John Wiley & Sons, Inc., 1967).
7. Langmuir, I. *J. Am. Chem. Soc.* 40: 75 (1918).
8. Dowd, J. E., and D. S. Riggs. *J. Biol. Chem.* 240: 863 (1965).
9. Morris, J. C., and W. J. Weber, Jr. "Removal of Biochemically-Resistant Compounds by Adsorption," Annual Technical Report, Division of Engineering and Applied Physics, Harvard University (May 15, 1962).
10. O'Connor, J. T., and C. E. Renn. *J. Am. Water Works Assoc.* 56: 1055 (1964).
11. Posselt, H. S., F. J. Anderson and W. J. Weber, Jr. *Environ. Sci. Technol.* 2: 1087 (1968).
12. Posselt, H. S., and W. J. Weber, Jr. in *Chemistry of Water Supply, Treatment, and Distribution,* A. J. Rubin, Ed. (Ann Arbor, MI: Ann Arbor Science Publishers, Inc., 1974).
13. Gadde, R. R., and H. A. Laitinen. *Anal. Chem.* 46: 2022 (1974).
14. Smith, S. B., et al. "Mercury Pollution Control by Activated Carbon: A Review of Field Experience," paper presented at 44th Annual Conference of the Water Pollution Control Federation, San Francisco, CA, October, 1971.
15. Nelson, F., H. O. Phillips and K. A. Kraus. *Proc. Purdue Ind. Waste Conf.* 29: 1076 (1974).
16. Huang, C. P., and M. H. Wu. *J. Water Poll. Control Fed.* 47: 2437 (1975).
17. O'Connor, J. T., D. L. Badorek and L. T. Thiem. Unpublished results.

18. Thiem, L., D. Badorek and J. T. O'Connor. *J. Am. Water Works Assoc.* 68: 447 (1976).
19. Barr, A. J., and J. H. Goodnight. *Statistical Analysis System,* Department of Statistics, North Carolina State University, Raleigh, NC (1972).
20. Dixon, W. J., Ed. *Biomedical Computer Programs* (Berkeley, CA: University of California Press, 1973).
21. Parks, G. A., and P. L. de Bruyn. *J. Phys. Chem.* 66: 967 (1962).
22. Kragteu, J. *Atlas of Metal-Ligand Equilibria in Aqueous Solution* (New York: Halsted Press, 1978).
23. Giles, C. H., D. Smith and A. Huitson. *J. Colloid Interface Sci.* 47: 755 (1974).
24. Giles, C. H., A. P. D'Silva and I. A. Easton. *J. Colloid Interface Sci.* 47: 766 (1974).
25. Snedecor, G. W., and W. G. Cochran. *Statistical Methods,* 6th ed. (Ames, IA: The Iowa State University Press, 1972).

CHAPTER 9

EXPECTATIONS AND LIMITATIONS FOR AQUEOUS ADSORPTION CHEMISTRY

Marc Anderson, Christopher Bauer, Douglas Hansmann,
Nicholas Loux and Robert Stanforth
  Water Chemistry Program
  University of Wisconsin
  Madison, Wisconsin

INTRODUCTION

Originally, we had intended to describe some of our own work on the adsorption of anions on metal oxide surfaces using electrophoretic mobility measurements. However, as the other chapters were received and reviewed, there was much discussion among the members of our research group as to the experimental and theoretical limitations in aqueous adsorption research. Although solving such problems is fundamental to the advancement of the science, these areas have not been clarified completely nor put into perspective in the other chapters. Consequently, they felt it important to focus on these concerns. In addition, this type of review should help scientists in other disciplines evaluate the state of the art of aqueous adsorption, which otherwise can be quite confusing because of the multiplicity of current adsorption models. Therefore, the intent is to illustrate some of the specific problems in adsorption experimentation and theory, and to make recommendations concerning the direction of future research. Although the scope of this discussion is not necessarily all-inclusive, the major problems that are readily apparent are addressed.

MATCHING THEORY WITH EXPERIMENTS

Experimental Shortcomings

Adsorption is a process by which a chemical species becomes bound to a surface. The study of adsorption chemistry thus requires the ability to measure accurately the amount of this adsorbed material. Then, by varying those parameters which control the extent of adsorption from solutions, e.g., pH, ionic strength, solid characteristics and adsorbate concentration, it is presumed that the reaction chemistry can be discerned, leading eventually to a generalized theory of aqueous adsorption.

There are two general approaches to quantitation: one based on separation of the solid phase from the supernatant solution and the other on in situ measurements. The predominant procedure at present is separation by centrifugation or filtration followed by analysis of the supernatant, or the solid, for the adsorbate. There is reason to believe, however, that the composition of the solid surface may not be the same as before the separation. It is well known that the efficiency of centrifugation and filtration is strongly dependent on particle size, and thus the distinction between solid and dissolved phase is usually defined operationally. In addition, the separation itself may perturb the electrical double layer with consequent alteration of equilibrium concentrations. Because these problems have not been resolved satisfactorily, in situ measurements are preferable because separation artifacts are avoided. However, if it can be determined by in situ analysis that such separations induce only negligible error, their use may be recommended. Applicable in situ methods are discussed later in this chapter.

The practical performance of adsorption experiments is subject to additional ambiguity. Most researchers specify pretreatment procedures and adsorption equilibration times preceding the measurement process. Adsorbent pretreatment is important because it is desirable to obtain a relatively homogeneous and stable (and consequently reproducible) substrate. Unfortunately, in most cases, the structure and composition of the surface have not been characterized. For this reason, theoretical treatments usually assume a uniform composition. May et al. [1], however, presented some evidence from studies on the solubility of gibbsite that the mineral phase on the surface is dependent on the solution pH and thus will not be uniform under all solution conditions. Freshly precipitated solids often display decreased adsorption capacity with aging time of the adsorbate-adsorbent system. Such behavior may, in part, be explained by the decrease in adsorbent specific surface area with aging or, alternatively, enhanced incorporation of the adsorbate into the bulk lattice with a freshly prepared substrate. To a large extent, the details of solid preparation and adsorbate equilibrium are matters of personal preference.

Even if surface structure is not characterized, it would seem desirable at least to standardize these equilibration practices or to determine some experimental indicator for sufficient "equilibration." It should be noted that current adsorption models assume that the system is at equilibrium, an assumption which is rarely verified experimentally.

These comments have referred mainly to preparation and standardization of the adsorbent. Another problem area involves adsorbate/adsorbent interactions, including precipitation. Precipitation and adsorption are intimately related, with adsorption a necessary first step for heterogeneous surface precipitation. Corey (Chapter 4) distinguishes between the two as follows: (1) adsorption is a two dimensional process (a surface layer) whereas precipitation is three dimensional (crystal buildup), and (2) solution adsorbate concentration is controlled by surface site concentration in adsorption and solution concentration in precipitation. Note that the definition of adsorption as a two-dimensional process is implicit in the Langmuir-Stern models and in surface complexation models such as those given by Morel (Chapter 7), Schindler (Chapter 1) and Hingston (Chapter 2).

Because the species controlling adsorbate removal from solution are different in adsorption and precipitation, there is no assurance that monolayer coverage will be complete at an adsorbate concentration less than that required for the onset of precipitation. It is important, therefore, in adsorption studies, to ensure that the observation of a decrease in the solution adsorbate concentration is actually caused by adsorption rather than precipitation.

An interesting illustration of adsorption-precipitation problems is found in the numerous studies of phosphate adsorption on aluminum oxides or hydroxides. Ferguson and King [2] calculated the solubility of aluminum phosphate in the presence of aluminum hydroxide. At a pH between 5 and 6, the soluble P concentration is less than 10^{-6} M. With P concentrations higher than 10^{-6} M in this pH range, the possibility is introduced that precipitation rather than adsorption is the P removal process. Phosphate concentrations well above 10^{-6} M have been used in most adsorption studies. Table I gives a survey of the P concentration used in several studies. Except for one study [9], the concentrations were above saturation for aluminum phosphate. This suggests strongly that precipitation may have been an unacknowledged interfering process in most of these studies. Indeed, Chen et al. [4] found aluminum phosphate crystals in suspension after 18 days from the reaction of 10^{-4} M P with Al_2O_3 at pH 4.3. The presence of observable crystals under conditions commonly used in adsorption experiments demonstrates that aluminum phosphate precipitation can occur. Chen et al. [4] also observed that P loss from solution was diphasic with time, with a rapid P loss in the first 24 hours followed by a slower loss from solution over a period of 40 days. The first process was ascribed to adsorption and the second to precipi-

Table I. Phosphorus Concentration Ranges Used in Studies of Adsorption on Aluminum Oxides at pH 5-6

Reference	Phosphorus Concentration (M)	Reported pH
3,4	$10^{-6} - 10^{-4}$	5-6
5	$10^{-4} - 10^{-1}$	5
6	$10^{-7} - 10^{-3}$	5.5
7	$10^{-6} - 10^{-4}$	5-6
8	$10^{-4} - 10^{-3}$	5-6
9	$10^{-9} - 10^{-7}$	5-6
10	$10^{-3} - 10^{-2}$	5

tation. This diphasic loss pattern has been observed, and similar adsorption-precipitation explanations postulated by others as well [6,11]. Thus, the experimental distinction between adsorption and precipitation is kinetic and is different from the theoretical distinctions given by Corey (Chap. 4). No one has demonstrated whether or not the two distinctions are the same, i.e., that the rapid 24-hour removal is confined to monolayer coverage. Furthermore, if the adsorption data are to be used in thermodynamic calculations, the adsorption reaction must be at equilibrium, or suitable modifications to the theory must be introduced. Even if the 24-hour removal is strictly adsorption, no one has shown that the reaction achieves equilibrium within a 24-hour period. It would seem prudent in adsorption studies, where precipitation is possible, to distinguish carefully between adsorption and precipitation reactions. A satisfactory approach would be to work in a concentration range below saturation.

A specific example of the confusion that can result from ignoring the possibility of precipitation can be found in studies of P adsorption on "Alon" (γ-Al_2O_3) by Huang [8] and Anderson and Malotky [9]. This example is chosen because the solid was "identical" in both studies. A major problem in comparing results from different adsorption studies is the difference in solid characteristics, particularly surface area. Because the solids in these two studies were "identical," and other experimental differences are relatively unimportant, the results should be comparable. However, Huang, using P concentrations up to 10^{-3} M, found a maximum adsorption over 1.5 times larger than Anderson and Malotky, who used P concentrations not greater than 10^{-6} M. The P concentration at pH 5 in Huang's study was three orders of magnitude above saturation (assuming Alon and gibbsite have about the same solubility). It would seem reasonable to assume that the higher maximum removal from solution in Huang's study was caused by precipitation rather than adsorption. Two other experimental differences that could be suggested as accounting for the different observed maximum adsorptions in the two studies were differences in washing procedure and the way pH was included in describing the adsorption envelope. Although later studies showed the washing technique to be unimportant [12], Anderson and Malotky [9] determined a pH-independent maximum adsorption which should be equal to or greater than the adsorption at any particular pH. On the other hand, the pH-dependent isotherms and linearizations (a double reciprocal plot) in Huang's study both resemble what one would expect from adsorption alone. If precipitation indeed occurred in Huang's study, the isotherms and linearizations are apparently insensitive in distinguishing adsorption from precipitation. The start of precipitation does not cause an obvious break in the "adsorption" isotherm, but rather gives a curve resembling a Langmuir plot. From this example, it is clear that before thermodynamic constants are calculated it is important for solute removal to be shown to be by adsorption rather than precipitation and that the system is at equilibrium.

Additionally, P release studies using both isotopic exchange [13,14] and desorption techniques [15,16], have demonstrated that phosphate is bound to aluminum and iron oxide surfaces in sites of differing lability (Hingston, Chapter 2). Explanations include different bonding mechanisms [15,16] or changing surface conditions [17]. Whatever the explanation, it is apparent that a model that postulates a uniform surface and a single surface reaction may be too simplistic to explain adequately solute release from these oxides.

Theoretical Shortcomings

To explain adsorption chemistry a number of theoretical models have been proposed at different levels of sophistication ranging from simple, nonspecific binding of the adsorbate, to complex models utilizing several different component energies of adsorbent-adsorbate interactions along with specified structures for the electrical double layer. In some degree, all of these models have been successful, the extent of success being determined by the purpose of each particular study. In cases where the chemical system is fairly narrowly defined (for example, in many engineering or agricultural applications) simple theoretical models suffice. However, to achieve a more universal applicability, there is a great interest in eliciting the fundamental nature of the adsorption process. It is apparent that simple models lack the generality to encompass more than a limited number of adsorbent-adsorbate systems.

Several intricate chemical models have been proposed and are apparently quite successful in describing experimental data. These models have been described in detail by Schindler in Chapter 1, Hingston in Chapter 2 and James in Chapter 6 of this book as well as by Bowden et al. [18] and James and Healy [19]. In Chapter 7, however, Morel et al. demonstrate that, in effect, the models are too flexible; that there are enough fitting parameters such that each model can be made to match any set of data.

It is instructive to review briefly the current means by which these models are applied. For instance, the Stern model has particularly been emphasized [9,18] for specific adsorption. However, the Stern equation does not readily incorporate all of the species which may be exchanged or bound at the surface and may unnecessarily complicate interpretation. Instead, surface complexation or ion exchange formulations, which utilize explicit equilibrium constant equations, are more direct. For this reason, the remainder of the discussion is focused on these models alone.

The total free energy for the adsorption reaction can be divided into several distinct terms [19]. In general:

$$\Delta G_{ads} = \Delta G_{chem} + \Delta G_{elect} + \Delta G_{sol}$$

and

$$K_{ads} = \exp[-\Delta G_{ads}/RT]$$

where ΔG_{ads} = free energy of the overall reaction
ΔG_{chem} = free energy of the chemical interactions
ΔG_{elect} = free energy of the electrostatic interactions
ΔG_{sol} = free energy of the solvation interactions
K_{ads} = an experimentally measured equilibrium constant
R = gas constant
T = absolute temperature

K_{ads} is a composite of the ΔG factors. The precise formulation of each energy term varies from model to model.

One goal of experimental work is to measure the so-called "intrinsic" adsorption constant due only to the chemical term. This term is assumed to be "constant" regardless of solution properties such as pH and ionic strength. Then by means of theoretical derivations for the electrostatic and solvation terms, the total adsorption constant can be calculated for varying conditions of pH and ionic strength.

Calculating Intrinsic Stability Constants

To isolate the intrinsic constant, the contribution from electrostatics and solvation must be eliminated. This is accomplished ostensibly by extrapolating data obtained at various values to solution conditions at which the net surface charge and surface potential are zero. Normally, this point is defined by the pH_{pzc} or by the pH_{iep}. Ionic strength also has a measurable effect and can apparently be eliminated by extrapolation to zero. It is less clear how ionic strength contributes to the adsorption equilibrium constant. James, in Chapter 6, has described the use of these extrapolation techniques and has proposed that it is the weak binding of the supporting electrolyte ions to the surface which accounts for the ionic strength behavior. Using this procedure, the intrinsic adsorption constant for the adsorbate is estimated by extrapolating to zero ionic strength, and that for the electrolyte ions by extrapolating to an ionic strength of $1\ M$.

Although weak binding of the electrolyte ions is an attractive concept for handling ionic strength effects, Schindler (Chapter 1) and Dousma [20] point out that some of the inherent assumptions may not be satisfied under all solution conditions. Despite uncertainties concerning the role of the electrolyte ions in adsorption reactions, these "inert" ions may be important species that have heretofore received little attention. In addition, the apparent presence of adsorbed electrolyte raises the possibility that the nonintegral stoichiometries, as observed for divalent metal adsorption (for example, see Schindler, Chapter 1), may be caused by variations in surface electrolyte ion concentrations. Because only pH and metal ion concentrations were monitored in this study, the possibly important role of the electrolyte ions was not considered. In any case, accurately discerning the exchange of electrolyte ions with the surface may be a difficult task because of the relatively high solution concentrations of these ions used to avoid other problems in measurement.

There are three additional problems which require attention. First, some caution is suggested in the utilization of the pH_{pzc} parameter. Here, the net surface charge is zero, but this does not mean that the surface is without charged sites. In fact, as much as 30% of the total surface acid sites may be charged. This conclusion necessarily follows if one considers the surface proton equilibrium to be described by:

$$-SOH_2^+ \underset{}{\overset{K_{a_1}}{\rightleftharpoons}} -SOH \underset{}{\overset{K_{a_2}}{\rightleftharpoons}} -SO$$

Application of the familiar species distribution equations for polyprotic acids to this surface acid leads to the above value. A greater difference between the two acidity constants means a greater fraction of sites are charged at the pH_{pzc}. Whether localized nonzero surface charge has significant effects has not been ascertained, nor have reacting models included this refinement.

Second, as pointed out previously from May et al. [1], there is reason to suspect that the chemical composition of the surface may vary with solution pH (for example, from one crystalline form to another). Thus the intrinsic or chemical free energy of adsorption may not be constant under variable solution conditions as it is currently assumed.

Finally, the effects of specific adsorption on acidic surface sites has yet to be discerned. For example, in the case of adsorption of phosphate on Alon [11], comparison of the maximum surface coverage with the number of acidic sites on the solid indicates that about nine acid sites lie beneath a phosphate molecule. Interpretation of acid titration curves is complicated then because it is not known whether these covered sites can still exchange protons and what effects, if any, the presence of the phosphate has on the surface acidity constants or, alternatively, of the surface on the phosphate acidity constants.

Modeling the Adsorption Manifold

Assuming that one has an accurate estimate of the intrinsic stability constant for a given solid and adsorbate, the next challenge is to predict, by means of a theoretical model for electrostatic and solvation energy contributions, the measured adsorption behavior as a function of various solution parameters.

Let us consider only the electrostatic contribution. Although the surface potential cannot be measured directly, it can be determined indirectly by summing the total amount of charge on the solid surface derived from adsorption of ions from the solution. This may, depending on one's model, include some or all of the following: H^+, OH^-, adsorbate species and "inert" ions. The distributions of these species on the solid surface are calculated from experimentally measured intrinsic stability constants. Using the Guoy-Chapman treatment of the electrical double layer, the surface potential is calculated.

In practice, some models separate the double layer into surface, Stern and diffuse regions as shown in Figure 1, which are occupied by potential-determining ions (H^+), specifically adsorbed ions and nonadsorbed counterions, respectively. The charges developed in the surface and Stern layers are calculated by summation of charged species, as described above. The charge in the diffuse layer is calculated assuming electroneutrality of the surface region.

The diffuse layer potential is calculated via the Guoy-Chapman equations, and finally the surface and Stern potentials are estimated using capacitance relationships (Chapter 6) [21,22]. The capacitance values required here are simply fitting parameters in the model.

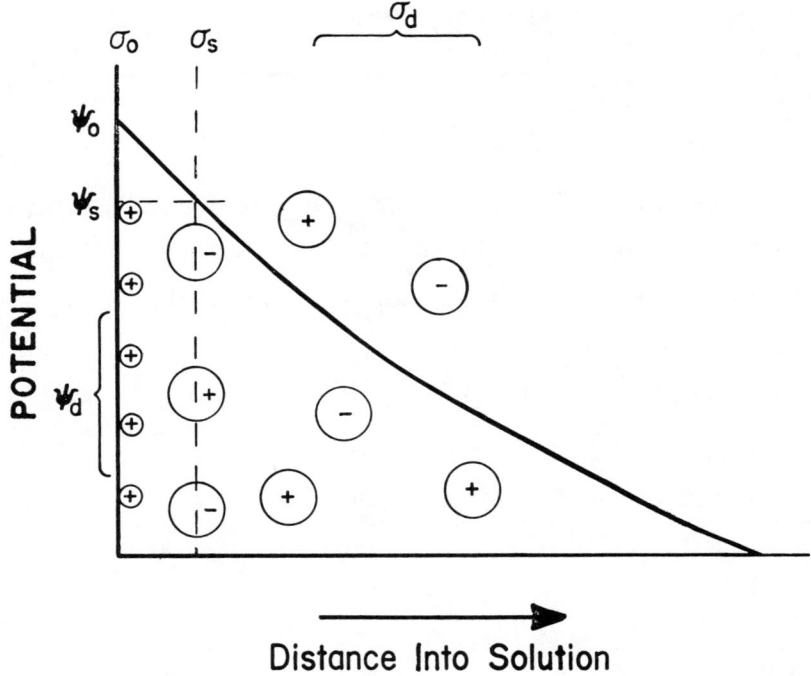

Figure 1. Representation of the electrical double layer showing potential determining H^+ ions, specifically adsorbed ions in the Stern layer and counterions in the diffuse layer.

One major limitation of these types of models has been discussed by Morel et al. in Chapter 7. In cases where the geometry or composition of the adsorbed species is unknown, these models select the form or forms which best fit the experimental adsorption envelope. However, the surface species obtained are very much dependent on how the model is formulated. For example, Schindler (Chapter 1) finds that divalent transition metals are bound as monodentate ($SO-M^+$) and bidentate ((SO_2)M) surface complexes. Davis and Leckie [21,22], however, find that similar metal species are bound as simple ions ($SO-M^+$) and hydroxy species (SO-MOH). Similar ambiguities abound in the literature (Table II), and it is apparent that too much "information" is being demanded from these models. An alternative way of formulating the problem is illustrated in Figure 2. In terms of free energy, for all models the initial state is a solid surface and a set of ions in solution and the final state is given by the experimental adsorption envelope. Each model assumes a theoretical representation for the electrostatic (and solvation) interactions. The electrostatic free energy is then compared with the free

Table II. Postulated Adsorption Reactions

Reaction[a]	Reference
Inorganic Adsorbates	
1. $-SOH + M^{z+} \rightleftharpoons SOM^{(z-1)+} + H^+$	23, 24
2. $2-SOH + M^{z+} \rightleftharpoons (-SO)_2 M^{(z-2)+} + 2H^+$	Chapter 1
3. $-SOH + H_x AO_y^- \rightleftharpoons -SOAO_{(y-1)} H^-_{(x-1)} + H_2 O$	8, 25
4. $2-SOH + H_x AO_y^- \rightleftharpoons (-SO)_2 AO_{(y-2)} H_{(x-2)} - 2H_2 O$	26
5. $-SOH + H^+ + A^- \rightleftharpoons -SOH_2^+ \ldots A^-$	21, 22
6. $-SOAO_x H_y^{z+} \rightleftharpoons -SAO_x H_{(y-1)}^{(x-1)+} + H^+$	9
7. $-SOH + AO_y^{-2} + M^{z+} \rightleftharpoons -SOAO_x M^{(z-2)+}$	27
8. $4-SOH + 2H_2 AO_x^- + M^{2+} \rightleftharpoons (-SOH)_2 (-SO)_2 (AO_{(x-1)})_2 M + 2H_2 O$	6
Organic Adsorbates	
Cations	
9. $R^+ + -SOM \rightleftharpoons -SOR + M^+$	28
10. $R^+ + -SOH \rightleftharpoons -SOR + H^+$	28
Bases	
11. $R + -SOM \rightleftharpoons -SOXR$	28
12. $R + -SOH \rightleftharpoons -SOHR$	28
13. $RH^+ + -SOM \rightleftharpoons -SOHR + M^+$	28
14. $RH^+ + -SOH \rightleftharpoons -SOHR + H^+$	28
Acids	
15. $RH + -SOM \rightleftharpoons -SOHR^- + M^+$	29
16. $RH + -SOH \rightleftharpoons -SOHR^- + H^+$	29
Bridging	
17. $R + H_2 O + SOM \rightleftharpoons -SOM\cdots H_2 O \cdots R$	30
18. $R + H_2 O + SOH \rightleftharpoons -SOH\cdots H_2 O \cdots R$	30

[a] $-SOH$ = surface exchange site; M^{z+} = cation; $H_x AO_y^-$ = anion; R = organic molecule.

EXPECTATIONS AND LIMITATIONS

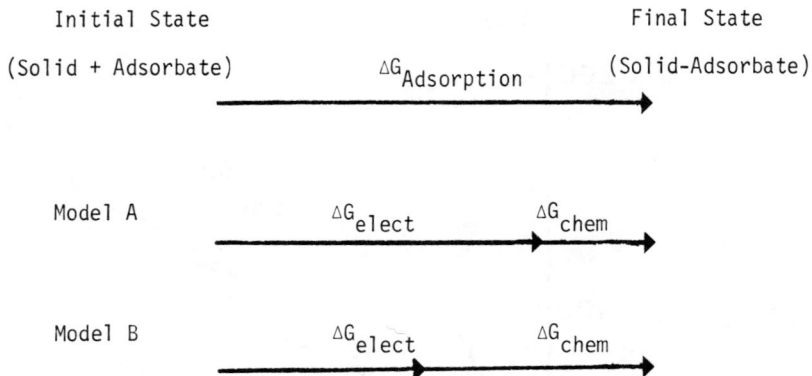

Figure 2. Representation of model differences from a thermodynamic point of view.

energy of the overall adsorption, i.e., the experimental data. The difference in energy is ascribed to chemical interactions, namely the type of surface bond. Thus, there may be as many different surface species "found" as there are models for the electrostatic properties.

In Chapter 7, Morel et al. suggested that further progress in adsorption research would benefit from forging a consensus model to eliminate the variability between results. However, unanimity is no advantage if the chemistry is incorrect. Although chemical models are sophisticated instruments, it is apparent that a fundamental limitation has been reached. Independent information is required about the extent of the electrostatic forces and about the chemical species on the surface to uncouple the interdependence of these terms in the models.

An independent measurement of the surface potential is desirable to compare theory with model prediction. Although direct measurement is impossible, an estimation might be obtained from the zeta potential, which in turn is estimated by means of electrophoretic mobility, streaming potential or electroosmosis measurements. The zeta potential is the potential at the plane of shear and cannot be exactly identified in practice with any of the three potential terms discussed previously (see Figure 1) because the distances of ψ_ζ from the surface varies with ionic strength and particle size. Theoretically, the zeta potential is roughly proportional to the diffuse layer potential (ψ_d) at low σ_d. As ψ_d increases, ψ_ζ levels off and becomes independent of ψ_d. The value of ψ_d at which this leveling off occurs increases as the ionic strength decreases [31]. A practical illustration is given in Figure 3 for the adsorption of phosphate on Alon [9]. The surface potential was calculated using a Guoy-Chapman approach and zeta potentials were calculated from electrophoretic mobilities. It is reasonable to expect that the zeta potential

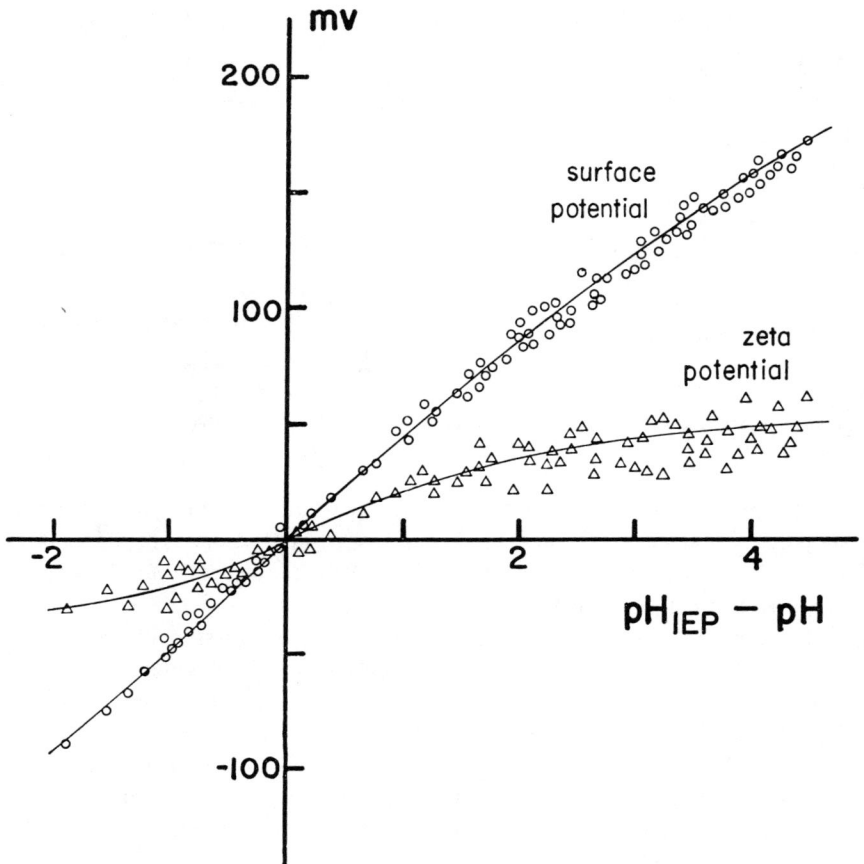

Figure 3. Variation of surface and zeta potentials with pH.

could be used to check surface potentials predicted by the models. Relatively little use has been made of electrophoretic mobility in adsorption studies, but it may prove to be a necessary complement to strictly chemical measurements.

Adsorption at High Relative Adsorbate Concentrations

Much of the literature concerns systems where the adsorbate concentration is a significant fraction of the ionic strength of the solution. Under such conditions an appreciable proportion of the adsorbate may be present in the diffuse layer and thus not strictly bound to surface sites as is required by complexation models. In addition, this situation will arise if the recommendation of observing the role of "inert" electrolyte ions at the surface is implemented.

In order to measure changes in the bulk concentration of electrolyte caused by sorption processes, the electrolyte may need to be more dilute than the 0.1 or 1 M levels currently being employed in many studies. Adsorption may therefore be expressed better as:

$$\Gamma_i = \Gamma_{i, \text{surface}} + \Gamma_{i, \text{edl}} \quad (1)$$

The $\Gamma_{i, \text{surface}}$ term is identical to that utilized in other models presented in this text, referring to chemically bound adsorbate. The Γ_{edl} term represents additional unbound adsorbate ions present in the diffuse layer necessary to maintain the electroneutrality of the system.

To estimate this new term, Γ_{edl}, a derivation parallel to the Debye-Huckel approximation for the double layer can be used [32]. One begins with a expression for the total charge in the diffuse layer:

$$\sigma_d = \int_0^\infty \rho \, dx \quad (2)$$

where ρ is the charge density at any point and the limits of integration are set from the beginning of the diffuse layer (see Figure 1), defined as $x=0$, to the bulk solution, where $x=\infty$ and $d\psi/dx=0$. In contrast, the original derivation integrates from the solid surface. Including appropriate substitutions, integration results in:

$$\sigma_d = \frac{\epsilon \psi_s}{4\pi \kappa^{-1}} \quad (3)$$

$$\kappa = \frac{8\pi e^2 I}{\epsilon kT}^{1/2} \quad (4)$$

where ψ_s = the potential at the Stern layer or beginning of diffuse layer,
ϵ = the dielectric constant,
e = the charge of the electron,
k and T = their usual thermodynamic values,
κ^{-1} = the Debye length,
I = the ionic strength

These equations demonstrate that ionic strength, diffuse layer charge and diffuse layer potential are interrelated. Assuming that the Boltzmann distribution describes the concentration gradient in the diffuse layer:

$$n_i = n_{i,o} \exp[-Ze\,\psi_d/kT] \tag{5}$$

where Z = the ion charge,
$n_{i,o}$ = its bulk solution concentration.

Integration of this equation over the entire diffuse layer gives the total amount of adsorbate in the diffuse layer:

$$\Gamma_{i,edl} = \int_{\psi_s}^{0} n_{i,o} \exp[-Ze\,\psi_d/kT]\,d\psi_d \tag{6}$$

$$= \frac{n_{i,o}\,kT}{Ze} [\exp(-Ze\,\psi_s/kT) - 1] \tag{7}$$

Thus, Equation 1 becomes:

$$\Gamma_i = \Gamma_{i,\text{surface}} + \frac{n_{i,o}\,kT}{Ze} [\exp(-Ze\,\psi_s/kT) - 1] \tag{8}$$

Solving Equation 3 for ψ_s and substituting gives

$$\Gamma_i = \Gamma_{i,\text{surface}} + \frac{n_{i,o}\,kT}{Ze} [\exp(-4Ze\,\sigma_d\,\kappa^{-1}/\epsilon\,kT) - 1] \tag{9}$$

These results are in contrast with the model of Davis et al. [21,22]. The latter model accounts for the effect of ionic strength solely on adsorbate uptake by means of the competitive adsorption of electrolyte ions. Such a formulation includes specific reactions at surface sites and is then amenable to multicomponent equilibria calculations.

In the derivation herein, ionic strength variations are accounted for in κ^{-1} as a phenomenon affecting double layer compression. It is still possible, if necessary, to include additional potential layers for specific adsorption, as in the models of Davis et al. [21,22] and Bowden et al. [25].

RECOMMENDATIONS FOR FUTURE WORK

After considering the many shortcomings of experimental procedures and adsorption theories, one begins to wonder whether anything can be known with certainty about the chemistry of aqueous adsorption. Although the state-of-the-art is certainly not as hopeless as this statement suggests, the previous discussion has built a strong case for the reevaluation of current experimental procedures and for the development of fresh ones.

In the Footsteps of Gas Adsorption

An interesting parallel can be drawn between aqueous and gas adsorption research, which illuminates the weak areas in the former. For gas adsorption, it is possible in many instances to create a given type of surface reproducibly, to identify it, to treat it with the adsorbate gas, and to study the results immediately in situ. Such control over the chemistry has been made possible by means of recent advances in high vacuum technology and, correspondingly, in surface spectroscopy [33-36]. More specifically, in the clean environment of a 10^{-9} torr vacuum, a solid can be cleaved and treated chemically and thermally to produce a particular surface. This surface can be identified, for instance, by low-energy electron diffraction to distinguish whether, for example, the 100 or the 110 crystal surface of a metal substrate is exposed. After adsorbing a certain gas, again within the same chamber, the adsorbed species can be studied directly on the surface by a handful of complementary methods; for instance, X-ray and ultraviolet photoelectron spectroscopy, Auger electron spectroscopy, secondary ion mass spectrometry, electron stimulated desorption and ion scattering spectroscopy.

Ideally, similar control is desirable in the case of aqueous adsorption; however, at present, all of the experimental steps are deficient. Creation of a colloid surface of a given type can be done only with considerable care; reproducibility can be achieved mainly by control of equilibration techniques, although these vary from lab to lab. The actual surface is rarely characterized to assure that the above methods succeed and that foreign adsorbents are absent. The methods of exposure to adsorbents varies and the duration of that exposure is another question. Finally, determination of the results of the adsorption have relied heavily on a few types of measurements, mainly pH and dissolved adsorbate concentration. It can be definitively stated that aqueous adsorption is well behind gas adsorption in the control of experimental parameters.

Where Do We Go From Here?

Now the question becomes: what can be done to rectify these faults and where do we go from here? It is too easy to maintain the status quo; that is, to continue using techniques and generating data that are applicable usually to only a few particular systems and to rely on theoretical models which are insensitive to the chemical surface species involved. In many cases, the information generated from such studies has been useful for specific systems or purposes. However, this limited outlook hinders development of procedures and theories sufficiently sophisticated to match the complexity of the chemistry. Alternatively, can surface spectroscopies be applied to aqueous adsorption just as successfully as they have been to gas adsorption? The primary difficulty in applying these techniques has been alluded to previously. Removal of the solid from its solution may induce unknown changes in the surface characteristics. Such changes may be anticipated for vacuum methods because of the inherent dehydration of the solid. Of course, if it can be shown that dehydration does not interfere with chemical interpretations, then there are great advantages to be gained by exploring these techniques. Obviously, the ideal condition under which to study aqueous adsorption is in situ, without separating the solid and supernatant. Consequently, analytical techniques capable of probing the surfaces of aqueous colloids are required. It is sobering to realize that fundamental information on the binding of gaseous adsorbates has required the investment of hundreds of thousands of dollars in sophisticated surface sensitive instruments, whereas to obtain similar information on aqueous adsorbates we have relied on the likes of atomic adsorption and a pH electrode. Can the study of surfaces in aquatic systems require any less effort or expense? (The surface spectroscopies have the obvious advantage of the marketability of their substrates—semiconductors and catalysts. We should be so fortunate as to make colloids in natural waters and soils profitable products.)

Probing Colloid Surfaces

With the aim of more accurately representing the structure and orientation of adsorbed ions and molecules several investigators have applied spectroscopic methods to both solids and solid-solution systems. Rouxhet et al. [37], and Scokart et al. [38] successfully applied infrared (IR) spectroscopy to the study of surface structure while Yoon and Salman [39] and Kieselev and Lygin [40] studied adsorbed molecules using this technique. However, IR is severely limited by the strong absorbance of water molecules, requiring the dehydration of the sample before analysis. The problem of surface alteration resulting from dehydration is the same as found in the application of high vacuum technology; thus IR is relegated to the characterization of dry surfaces before and after aqueous adsorption experiments.

Information gained by infrared spectroscopy might be complemented by Raman spectroscopy. Raman has two primary adantages: (1) water is only weakly Raman active and therefore does not interfere when used as the solvent, and (2) signal enhancing phenomena, such as resonance Raman and coherent Raman effects, can be utilized to provide adequate sensitivity to detect the adsorbed species. To date, emphasis has been placed on organic adsorbates both in the vapor phase and in solution [41-43], presumably because many organics are highly Raman active. The potential for studying inorganic adsorbates in aqueous solution exists [44], and several anions of particular interest such as phosphate [45], arsenite [46] and silicate [47] have been studied in homogeneous solution. Furthermore, growing interest in the application of Raman spectroscopy for surfaces is demonstrated by studies of the orientation and binding of organic and inorganic molecules to electrode surfaces [41,42,48] and by spectroelectrochemistry [49], which allows electroactive species to be studied in close proximity to the electrode surface using properties independent of their electrochemical behavior. One potentially serious problem, however, may be that scattering caused by the solid particles may obscure the Raman signals from the adsorbed species.

Recently, electron spin resonance spectroscopy (ESR) of spin-labeled organic molecules [50], ionic organic radicals [51,52] and paramagnetic ions such as Cu(II) [53] and Mn(II) [54,55] have provided information regarding the binding and mode of adsorption in aqueous systems. Signal frequencies reflect the electronic environment of the probe species. Line shape analysis gives information related to relaxation behavior of the probe species, which is related to the solvent environment (viscosity, molecular mobility) [53].

Nuclear magnetic resonance (NMR) may be a particularly powerful tool for the study of surface species for two reasons: (1) specialized techniques developed recently permit the study of solid state materials; this advance opens the way to investigation of surfaces as well [56]; and (2) NMR has multinuclear capabilities. Many adsorbates of interest contain, for instance, the nuclei C, F, H, N or P, all of which can be observed and all of which are amenable to solid-state techniques. The primary limitation of NMR may be its detection limits. Assuming an adsorbent with a surface area of several hundred m^2/g, an adsorbate at a surface coverage of about 10^{-6}-10^{-5} mol/m^2, and a slurry of 1 g solid per liter, the concentration of adsorbed material is only 10^{-4}-10^{-3} M. This may be just within the detectabilities of NMR.

Where Raman spectroscopy, ESR and NMR give direct information about surface binding, electrochemical measurements, by ion-selective electrodes for instance, can monitor all the ions suspected of entering and leaving the surface. When coupled with an acidity titration, such measurements give indirect evidence about surface reaction stoichiometrics and binding energies by means of a mass balance.

Of course, additional degrees of freedom in experimental design, such as the use of radioactively labeled materials and variations in solvent type, temperature and equilibrium times are to be recommended to provide as many independent descriptions of a given adsorbent/adsorbate system as possible. In this manner, whatever models are developed can be tested rigorously and with minimal experimental bias.

Return to Titration

After suggesting that currently measured solution parameters such as dissolved adsorbate concentration, pH and ionic strength were not sufficient to address the problems associated with adsorption at the solid-liquid interface, and after suggesting that more elaborate techniques such as ESR, NMR and Raman spectroscopy be used to probe these problems, it may seem strange to suggest a return to titration. However, at this time it should be pointed out that some very interesting results are being obtained in the laboratories of DeBruyn using simple but very well controlled titration techniques [20]. When correlated with in situ absorbance measurements, these titrations have done much to help understand the very complex polymerization and microprecipitation of hydrolyzable cations. Similar, carefully controlled titrations of well-characterized solids with both cationic and anionic adsorbates at fixed pH and ionic strength under nitrogen could be performed to improve our understanding of adsorption and surface precipitation.

Experimental Standardization

Another aspect of concern is transferability of results from one laboratory to those in another. Interlaboratory comparison of several types are used routinely in analytical chemistry to compare measuring capabilities between laboratories. Because of the difficulties of adsorption measurements, it would indeed be interesting to compare results of a number of currently used procedures among experimenters. This can be performed on several levels: (1) measurement techniques alone, using well-characterized and well-behaved solids, such as polystyrene latex particles for inorganic adsorbates and activated carbon for organic adsorbates; (2) measurements on natural colloids, prepared at one site and distributed to participating laboratories; and (3) measurements on a colloid prepared locally—portions of a batch of dry solid are distributed and each laboratory prepares the colloid as it does normally.

Experience from interlaboratory testing in analytical chemistry has shown that the state-of-the-art is not as good as it may seem within a single laboratory [57]. Would such comparisons in adsorption systems be just as revealing?

CONCLUSION

Depending on the degree of resolution, there exist models that are capable of predicting the concentration of adsorbate ions in solution for a given system under given conditions. Unfortunately, these models are not as yet universally applicable. The present danger that is becoming increasingly apparent is that we are expecting too much from our models. Modeling efforts are expanding at a greater rate than are experimental methods of verification. More knowledge is needed about the chemistry of surface reactions before modeling efforts are to advance. There are many problems, many possible techniques, and a wealth of work remaining for the experimental scientist.

REFERENCES

1. May, H. M., P. A. Helmke and M. L. Jackson. *Geochim. Cosmochim. Acta* 43: 861 (1979).
2. Ferguson, J. F., and T. King. *J. Water Poll. Control Fed.* 49: 646 (1977).
3. Chen, Y. S. R., J. N. Butler and W. Stumm. *J. Colloid Interface Sci.* 43: 421 (1973).
4. Chen, Y. S. R., J. N. Butler and W. Stumm. *Environ. Sci. Technol.* 7: 327 (1973).
5. Muljadi, D., A. M. Posner and J. P. Quirk. *J. Soil Sci.* 17: 212 (1966).
6. Helyar, K. R., D. N. Munns and R. G. Buraw. *J. Soil Sci.* 27: 307 (1976).
7. Hsu, P. H., and D. A. Rennie. *Can. J. Soil Sci.* 42: 197 (1962).
8. Huang, C. P. *J. Colloid Interface Sci.* 53: 178 (1975).
9. Anderson, M. A., and D. T. Malotky. *J. Colloid Interface Sci.* 72: 413 (1979).
10. Bache, B. W. *J. Soil Sci.* 15: 110 (1964).
11. Malotky, D. T. Ph.D. Thesis, University of Wisconsin, Madison (1978).
12. Stanforth, R. R., and M. A. Anderson, University of Wisconsin, Madison. Unpublished results.
13. Atkinson, R. J., A. M. Posner and J. P. Quirk. *J. Inorg. Nucl. Chem.* 34: 2201 (1972).
14. Kyle, J. H., A. M. Posner and J. P. Quirk. *J. Soil Sci.* 26: 32 (1975).
15. Kafkafi, V., A. M. Posner and J. P. Quirk. *Soil Sci. Soc. Am. Proc.* 31: 348 (1967).
16. Scholten, A. G. Ph.D. Thesis, University of Wisconsin, Madison (1965).
17. Hingston, F. J., A. M. Posner and J. P. Quirk. *J. Soil Sci.* 25: 19 (1974).
18. Bowden, J. W., A. M. Posner and J. P. Quirk. *Aust. J. Soil Res.* 15: 121 (1977).
19. James, R. O., and T. W. Healy. *J. Colloid Interface Sci.* 40: 65 (1972).
20. Dousma, J. Ph.D. Thesis, University of Utrecht, The Netherlands (1979).
21. Davis, J. A., R. O. James and J. O. Leckie. *J. Colloid Interface Sci.* 63: 480 (1978).
22. Davis, J. A., and J. O. Leckie. *J. Colloid Interface Sci.* 67: 90 (1978).

23. Schindler, P. W., B. Furst, R. Dick and P. U. Wolf. *J. Colloid Interface Sci.* 55: 469 (1976).
24. Hohl, H., and W. Stumm. *J. Colloid Interface Sci.* 55: 281 (1976).
25. Bowden, J. W., A. M. Posner and J. P. Quirk. *Aust. J. Soil Res.* 15: 121 (1977).
26. Hingston, F. J., A. M. Posner and J. P. Quirk. *Disc. Faraday Soc.* 52: 334 (1972).
27. Davis, J. A., and J. O. Leckie. *Environ. Sci. Technol.* 12: 1309 (1978).
28. Weber, J. B. "Interaction of Organic Pesticides with Particulate Matter in Aquatic and Soil Systems," in *Fate of Organic Pesticides in the Aquatic Environment*, R. F. Gould, Ed. (Washington, DC: American Chemical Society, 1972).
29. Bailey, G. W., and J. L. White. *Residue Rev.* 32: 29 (1970).
30. Theng, B. K. G. *The Chemistry of Clay-Organic Reactions* (London: Adam Hilger, Ltd., 1974).
31. Lyklema, J., and J. T. G. Overbeek. *J. Colloid Interface Sci.* 16: 501 (1961).
32. Hiemenz, P. C. *Principles of Colloid and Surface Chemistry* (New York: Marcel Dekker, Inc., 1977).
33. Somorjai, G. A. *Principles of Surface Chemistry* (Englewood Cliffs, NJ: Prentice Hall, 1972).
34. Evans, C. A., Jr. *Anal. Chem.* 47: 818A (1975).
35. Evans, C. A., Jr. *Anal. Chem.* 47: 855A (1975).
36. Morrison, S. R. *The Chemical Physics of Surfaces* (New York: Plenum Publishing Corporation, 1977).
37. Rouxhet, P. G., P. O. Scokart, P. Canesson, C. DeFosse, L. Ridrique, F. D. Declerck, A. J. Leonard, B. Delmon and J. P. Damon. In *Colloid and Interface Science, Vol. III,* M. Kerker, Ed. (New York: Academic Press, Inc., 1976).
38. Scokort, P. O., S. A. Selin, J. P. Damon and P. G. Rouxhet. *J. Colloid Interface Sci.* 70: 209 (1979).
39. Yoon, R. H., and T. Salman. In *Colloid and Interface Science, Vol. III,* M. Kerker, Ed. (New York: Academic Press, Inc., 1976).
40. Kiselev, A. V., and V. I. Lygin. *Infrared Spectra of Surface Compounds* (New York: John Wiley & Sons, Inc., 1975).
41. King, F. W., R. P. Van Duyne and G. C. Schatz. *J. Chem. Phys.* 69: 4472 (1978).
42. Gardiner, D. J. *Anal. Chem.* 50: 131R (1978).
43. Craver, C. D. In *Infrared and Raman Spectroscopy, Part C,* E. G. Brame, Jr. and J. G. Grasselli, Eds. (New York: Marcel Dekker, Inc., 1977).
44. Nakamoto, K. *Infrared and Raman Spectra of Inorganic and Coordination Compounds,* 3rd ed. (New York: John Wiley & Sons, Inc., 1978).
45. Pinches, S., and D. Sadeh. *J. Inorg. Nucl. Chem.* 30: 1785 (1968).
46. Loehr, T. M., and R. A. Plane. *Inorg. Chem.* 7: 1708 (1968).
47. Fortnum, P., and J. O. Edwards. *J. Inorg. Nucl. Chem.* 2: 264 (1956).
48. Gold, H. S., and R. P. Buck. *J. Raman Spectros.* (1979).
49. Heineman, W. R. *Anal. Chem.* 50: 390A (1978).
50. Smith, I., G. W. Stockton and A. P. Tulloch. In *Colloid and Interface Science, Vol. I,* M. Kerker, Ed. (New York: Academic Press, Inc., 1977).
51. Muha, G. M. *J. Catalysis* 58: 470 (1979).
52. Flockhart, B. D., and R. C. Pink. *J. Catalysis* 61: 291 (1980).

53. Bassetti, U., L. Burlamacchi and G. Martini. *J. Am. Chem. Soc.* 101: 5471 (1979).
54. Burlamacchi, L. *J. Chem. Soc. Faraday Trans.* 71: 54 (1975).
55. Burlamacchi, L., G. Martini and M. F. Ottavianni. *J. Chem. Soc. Faraday Trans.* 72: 324 (1976).
56. Griffin, R. G. *Anal. Chem.* 49: 951A (1977).
57. Hertz, H. S., W. E. May, S. A. Wise and S. N. Chesler. *Anal. Chem.* 50: 428A (1978).

GENERAL INDEX

activation energy 165-166,168
activity of solids 163-165,171
activity of surface species 7,61, 268-269
aging
 of hydroxides 141
 of precipitates 164,328
 See also modeling, effects of
Auger electron spectroscopy 341

BET
 See surface area
Boltzmann factor 61,197,226,237, 339
bonding concepts 20-22,66,174,179
Bronstead acid 183,214
buffer capacity 77-78,287

capacitance, interfacial 53,59,221- 222,239-240,245,247-249,253, 271-278,281,283,287,289-290, 292,334
cation
 chelated metals and carbon absorption 295,301,310,321-324
 exchange and exchange capacity (CEC) 124-137,139-141
 polynuclear 173-176
 selectivity 109,110-114,129-133
competition, ionic 40,64-65,73-76, 80
complexes
 agents in carbon absorption 295- 296,300,310
 bonding on surface 20-22

formation of surface 5,121-122, 206,214,220,264-265,267
stoichiometry of surface 11,55,264, 269-270,301
structure of surface 20-22
ternary type I, II 26-30
 See also modeling, effects of; stability constants
concentration, role in adsorption 116-121,297-298
conductivity, specific 255-257
conductometric titration 224,246
conjugate acids, effect of pKa on adsorption 59-60
coordination 2-4,210
coprecipitation
 See precipitation
coulombic energy
 See modeling, effects of
crystal growth 161-169,178

Debye-Huckel approximation 339
desorption 104
diagenesis 139
dielectric constant, interfacial 239, 245,248
diffuse layer potential and charge
 See electrical double layer
distribution constant 32-33,73-74, 171-174
double hydroxides 114-116

EDTA 39,122,301-303,306-309, 322-323

349

electrical double layer (EDL)
  concepts related to uptake and release of ions at interfaces 52-54,58,66,219-225,227-230,233-235,238-241,245,247-250,254,266,328,334,339-340
  See also ionic strength; zeta potential
electron spin resonance spectroscopy (ESR,EPR) 95,110,140,343
electron stimulated desorption spectroscopy 341
Elovich equation 68,70,71
equilibration time 328-330
exchange
  See cation; ligands; modeling, effects of competing ions

fertilizers 78-79,177
free energy 52-54,102,117-118,163,165,167,172-173,264-265,297,332,336
Frenkel defect 170
Freundlich equation
  See modeling

Gibbs-Duhem equation 56
Gouy-Chapman equation 52,59,238,334
Gran plot 303

heterocoagulation 220
heterogeneity of surface sites 118,232-235,297-300,331
heterogeneous nucleation 162,166-167
homogeneous nucleation 164-165
hydrolysis
  with respect to adsorption 20,53,106-110,220-235,264-266
  with respect to precipitation 173
hydrogen electrode 185-189

hydroxylated surface
  acidity sequence of oxides 204
  complex formation with 5
  exchange with ligands 5-7
  properties of 10,66
  protolytic behavior of 2,4,6,7-10,183-184,195,220,224-234,236-244,246-255,264-292,333,336

impurities in precipitation processes 168
incongruent dissolution 173
infrared spectroscopy 2,62,66,76,116,140,172,239,342
interfacial tension and energy 164-166,169
ionic strength 8,53-54,59,63-64,220,222,225,227-235,238,240-242,244,250-251,254,265-266,274-275,292,333,339-340
ion pair formation with surfaces 53,237-240,248,266,274-277,281-283,290-291,333
ion scattering spectroscopy 341
isoelectric pH (IEP) 40,98-99,222,233,278,285,310-311,321-333
isomorphic substitution 125-126,141,172
isotherms
  See modeling
isotopic exchange 70-73,95,140,239

kinetics
  adsorption-desorption of anions 66-70,329-330
  adsorption-desorption of cations 93-95
  isotopic exchange 72,80,95
  nutrient uptake 77-79

GENERAL INDEX

rate of diffusion controlled crystal growth 167
rate of nuclei formation 165
titration 205
Kurbatov plot 11,15,34,101-103

Langmuir equation
See modeling
lattices 232-234,237,245,252-256
Lewis acid 2,5,22,30,41,183
Lewis base 5,41
ligands
  dissolved, effect on adsorption 39-40,301,303,306-308,321-324
  exchange of 2,5,53,73-76,101-103,110-114,220,287
  hydroxyl, properties of surface 10,183-184,265-268
  nature of surface 2-4
  organic complex forming 39,122,295,296,301,303,306-307,309,322
  polydentate 6,269-270
  structure and bonding surface 20-22,267
  See also modeling, effect of competing ions; specific adsorption

Michaelis-Menten equation 297
MINEQL
  See programming, equilibrium modeling
  Freundlich equation 62,118,296-297,305-306,308-309,324
  isotherms 37-39,61-63,264
  Kurbatov plots 11,34,101-103
  Langmuir equation 24,37,54,57,60,62,73,116-118,175,184,189,202,204,206,214,278,284,296-301,303,305,306,308,310,312,321,331
  nutrient uptake 77-78

Stern equation 52,60,197,268,332,334
Stern-Grahame model 224,235
  See also ion pair formation with surfaces
modeling, effects of
  aging 66,99,141,328
  charged species 60-61
  competing ions 40,64-65,73-76,80
  complexation 121,267
  concentration 116-121,297-298
  dissolved ligands 39-40,301,303,306-308,321-324
  equilibration time 328-330
  pH 7-20,23-28,34-37,54-58,139-140,220-224,234,236-244,246-255,267-268,271-273,275,277-292,300,304-306,311,321,336
  pK_a, pK_b 54-60,139-140
  potential 7-10,58-63,73,220,224-234,236-244,246-255,270-277,281-283,292
  surface charge 7,57-66,73,119,140-141,220-234,236-244,246-255,270-272
  temperature 68-69,122-123
  See also ionic strength
Mossbauer spectroscopy 116

Nernst equation 208,210,213-214,223,225-226,234-235,271
nucleation and crystal growth 164-167
nutrients 77-81

Ostwold-Freundlich equation 164

particle size 163-164
pH
  dependence of cation speciation on 95-104

dependence of charge on clays 127
effect on solid solution 174
See also hydroxylated surfaces,
  protolytic behavior of; modeling,
  effects of; stability constants
phyllosilicates
  cation exchange in 125-127
  coatings on 126
  dissolution, precipitation in 177
  solid solutions in 172
  structure of 124-125
point of zero charge (PZC) 12,63,
  98-99,139,189,192,205-206,208-
  209,214,220-226,228,233,236,240-
  242,244,246-247,250-253,271,287,
  333-334
polarity of solids 112
polymorphism 168-169,179
polynuclear ions 64,104,173,176
potassium exchange 135-137
potential determining ions 53-54,220-
  234,236-244,246-255,271
potentiometric titration 183-191,224,
  236,240,246-247,254-255,278-283
precipitation
  coprecipitation of double hydroxides
    114-116
  relation to adsorption and
    hydrolysis 107,174,329-330
  relation to specific adsorption 161
programming, equilibrium 240,257-
  258,290-293
protolysis, surface hydroxyl
  See hydroxylated surface

radio colloids 104
Raman spectroscopy 243
reversibility 65-66,141,173

Schottky defect 170
secondary ion mass spectrometry 341
  310-311

selectivity sequences 109-114,129-
  133,140
separation processes, effect on
  adsorption 328
silver-silver chloride electrode 185
solid solution 170-175
solubility product 162-164,171,
  175-179
specific adsorption
  anions 22-26,52-54,58,63-64,73-76,
    173,238,266-267,272,283,332,334
  cations 53-54,99-101,107-137,139,
    206,238,266,272,283,332,334
  relation to potentiometric titration
    192,214
  relation to precipitation 161,174
specific conductance
  See conductivity
stability constants
  comparison of 213
  effect of pH on 117
  evaluation of 11-20,171-174,220-
    240,243,251-253,233-234,282-
    283
  intrinsic 7-10,14-20,24-25,73-74
    196,207-209,213,240-245,248-
    254,285-286,273,281,332
  microscopic acidity 26,196,198-
    202,219,225-237,248-254,266,
    285-286,292
  pair formation (ion pair formation
    with surface) 8,237-240,248-254,
    266
  See also distribution constant
steric hindrance 29,76
stoichiometry
  See complexes
supersaturation 164-167,178
surface
  acidity 183,194,202,214
  area 2,76,167,204,268-269,295,
    310-311

coverage 119,140-141,295,308-312,319
porosity 53,204
site density 2,12-13,18,23-25,33-34,58-62,64,76,120,140,183,197,204,220,254,265-270,297
solvation 195,332,334
See also capacitance; complexes; heterogeneity of surface sites; hydroxylated surface; ion pair formation; modeling, effects of; stability constants

temperature, effect on adsorption 68-69,122-123
temperature, effect on precipitation 167,170-171,177
ternary surface complexes 26-30,39

transport
 coefficient for crystal growth 167
 of anions in the environment 81-82
 of solutes to surface 167

ultraviolet photoelectron spectoscopy 341

volta potential 223

water, natural 264
water supply 295
weathering reactions 134-135

X-ray diffraction 116,341

zeta potential 220,222,224,234,246,254,257,338
Zwitter ions 224,230,233

INDEX TO ELEMENTS, IONS AND MINERALS

allophane 127
aluminum 2-4,9,12,18-19,22,30,51-52,55-57,64-66,68-72,74,78,80,95-102,105,109-117,120,122-127,133-135,138-141,172-173,176-178,183-184,189-192,196,198-199,201-214,249,258,279-280,283-285,287-290,329-330.
ammonium 119,121-124,130,133,170
anatase
  See titanium
apatite 169,172,177,186,203-204
aragonite 168
arsenic 57,63,76,81-82,258

barium 18,20,94,105-106,112-113,118,131,133,135,206,208-210,213
barrandite 172
benzoate 66
biotite 127,133,135,136
boron 51,80
bromide 66,99,170,174
brucite 124,125

cadmium 12,16,18,20,38-39,41-45,65,93-94,102,105,110,113,114,122-123,137-138,140,213,296,300,301,302,303-312,320-323
calcite 168,177
calcium 5,18,20,64-65,78,94,96,105,108,112,115-117,119-121,128,130-131,133,137,140,169-170,172,176-178,184,206,208-210,213-216,302-304
carbon (activated) 295,299,303-304,306-307,310,321,323
carbonate-bicarbonate 5,66,76-81,115-116,134,162,170-172,176,184
cerium 4,103
cesium 111,112,127-131,137,139,141
chloride 39,55,63,65,73,76,81,98-99,115,121-123,127,135,137,138,170,174,176,185-194,200-201,203,205,206,208,211,236-240,242-244,248-249,257,266,271,274-277,290-292
chlorite 125,127,135,139
chromium 104,115-117,302
citrate 122
cobalt 12,21-22,41-43,93-94,102,105-106,113-114,118,137,140,172,176,211-214
copper 16-18,20-21,29-31,35-38,42,94,102,105,114,119-120,122,131,133,137-140,172,210-214,343
cyanide 122
cysteine 122

deuterium 2,26,66
dolomite 168

fluoride 55-57,60,63,80,82,136,172,343

gibbsite
  See aluminum
glutamate 122
goethite
  See iron
greenalite 177

INDEX TO ELEMENTS, IONS AND MINERALS

halloysite 125,135
hematite
 See iron
humic acid 30

illite 36,138
imogolite 127
iodide 2,66,167,221-223,230-253
iron 2,4-5,9,12,18-20,22-27,30,51-
 52,55,57,64-66,72,74-76,78,80,98,
 100-105,111-116,119-126,134-136,
 140-141,172-175,177,184,201,203-
 205,213-214,249,300

kaolinite 76,124-125,131,138,174,
 177

lead 12,18,20,94,102,105,114,131,
 140,213,264,266-274,276-277,284-
 287,289-292,296,300,310-311
lithium 111-113,129,131,139,141,
 242,244-246,249

magnesium 5,18,20,63,94,102,105,
 112-113,115-116,124-125,128,131,
 133-137,172,206,208-210,213-214,
 249
mangangese 4,12,18,93-99,101,105-
 106,112-118,126,140,171-172,177,
 300,320,340
mercury 39,53,114,122,140,221,
 225,245,253,301
mica 124-125,128,131,135-137,139
 167
molybdenum 51,57,63,76-82
montmorillonite 30,124,127,131,
 133,138-139
muscovite 127,133,136

neptunium 104
nickel 12,22,94,102,105,114-115,
 140,172,210-214
nitrate 19,66,73,76,80-81,96-103,
 106,108,110,119-120,170,203,
 242,245-247

oxides
 See element of interest
oxalate 66,79,169-170

palygorskite 125
perchlorate 2,4-5,17,39,42-43,93,98,
 203,244,251-252,254-255,266,279,
 281,284-285,288
phenoxyacetic acid (2,4-D) 64
phlogopite 137
phosphorus 2.24-27,51,57,63-66,72-
 81,169,172-175,177-178,268,329-
 330,343
platinum 185-186
potassium 63,98-100,106,111,119,
 123,125,127-131,135-137,139,141,
 172,177,186-187,190,203,236,247,
 249

rubidium 111,127-129,139,141
ruthenium 103
rutile
 See titanium

selenium 22-23,51,57,63,65,72-76,
 81-82
sepiolite 125
silica 2,4,5,9,12,16,18,20-22,29-30,
 35-37,41-45,52,57,60,63-64,68-71,
 74,80,98-99,104-106,109,111-114,
 118,120,122,125-126,129,134-136,
 140,141,172-174,176-177,184,202-
 204,213,232,241,249
silver 30,112,122,167,170,185-187,
 213,221-223,230,235,245,253
smectite 127,135

INDEX TO ELEMENTS, IONS AND MINERALS

sodium 4,5,17,19,39,41-43,55,63-65,93,96-100,102-103,111-112,119-120,129-130,131,133,137-139,172,177,185-194,200-201,203,205-206,208,211,236-240,241-244,248,256,266,271,274,276-277,279,281-285,288
strengite 172
strontium 65,94,102-103,105,110,112-113,118,131,133,171,206,208-210
sulfur 30,51,63,65,76,79-80,99,106,134,186

thorium 2,9,118,184,203-204
tin 4,114
titanium 9,16-18,22,38-39,98-100,105,111-112,117-118,126,184,201,203-205,213,241,244-247,249

uranium 12
urea 170

variscite 172
vermiculite 124,127-128,133,135,136

yttrium 103

zinc 4,12,18,65,94,96-97,100,102,105-106,113-114,119-120,122,137,138,205,210-213,296,300,302,310-311
zirconium 22,103,111,184,203-204